ORGANIC PHOTOCHEMISTRY:
A VISUAL APPROACH

ORGANIC PHOTOCHEMISTRY: A VISUAL APPROACH

JAN KOPECKÝ

Prof. Jan Kopecký
Department of Chemistry
C.W. Post Campus
Long Island University
Brookville, NY 11548

Library of Congress Cataloging-in-Publication Data

Kopecký, Jan.
[Organická fotochemie v obrazech, schématech a tabulkách. English]
Organic photochemistry; a visual approach / Jan Kopecký.
p. cm.
Translation of: Organická fotochemie v obrazech, schématech a tabulkách.
Includes bibliographical references and index.
ISBN 0-89573-296-3
1. Organic photochemistry. I. Title.
QD275.K6613 1991
547.1′35—dc20 91-36270
CIP

Printed in the United States of America.

ISBN 0-89573-296-3 VCH Publishers
ISBN 3-527-26952-5 VCH Verlagsgesellschaft

Printing History:
10 9 8 7 6 5 4 3 2 1

VCH Publishers, Inc.
220 East 23rd Street
Suite 909
New York, New York

VCH Verlagsgesellschaft mbH
P.O. Box 10 11 61

VCH Publishers (UK) Ltd.
8 Wellington Court
CB1 1HW
dom

FOREWORD

Professor Kopecky has written - or rather, drawn - a delightful book. The basic concept, using the language of drawings rather than words, is refreshingly novel for a textbook, and remarkably effective. Students of organic chemistry and old pros alike are certain to find the reading quite easy going and are sure to derive much pleasure from the perusal of this volume.

I had the good fortune to meet Professor Kopecky when I was still an undergraduate student. He attempted to teach me organic chemistry, and directed me towards stereochemistry and quantum chemistry as the two upcoming disciplines that would transform organic chemistry as it then existed. Not at all bad as foresight, in the Fifties!

As a summer research student in his laboratory, I participated in the synthesis of heterocycles for which we hoped to find biological activity, without much of a rational basis. In free moments, we dreamt of the structure of molecules that were heretical in those days, and for which any self-respecting professor of organic chemistry would have immediately failed a student during an exam.

It is astonishing how many of these "impossible" structures have since become commonplace, and it is remarkable how many of the original preparations involved a photochemical reaction as a key step. Although organic photochemistry is sometimes considered merely marginal to the field of organic synthesis, it is clear that it has affected organic chemistry profoundly in many ways, both directly and indirectly. Professor Kopecky's book demonstrates this fact in a particularly convincing fashion, and it is to be hoped that a variety of organic chemists, and not only photochemists, will find the time to browse in it.

As the last chapter of the book clearly demonstrates, photochemistry has affected more than the thinking of chemists. For better or for worse, it plays a key role in the environment around us, and it would indeed not be a bad idea for all persons with an interest in chemistry, and in natural science in general, to read a few pages in Professor Kopecky's book.

There were periods in the history of organic photochemistry when the discipline appeared to have passed its zenith and to have exhausted its intellectual potential. However, as Professor Kopecky's book amply demonstrates, photochemists are resilient folk, and their discipline is apparently inexhaustible. Repeatedly, novel directions were uncovered, such as the electron-transfer based reactions of the last decade, and one must agree with Turro that "the fun has just begun." All those already enjoying the fun, and those thinking about joining them, now have an opportunity to spend numerous pleasant moments with Professor Kopecky's book.

JOSEF MICHL

DEPARTMENT OF CHEMISTRY AND BIOCHEMISTRY

UNIVERSITY OF COLORADO

To my wife, Joan

CONTENTS

PREFACE

This book is an introduction to organic photochemistry, the study of electronically excited states of organic molecules. It treats their formation, physical deactivation pathways, and chemical reactions.

The format of this book, as indicated by its title, is unique. The author has attempted to present a visualization of photochemistry by means of reaction schemes, graphs, tables, and diagrams. The figures have been designed to be self-explanatory, and the corresponding text is meant to supplement these figures when additional explanations are needed.

Chapters 1 through 4 are concerned with the theoretical basis of photochemistry. Photochemical reactions, classified by chromophore type, especially those important in the synthesis of organic compounds are described in Chapters 5 through 13. Since photochemistry requires the use of unique experimental techniques, these are detailed in Chapter 14. Factors influencing photochemical reactions, and the applications of these reactions in various fields, are covered in Chapters 15 and 16.

This book is designed to serve as an introduction to photochemistry for both graduate and upper-level undergraduate students. It is my belief that it will also prove useful for synthetic organic chemists who need to quickly grasp the rudiments of organic photochemistry. The inclusion of extensive references to the original literature and review articles may also make it a useful source for practicing photochemists.

The examples used in this book were chosen to appropriately illustrate the principles of organic photochemistry and synthetic routes to organic compounds and, as a result, may not always be extracted from the most current literature. The reader is encouraged to check the references for additional, and more recent, examples. I chose to present photochemical synthetic methods that have been experimentally checked for accuracy, especially those that appeared in *"Organic Photochemical Synthesis."* The number of examples used in this book is limited by its unusual format; however, I hope that the reader will find the additional references at the end of each chapter sufficient.

Organic Photochemistry: A Visual Approach would not have been realized without the help of the co-workers mentioned in Part II of the Introduction. In addition, I would like to extend my gratitude to Dr. Jiri Pancir for his contributions to some of the theoretical discussions in Chapters 1 through 4, and especially to Dr. Josef Michl for his expert criticism and many valuable suggestions. If any errors remain, it is because I did not follow Dr. Michl's advice closely enough. Finally, I would like to thank my wife Joan for her patience, support, and practical assistance during the preparation of the manuscript. As a small token of my appreciation for all her help, I have dedicated this book to her.

JAN KOPECKÝ
Long Island University

March 1991

INTRODUCTION

I.

Structural formulas are concrete expressions of very precise ideas. They do not need any explanation for those familiar with the principles of the language of formulas and are easily understood in all details, so that it is absolutely unnecessary to repeat by words the ideas which they convey. They satisfy all the requirements of formulas; the goal of the formula language is to provide in a small space a large number of concepts in an unambiguous way.

Friedrich A. Kekule
(1883)

II.

COMPUTER-ASSISTED DRAWING OF FIGURES:

Jingqi Wang
Kinping Tsoi

MANUSCRIPT EDITING:

Joan E. Shields

PROOFREADING:

Robert E. Schaffrath

CONCEPT, SCREENPLAY, AND DIRECTION:

Jan Kopecky'

I. This statement by one of the world's foremost chemists prompted the author to experiment with the production of this book, whose contents are primarily expressed by formulas, reaction schemes, definitions, graphs, and tables. The role of the text is merely supplementary and is meant to explain only those aspects of the subject that are difficult to describe in graphical form. The author has attempted to construct the individual figures to provide the largest possible amount of information on a subject. Therefore, it is expected that the reader will be able to follow the content of most of the figures without the supplementary text, at least at a certain level of understanding. Whether this goal has been accomplished can be verified simply by "reading" the figure and comparing it with the corresponding text. Undoubtedly, this test will occasionally reveal that the author did not do justice to Kekule's statement and repeated once more, in words, ideas that were already expressed by the formulas. The reader can then judge to what degree the attempt to visualize the subject of organic photochemistry has actually succeeded.

II. This book could not have been realized without the help of my colleagues, students, and friends. The titles in the figure are a mere list of names and areas in which various individuals converted the author's ideas into reality, and do not adequately express my gratitude for their generous and efficient help. Last but not least, I would like to thank the staff of VCH Publishers for their assistance, and especially for their courage in agreeing to perform this experiment.

Chapter 1
INTRODUCTION TO PHOTOCHEMISTRY

INTRODUCTION TO PHOTOCHEMISTRY

1.1

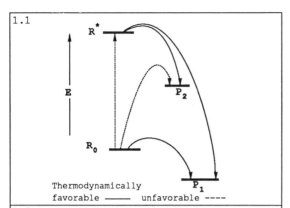

Thermodynamically
favorable ——— unfavorable ----

1.2

GROTTHUS-DRAPER LAW

(First Law of Photochemistry)

Only radiation absorbed in a system can produce a chemical change.

STARK-EINSTEIN LAW OF PHOTOCHEMICAL EQUIVALENCE

Number of activated molecules = number of quanta of radiation absorbed.

$$1h\nu + 1R_0 = 1R*$$

(The amount of radiation absorbed is limited to one quantum per molecule taking part in the reaction - exceptions to the law have been observed in two-photon absorption processes.)

1.3

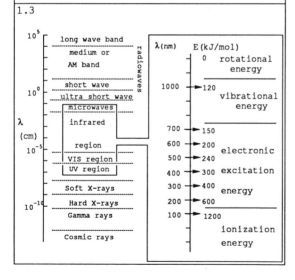

1.1. Quantum chemistry predicts that molecules exist in a variety of electronic states that differ in both electronic energies and wave functions. Classical chemistry involves reactions of molecules in their *electronic ground state* (lowest electronic energy state). The reactivity in states with higher electronic energy (*electronic excited states*) is the subject of photochemistry.[1] Ground and excited states are terms used to describe ground electronic and electronically excited states, respectively. Organic photochemistry is the study of the reactions that occur through excited states of organic molecules. One advantage of photochemistry is that reactions that are thermodynamically unfavorable when the reactants are in the ground state (R_0), e.g., $R_0 \rightarrow P_2$, may occur from an excited state ($R*$), e.g., $R* \rightarrow P_2$.

1.2. Two fundamental laws were formulated in the early years of photochemistry. Although the *Grotthus-Draper law* does not apply to chemi-excitation processes (see 13.36) and the *Stark-Einstein law* fails in multiphoton excitations with laser light,[2] these laws govern most photophysical and photochemical processes.

1.3. Generally, electronic excitations that produce photochemical reactions are induced by the absorption of ultraviolet (UV) or visible (VIS) electromagnetic radiation by a molecule. This radiation energy corresponds to the excitation energies of organic molecules ($\lambda = 200-700$ nm). Absorption occurs in multiples of a quantum of radiation, i.e., *a photon*, whose energy is a function of its frequency (ν) (Planck). Upon absorption of a photon, an excited state is produced. Electronic excitation of a molecule can also occur by interaction with another molecule with high internal energy.

4

1.4

$$E_{exc} = E^* - E_0 = h\nu = hc/\lambda$$

E_{exc} = electronic excitation energy

E^* = energy of the excited state

E_0 = energy of the ground state

h = Planck's constant

ν = frequency of absorbed radiation

c = speed of light *in vacuo*

λ = wavelength of absorbed radiation

1.5
ENERGY CONVERSION TABLE

Wavelength nm	Wave number cm^{-1}	Energy kJ/mol	kcal/mol	eV
200	50,000	599	143	6.20
250	40,000	479	114	4.96
300	33,333	399	95	4.13
350	28,571	342	82	3.54
400	25,000	300	71	3.10
450	22,222	266	64	2.75
500	20,000	240	57	2.48
550	18,182	218	52	2.25
600	16,666	200	48	2.07

1.6
BORN-OPPENHEIMER APPROXIMATION

The wave function and the energy of a molecule can be divided into nuclear and electronic components and the two can be treated separately.

Schrödinger Equation of the electronic wave function:

$$H_e \Psi_e(r,R) = E_e \Psi_e(r,R)$$

Molecular energy:

$$E_{mol} = E_{elec} + E_{vib} + E_{rot} + E_{trans}$$

where $E_{elec} \gg E_{vib} > E_{rot}$

1.4. The energy required for electronic excitation is the *excitation energy* (E_{exc}), which is inversely proportional to the *wavelength* (λ). The absorption of 1 mol (6.023×10^{23}) of photons (1 einstein) causes a molecule to acquire 1.198×10^5 kJ mol$^{-1}/\lambda$ (nm) or 2.864×10^4 kcal mol$^{-1}/\lambda$ (nm) of energy (molar E_{exc}). E_{exc} for the wavelength range 254–400 nm, most commonly used in organic photochemistry, is 470–200 kJ mol^{-1} (113–48 kcal mol^{-1}). This large energy gap between the excited and ground states explains the thermodynamics of photochemical reactions. The relation shown permits a facile calculation of the E_{exc} acquired by a chemical system upon absorption of radiation of known wavelength. It is useful to remember that radiation of λ = 200, 300, 400, and 600 nm corresponds to molar E_{exc} = 600, 400, 300, and 200 kJ mol^{-1}, respectively.

1.5. Various interrelated units describe radiation, excitation energy, and electronic transitions. In experimental photochemistry, wavelength (λ) in nanometers (nm) and energy in kJ or kcal are preferred, while *wavenumber* $\bar{\nu} = 1/\lambda$ (cm^{-1}) and *electronvolt* (eV) are more commonly used in spectroscopic and theoretical studies.

1.6. Because of the large difference in the masses of electrons and nuclei and the fact that nuclei move much more slowly, their motions can be treated separately (*Born-Oppenheimer approximation*). By solving the corresponding Schrödinger equation for each configuration of nuclei, R (position vector of nuclei), E_e (the electronic energy), and $\Psi_e(r, R)$, (the electronic wave function) where r is the position vector of the electrons, are obtained. The approximation can be visualized as electrons moving in a potential field of static nuclei. This permits all energy modes, electronic (E_{elec}), vibrational (E_{vib}), rotational (E_{rot}), and translational (E_{trans}) to be considered independently. All energy modes except E_{trans} are quantized.

1.7

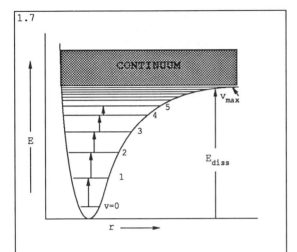

1.7. The solution of the electronic part of the Schrödinger equation, within the framework of the Born-Oppenheimer approximation, as a function of nuclear coordinates gives a *potential energy diagram* (often approximated by a *Morse curve*) for a given electronic state (ground or excited) of a diatomic molecule. The solution of the vibrational part of an energy curve gives the energy levels of *vibrational states*, indicated by horizontal lines. The long arrow in the figure represents the dissociation energy (E_{diss}) of the chemical bond; the short arrows represent transitions between corresponding vibrational states.

1.8

BOLTZMANN DISTRIBUTION LAW

$$n_j/n_i = e^{-(E_j - E_i)/kT}$$

n_j/n_i ... relative population of a molecule into any two energy levels, E_j and E_i

(J/mol)

k ... Boltzmann constant

T ... absolute temperature

1.8. Although electronically excited states can be produced by thermal activation, this method is useless for the formation of excited states of organic molecules. The *Boltzmann distribution law* shows that if an excited state has an $E_{exc} = 400$ kJ mol^{-1}, the relative population of the excited and ground states, n_{exc}/n_0, at 25°C is 8×10^{-71} and at 500°C is 7×10^{-5}. Even at the extreme temperatures achieved in a flame (at which most molecules would undergo rapid thermal decomposition), only 0.03% of the molecules would be electronically excited.

1.9

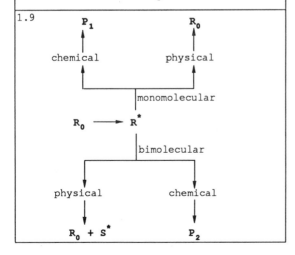

1.9. Photoprocesses involve two steps: (1) production of an excited state and (2) events that determine the fate of the excited state. The absorption of a quantum of radiation in the UV-VIS region produces a highly energetic species that dissipates its excess energy by two basic deactivation processes: (1) a chemical reaction leading to photoproduct(s) P_1, P_2 and (2) a physical process (see Chapter 3) that regenerates R_0 (and produces another excited species S* in a bimolecular reaction). Both pathways can proceed either monomolecularly or bimolecularly. The figure is oversimplified and the individual processes will be discussed in greater detail in subsequent chapters.

1.10

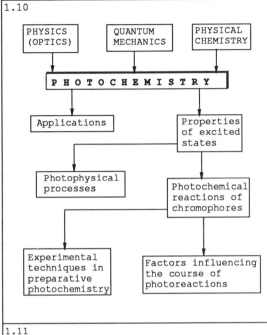

1.11

1.12

PHOTOCHEMICAL REACTIONS

Advantages	*Disadvantages*
-selective activation of particular reactant or part of a molecule	-only absorbed radiation can be used for chemical transformations
-specific reactivity of the electronically excited states	-quenching of the photochemical reaction by the product
-low activation energy of most photochemical reactions (low temperature photochemistry)	-unfavorable economy in the transformation of electrical energy into usable radiation energy

1.10. Organic photochemistry (Chapter 1), based on the laws of physics, physical chemistry, and quantum mechanics, is a description of the origin (Chapter 2) and deactivation of electronically excited states of organic molecules. Deactivation of these states occurs by either photophysical processes (Chapter 3) or photochemical reactions whose mechanisms are based on quantum mechanical models (Chapter 4) and that can be systemized according to the chromophores present (Chapters 5 through 13). Preparative photoreactions are normally performed in specialized equipment (Chapter 14) and their course can be influenced by various factors (Chapter 15). Theoretical and experimental photochemistry has been applied to various aspects of human activity (Chapter 16).

1.11. Two aspects of the study of photochemistry are: (1) theoretical, primarily of interest to theoretical and physical chemists and (2) the application of photochemical methods to the preparation of complex organic compounds. The focus of this book is the study of organic photoreactions with special emphasis on their use in the synthesis of complex molecules. Although its role in organic synthesis is not as universal as that of thermal reactions, organic photochemistry is a mature discipline. It offers a variety of useful transformations whose scope and limitations are well established and which are not possible by thermal methods. Photoreactions are valuable tools in the strategy of the synthesis of unusual molecules, natural products, etc.

1.12. In addition to the disadvantages shown, preparative photochemistry requires specialized equipment not always available in organic laboratories. In addition to the advantages mentioned, photoreactions provide considerable insight into the nature and mechanism of organic reactions and play a significant role in the design of a unified theory of chemical reactivity.

REFERENCES

(1) For terms recommended by IUPAC for use in photochemistry, see Braslavsky, S. E. and Houk, K. N. *Pure Applied Chemistry* **1988,** *61*, 1055.

(2) For a review of two-photon laser excitation, see Scaiano, J. C., Johnston, L. J., McGimpsey, W. G., and Weir, D. *Acc. Chem. Res.* **1988,** *21*, 22.

GENERAL READING IN ORGANIC PHOTOCHEMISTRY

Arnold, D. R. *Photochemistry*, Academic Press: New York, 1974.

Barltrop, J. A. and Coyle, J. D. *Principles of Photochemistry*. Wiley: New York, 1978.

Becker, H. G. O. *Einführung in die Photochemie*. Theime Verlag: Stuttgart, New York, 1983.

Buchardt, O., ed. *Photochemistry of Heterocyclic Compounds*. Wiley: New York, 1976.

Calvert, J. G. and Pitts, Jr., J. N. *Photochemistry*. Wiley: New York, 1966.

Cowan, D. O. and Drisko, R. L. *Elements of Organic Photochemistry*. Plenum: New York, 1976.

Coxon, J. M. and Halton, B. *Organic Photochemistry*. Cambridge University Press: Cambridge, 1974.

deMayo, P., ed. *Rearrangements in Ground and Excited States*. Academic Press: New York, 1980, Vols. 1–3.

Horspool, W. M. *Aspects of Organic Photochemistry*. Academic Press: London, 1976.

Klessinger, M. and Michl, J. *Lichtabsorption und Photochemie organischer Moleküle*. Verlag Chemie: Weinheim, 1989.

Lamola, A. A. *Creation and Detection of the Excited State*. Marcel Dekker: New York, 1971.

Michl, J. and Bonačić-Koutecký, V. *Electronic Aspects of Organic Photochemistry*. Wiley: New York, 1990.

Scaiano, J. C., ed. *CRC Handbook of Organic Photochemistry*. CRC Press: Boca Raton, Florida, 1989, Vols. 1 and 2.

Turro, N. J. *Modern Molecular Photochemistry*. Benjamin/Cummings: Menlo Park, California, 1978.

vonBünau, G. and Wolff, T. *Photochemie-Eine Einführung*. Verlag Chemie: Weinheim, 1985.

Wayne, R. P. *Principles and Applications of Photochemistry*. Oxford University Press: Oxford, 1988.

GENERAL READING IN ORGANIC PHOTOCHEMICAL SYNTHESIS

Coyle, J. E., ed. *Photochemistry in Organic Synthesis*. The Royal Society of Chemistry: Burlington House, London, 1986.

Horspool, W. M., ed. *Synthetic Organic Photochemistry*. Plenum: New York, 1984.

Margaretha, P. *Preparative Organic Photochemistry*. Academie-Verlag: Berlin, 1982.

Müller, E., ed. *Photochemistry*. In *Houben-Weyl: Methoden der organischen Chemie*. Theime: Stuttgart, 1975, Vols. IV/5a and 5b.

Ninomiya, I. and Naito, T. *Photochemical Synthesis*. Academic Press: San Diego, 1989.

Schönberg, A. *Preparative Organic Photochemistry*. Springer Verlag: New York, 1968.

Srinivasan, R., ed. *Organic Photochemical Synthesis (Org. Photochem. Synth.)*. Wiley–Interscience: New York, 1971 and 1976, Vols. 1 and 2.

PHOTOCHEMICAL JOURNALS

Journal of Photochemistry (J. Photochem.)

Molecular Photochemistry (Mol. Photochem.). **1969,** *1*.

Photochemistry and Photobiology (Photochem. Photobiol.). **1960,** *1*.

ORGANIC PHOTOCHEMISTRY SERIES

Advances in Photochemistry (Adv. Photochem.) Wiley: New York.

Organic Photochemistry (Org. Photochem.) Marcel Dekker: New York, Vol. 1–10.

Photochemistry: Specialist Periodical Reports. The Royal Society of Chemistry: Burlington House, London.

Chapter 2
ELECTRONICALLY EXCITED STATES

ELECTRONICALLY EXCITED STATES

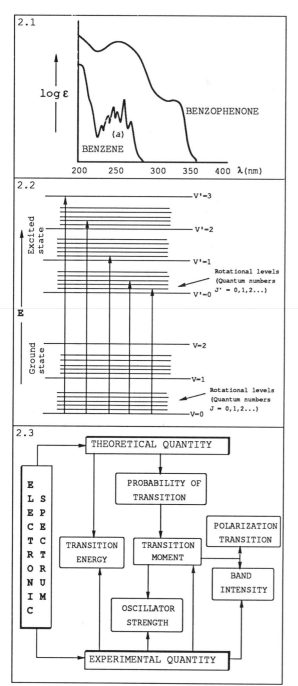

2.1. The Grotthus-Draper law requires an overlap of incident light with that absorbed by the chemical system to produce an excited state. *Molecular (electronic) spectroscopy* is a generally accepted method for recording excited states. In *molecular absorption spectroscopy*, electronic transitions are presented as a plot of the amount of monochromatic radiation absorbed [expressed as *absorbance* (*A*) or *molar absorption coefficient* (ϵ)] *vs.* a quantity related to the photon energy (wavelength or wavenumber). The UV-VIS spectrum displays several absorption bands corresponding to different excited states. The longest wavelength band with vibrational structure (*a*) is associated with the lowest excited state. The electronic transition responsible for a given spectral band is approximately localized in a part of the molecule (an atom or group of atoms) called a *chromophore* (see 2.48).

2.2. Pure electronic transitions should produce a line spectrum. However, since molecules are also vibrationally and rotationally excited, broad, structureless bands are usually observed, especially in solution. Vibrational structure (see (a), 2.1) is exhibited in the gas phase or in nonpolar solvents. The absorption band maxima (ϵ_{max} at λ_{max}) correspond approximately to energies of various excited states (E_{exc}).

2.3. The most important experimental spectral quantities are the *transition energy* and the *intensity* and *polarization* of the bands that are related to the energy and amount of radiation absorbed. The theoretical terms to describe a UV-VIS spectrum are as follows: (1) *transition moment* (*M*), the square of which determines (2) the *oscillator strength* (*f*), a measure of the band intensity, and (3) *transition polarization* (see 2.6).

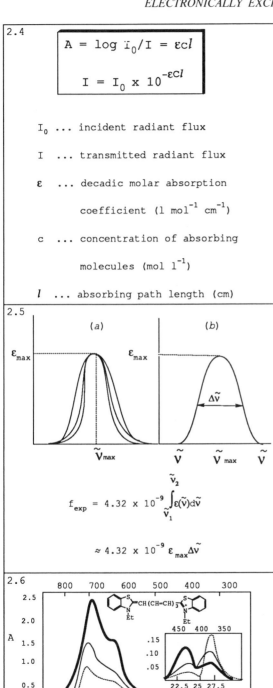

2.4. According to the *Beer-Lambert law* absorbance (*A*) of radiation by a homogeneous isotropic medium is proportional to the concentration (*c*) (pressure in the gas phase) of the absorbing species and to the absorption pathlength (*l*) (see 14.3). The proportionality constant (ϵ) is the *molar (decadic) absorption coefficient* at a given wavelength. ϵ_{max} is a measure of the intensity (magnitude) of the absorption band and an indicator of the "*allowedness*" of an electronic transition. The use of log ϵ provides the opportunity to plot widely different absorption intensities on the same graph.

2.5. The disadvantage of using ϵ_{max} as a measure of band intensity is illustrated by the three bands of different shape with the same ϵ_{max} (*a*). The intensity is more correctly expressed as the *integrated absorption intensity*, where ϵ is integrated within the wavenumber limits of the band. It is proportional to the *experimental oscillator strength* (f_{exp}), a dimensionless quantity. (For theoretical oscillator strength, see 2.27). With symmetrical bands (of Gaussian shape), the integral can be replaced by the difference in wavenumbers ($\tilde{\nu}$) at *half-width*, i.e., at a height equal to half the height at the maximum of the absorption band, $\frac{1}{2}\epsilon_{max}(b)$.

2.6. An important goal in the interpretation of an absorption spectrum is the assignment of individual bands to the corresponding electronic transitions. Valuable information is obtained from *linear polarized spectroscopy* in which spectra of oriented molecules, e.g., in a stretched polyethylene sheet, are recorded in polarized light. Among the internal axes of a molecule with different angles to the direction of stretching, the (*effective*) *orientation axis* ("long axis") is the one in which the electronic transitions polarized along this axis have the maximum ratio of "parallel" [$A_{\parallel}(\lambda)$, bold line] to perpendicular [$A_{\perp}(\lambda)$, dotted line] absorbances at a given wavelength (*dichroism*).[1]

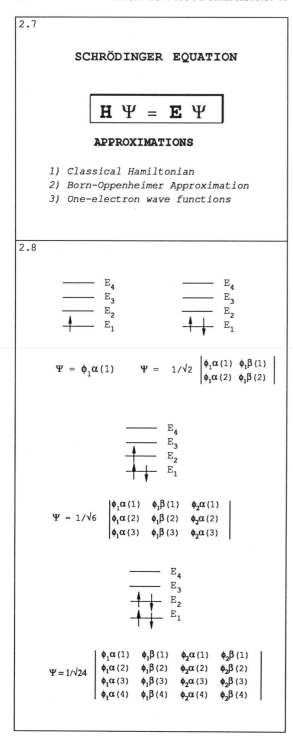

2.7

SCHRÖDINGER EQUATION

$$\mathbf{H}\,\Psi = \mathbf{E}\,\Psi$$

APPROXIMATIONS

1) *Classical Hamiltonian*
2) *Born-Oppenheimer Approximation*
3) *One-electron wave functions*

2.8

$$\Psi = \phi_1\alpha(1) \qquad \Psi = 1/\sqrt{2}\begin{vmatrix} \phi_1\alpha(1) & \phi_1\beta(1) \\ \phi_1\alpha(2) & \phi_1\beta(2) \end{vmatrix}$$

$$\Psi = 1/\sqrt{6}\begin{vmatrix} \phi_1\alpha(1) & \phi_1\beta(1) & \phi_2\alpha(1) \\ \phi_1\alpha(2) & \phi_1\beta(2) & \phi_2\alpha(2) \\ \phi_1\alpha(3) & \phi_1\beta(3) & \phi_2\alpha(3) \end{vmatrix}$$

$$\Psi = 1/\sqrt{24}\begin{vmatrix} \phi_1\alpha(1) & \phi_1\beta(1) & \phi_2\alpha(1) & \phi_2\beta(1) \\ \phi_1\alpha(2) & \phi_1\beta(2) & \phi_2\alpha(2) & \phi_2\beta(2) \\ \phi_1\alpha(3) & \phi_1\beta(3) & \phi_2\alpha(3) & \phi_2\beta(3) \\ \phi_1\alpha(4) & \phi_1\beta(4) & \phi_2\alpha(4) & \phi_2\beta(4) \end{vmatrix}$$

2.7. Quantum chemistry is a versatile framework for the interpretation of physical and chemical properties of atoms and molecules in both electronic ground and excited states. Consequently, a quantum chemical interpretation provides a useful description of a photochemical process. However, the results are dependent on the quantum chemical method used. These methods are characterized by the approximations imposed on the Hamiltonian H and the wave function Ψ. Most theoretical studies are based on: (1) the use of the classical nonrelativistic Hamiltonian; (2) the separation of translational, rotational, vibrational, and electronic motion (*Born-Oppenheimer approximation*) (see 1.6); and (3) the *limited basis set configuration interaction (CI)*.

2.8. Assuming that an electron moves both in the field of an atomic nucleus and the average field of other electrons (SCF method), the solution to this equation is a series of one electron functions (ϕ_i) or *molecular orbitals* (MO), each of which is associated with an energy E_i. Each MO, e.g., ϕ_1, $\alpha(1)$ and ϕ_1, $\beta(1)$, is usually approximated by a *linear combination of atomic orbitals* (LCAO) in the SCF approximation (LCAO-MO-SCF method). The *electronic wave function* of the entire molecule can be constructed from the products of the MOs with spin functions α or β in which two *spin orbitals* are formed from each MO, $\phi_1\alpha(1)$ and $\phi_1\beta(1)$. The overall wave function (*electronic configuration*) of an electronic state is expressed in the form of a *Slater determinant*, formed from spin orbitals, such that the exchange of two electrons (the exchange of two rows in the determinant) results in a change in the sign of the wave function (*Fermi–Dirac condition*). If two electrons have identical spin and orbital functions, i.e., if two columns of the determinant are identical, Ψ vanishes (*Pauli exclusion principle*).

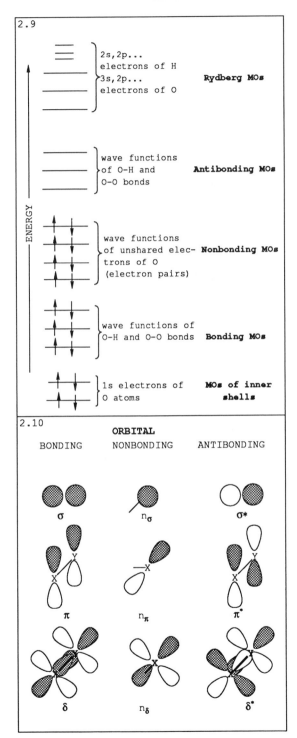

2.9

2s,2p...
electrons of H
3s,2p...   **Rydberg MOs**
electrons of O

wave functions
of O–H and   **Antibonding MOs**
O–O bonds

wave functions
of unshared elec-   **Nonbonding MOs**
trons of O
(electron pairs)

wave functions of
O–H and O–O bonds   **Bonding MOs**

1s electrons of   **MOs of inner**
O atoms   **shells**

2.10

ORBITAL

BONDING   NONBONDING   ANTIBONDING

σ n_σ σ^*

π n_π π^*

δ n_δ δ^*

2.9. This figure shows an orbital energy level diagram of the hydrogen peroxide molecule in its electronic ground state. The electrons in the inner shells make a negligible contribution to chemical bonding and their functions are, for all practical purposes, equal to atomic functions. MOs of valence electrons, i.e., $1s$ in the first row and $2s$ and $2p$ in the second row of the periodic table, are formally classified as *nonbonding* (atomic wave functions, e.g., *lone pairs*), *bonding*, and *antibonding*. Electrons in bonding orbitals stabilize a molecule; those in antibonding orbitals are destabilizing. The corresponding wave functions are quite different from the wave functions of the constituent atoms.

2.10. MOs constructed from wave functions of valence electrons (*valence-shell MOs*) are further classified according to the number and location of *nodal planes* (positions where the wave functions vanish). MOs localized on one of the atoms are nonbonding. Orbitals that have no nodal plane perpendicular to the bond axis are bonding orbitals (*in-phase* overlap); those with one nodal plane are antibonding (*out-of-phase* overlap). Orbitals having 0, 1, or 2 nodal planes passing through the bond axis are σ, π, or δ orbitals. π MOs exist only in a planar segment of a molecule, while δ orbitals occur on a linear segment. Antibonding orbitals are indicated by asterisks (σ^*, π^*, or δ^*). Both σ and σ^* MOs are cylindrically symmetrical about the internuclear (bond) axis; π and π^* MOs are antisymmetric about this axis. σ^* and π^* MOs possess a nodal plane perpendicular to the internuclear axis, and the π^* MO has an additional nodal plane passing through the bond axis. It follows that the nonbonding (n) MO has no antibonding counterpart. MOs constructed from higher atomic functions, e.g., the $2s$ orbital of hydrogen or the $3s$ and $3p$ orbitals of carbon, are known as *Rydberg orbitals* (see 2.21).

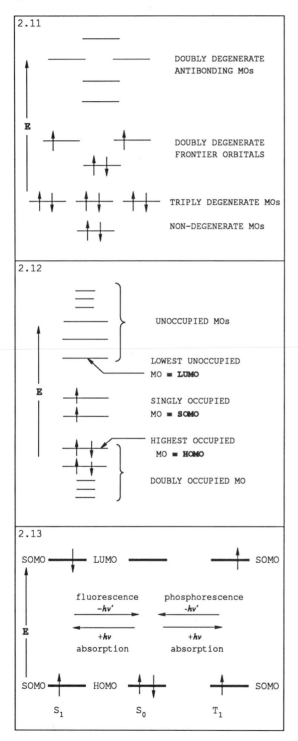

2.11. Two or more MOs having the same energy are doubly degenerate, triply degenerate, etc. (the term doubly is usually omitted). This degeneracy can have an accidental origin that often follows from approximations in the theoretical treatment or can result from a high degree of molecular symmetry. The appearance of degeneracy presents considerable difficulty in the interpretation of results at the MO level, especially for incompletely occupied frontier orbitals.

2.12. Molecular orbitals can be classified as occupied ("doubly"), singly occupied, or unoccupied. HOMO, SOMO, and LUMO are acronyms for *highest occupied molecular orbital, singly occupied molecular orbital,* and *lowest unoccupied molecular orbital.* HOMO and LUMO are frequently referred to as *frontier orbitals.* The wave functions of frontier and singly occupied orbitals play a key role in the theoretical understanding and analysis of the reactivity of excited states.

2.13. The pure spin electronic state approximated by the configuration in 2.9 is a *ground state,* i.e., the lowest possible energy state. A monoexcited configuration is formally produced by promoting an electron from an orbital to an unoccupied orbital in a ground state configuration (electronic transition). This can occur in two ways. A configuration with two SOMOs having identical spin functions in both SOMOs, i.e., either both α or both β, is a configuration with parallel spins (with total spin quantum number $S = 1$). The opposite configuration has antiparallel spins, one α and one β. An electronic configuration with antiparallel spins is, by convention, a *singlet state,* S_x, where x is 0.1, 2. . . ; S_0 is the singlet ground state; S_1, the *first excited singlet state,* etc. A configuration with parallel spins is a *triplet state,* T_x. Accordingly, the configuration in 2.9 corresponds to an S_0 state. For a more correct description of S_x and T_x states, see 2.15.

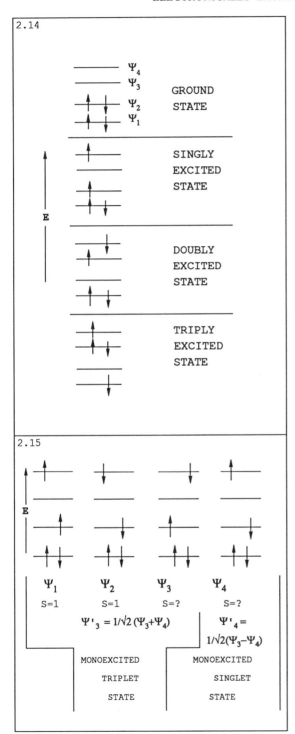

2.14. *Electronic configurations* are further classified according to the number of excited electrons relative to the ground state electronic configuration, e.g., singly excited, doubly excited, triply excited, etc. A *singly excited configuration* with parallel spins, as well as doubly and triply excited configurations with antiparallel spins, is shown. A configuration with no SOMOs is a *closed-shell configuration*, while one with at least one SOMO is an *open-shell configuration*. A configuration with an odd number of SOMOs represents a radical state. It should be noted that electronic spectra reflect singlet → singlet transitions, $S_0 → S_x$, and the individual spectral bands correspond to different singlet states. Singlet-triplet transitions, $S_0 → T_x$, are not generally observed in electronic spectra since these transitions are strongly forbidden by selection rules (see 2.28); however, this prohibition can be circumvented (see 3.13).

2.15. All possible assignments of spin functions to molecular orbitals that describe a singly excited configuration are shown. The functions Ψ_1 and Ψ_2 are the *eigenfunctions* of total spin (*pure spin functions*), while Ψ_3 and Ψ_4 are not. Pure spin functions (Ψ'_3 and Ψ'_4) can be produced from Ψ_3 and Ψ_4 by a linear combination. The new function Ψ'_4 has a total spin *eigenvalue* of 0. The function Ψ'_3 has a total spin eigenfunction of 1 and this value, as well as its energy, is identical to that of Ψ_1 and Ψ_2. The functions Ψ_1, Ψ_2, and Ψ'_3 are triply degenerate in their spins and together represent the triplet state. The nondegenerate function Ψ'_4 is a singlet state. Pure spin functions produced in this manner can be used as a simple description of excited states. In the absence of external fields, the functions Ψ_1, Ψ_2, and Ψ_3 are approximately degenerate (*zero field splitting*); in the presence of an external magnetic field the degeneracy disappears and the splitting is increased (the *Zeeman effect*).

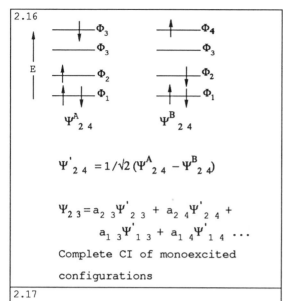

2.16. Pure spin wave functions (electronic states) can be produced by a single configuration, e.g., Ψ_1 and Ψ_2 in 2.15 or from a larger number of configurations, e.g., Ψ_3. Typically, when an electronic state is expressed as a combination of many configurations, it is known as *configuration interaction* (*CI*), a method which improves the accuracy of the description. The energy difference between a calculation with and without CI is the *correlation energy*. The calculation can be improved further by concurrently optimizing the MO with CI and by a method known as *multiconfigurational interaction* (*MC-SCF*).

2.17. The contributions of four singly excited configurations in *trans*-1,3-butadiene to the wave functions of the individual singly excited states are shown. In the *Pariser–Parr–Pople* (PPP) approximation the results are given in terms of pure spin functions (see 2.15) and the numbers represent the SOMOs. The states Ψ_A and Ψ_D are "pure" since they are predominantly represented by a single configuration. Such states are described qualitatively by a single MO configuration. On the other hand, the states Ψ_B and Ψ_C cannot be described accurately by a single configuration wave function.

2.17

	Ψ'_{23} (%)	Ψ'_{24} (%)	Ψ'_{13} (%)	Ψ'_{14} (%)
Ψ_A	86	0	0	14
Ψ_B	0	50	50	0
Ψ_C	0	50	50	0
Ψ_D	14	0	0	86

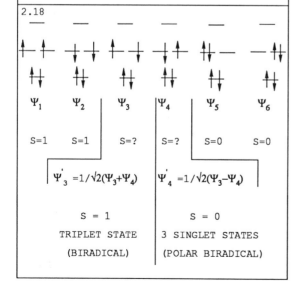

2.18. Systems with degenerate frontier orbitals are more complex. When spin is considered, two electrons can be placed in the degenerate orbitals in six different ways. After the transformation to pure spin wave functions, a triply degenerate triplet and three singlet states result. The latter are *biradical* or *zwitterion* and the triplet state is a *biradical*. The term zwitterion originates from the redistribution of charges leading to polar structures that frequently result in these states. For the importance of these configurations in quantum mechanical models of excited states and consequently of photochemical reactions, see 4.20.

2.19

SYSTEM	STATE SYMBOLS	EXAMPLES OF TRANSITIONS
ENUMERATIVE	S_0, S_1, S_2...S_x[a] T_1, T_2,...T_x	S_0->S_1; S_0->S_x S_1->T_1
KASHA	$^1(\pi,\pi^*)$; $^3(n,\pi^*)$[b] $^1(n,\sigma^*)$; $^1(\sigma,\pi^*)$	n->π^*; π->π^*; n->σ^*
GROUP THEORY	$^1A_{1g}$; $^1B_{2u}$; $^3E_{1u}$[c]	$^1A_{1g}$->$^1B_{2u}$

[a]Subscript zero represents the electronic ground state 1,2...; the first, second,...) singlet or triplet states, respectively.

[b]Superscript 1 or 3 represents the multiplicity of the state (singlet or triplet).

[c]For an explanation of the symbols, see 2.29 and 2.33.

2.20

LOWEST ELECTRONIC STATES OF THE >C=O
(without inner electrons)

ELECTRONIC STATE	EXCITED STATE	ELECTRON-SPIN CONFIGURATION
S_0	---	$(\pi_{CO}\uparrow\downarrow)^2 (n_0 \uparrow\downarrow)^2 (\pi^*_{CO})^0$
T_1	$^3(n,\pi^*)$	$(\pi_{CO}\uparrow\downarrow)^2 (n_0 \uparrow)^1 (\pi^*_{CO}\uparrow)^1$
S_1	$^1(n,\pi^*)$	$(\pi_{CO}\uparrow\downarrow)^2 (n_0)^1 \uparrow^1 (\pi^*_{CO}\downarrow)^1$
T_2	$^3(\pi,\pi^*)$	$(\pi_{CO}\uparrow)^1 (n_0 \uparrow\downarrow)^2 (\pi^*_{CO}\uparrow)^1$
S_2	$^1(\pi,\pi^*)$	$(\pi_{CO}\uparrow)^1 (n_0\uparrow\downarrow)^2 (\pi^*_{CO}\downarrow)^1$

2.21

2.19. Several different systems are used for the assignment of electronic states, depending on the type of information available. If the electronic configuration of the state is unknown, the *enumerative system*, in which the excited states are labeled according to their spin multiplicity and increasing energy, is employed. In the simple *MO representation* (*Kasha*), the pure spin excited states are labeled by the type of MO (see 2.10) involved in the dominant configuration. When symmetry considerations are important, a system based on group theory is required.

2.20. Group theory is used most often in theoretical and spectroscopic studies. However, in organic photochemistry the enumerative and Kasha systems are commonly used. The designation of an electronic transition, using the MO representation, is based on the type of ground state orbitals (σ, π, n, π^*, σ^*) involved in the process and an arrow directed from the symbol for the initially occupied MO to the symbol for the originally unoccupied MO. Indication of the electronic states of a simple carbonyl compound, shown for the ground state and two lowest singlet and triplet states, can be based on either the enumerative system, in which the order of increasing energy is easily understood, or Kasha's assignment, in which the spin multiplicity is indicated by a superscript to the left of the parenthesis.

2.21. In contrast to valence orbitals (see 2.10), Rydberg orbitals are diffuse, nonbonding, and thinly spread in space. Rydberg states are produced by excitation of electrons into Rydberg orbitals, i.e., orbitals with a large principal quantum number, e.g., a $\pi \rightarrow 3s$ transition generates the $^1(\pi, 3s)$ state. Thus, the chemical behavior of Rydberg states is that of radical cations (see 5.5). High energy transitions, e.g., $\sigma \rightarrow \sigma^*$, $\pi \rightarrow \sigma^*$, and n $\rightarrow \sigma^*$, are close to or mix with Rydberg transitions. Therefore, pure Rydberg states are rare.

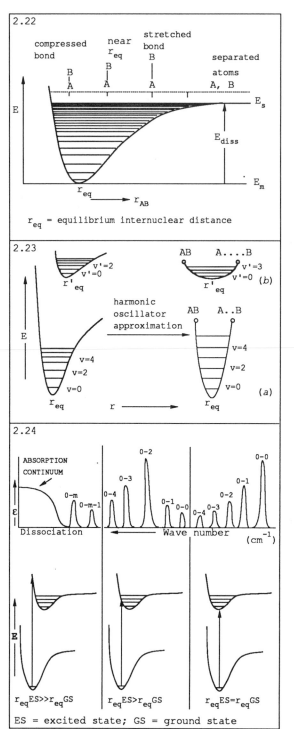

2.22

compressed bond · near r_{eq} · stretched bond · separated atoms A, B

E

E_s

E_{diss}

E_m

r_{eq}

r_{AB}

r_{eq} = equilibrium internuclear distance

2.23

$v'=2$ · $v'=0$ · r'_{eq}

AB A....B

$v'=3$ · $v'=0$ (b) · r'_{eq}

harmonic oscillator approximation

E

$v=4$ · $v=2$ · $v=0$ · r_{eq} · r

AB A..B

$v=4$ · $v=2$ · $v=0$ (a) · r_{eq}

2.24

ABSORPTION CONTINUUM

0-2 · 0-3 · 0-4 · 0-m · 0-m-1

0-0

0-1 · 0-0 · 0-4 · 0-3

0-0 · 0-1 · 0-2

ε

Dissociation   Wave number (cm^{-1})

E

$r_{eq}ES \gg r_{eq}GS$ $r_{eq}ES > r_{eq}GS$ $r_{eq}ES = r_{eq}GS$

ES = excited state; GS = ground state

2.22. The energy of electronic states as a function of nuclear coordinates is an *energy hypersurface*. The only variable in a diatomic molecule is the distance between the nuclei (r). The energy approaches infinity as the nuclei approach one another, pass through a minimum (E_m) at an equilibrium nuclear separation (r_{eq}), and approach a limit at E_s, as r approaches infinity. $E_s - E_m$ is the *dissociation energy* (E_{diss}).

2.23. A solution of the vibrational problem for an energy curve gives the energy levels of the vibrational states as horizontal lines. For several of the low energy vibration levels, the Morse curve can be replaced approximately by a parabola with equidistant vibrational levels (*harmonic oscillator approximation*). The parabola is steep for strong bonds and the distances between the individual vibrational levels is large; the opposite is true for weak bonds. A similar relationship exists between the electronic ground and excited states. States characterized by a *vibr*ational and an electro*nic* (but not a rotational) wave function are *vibronic states*. Thus, the energy and the overall wave function can be divided approximately into vibrational and electronic contributions.

2.24. For simplicity, transitions between any two electronic states are considered transitions between vibronic states. The relative intensities of bands in the spectrum are governed by the *Franck–Condon principle*, which states that the electronic transition of highest probability (intensity) is that from the ground vibronic state of the lowest electronic state (labeled 0) to the vibronic state of the higher electronic state lying vertically above it (*vertical transition*), labeled 0', 1', 2' . . . , etc. Transitions to other vibrational levels occur also, but with lower probability (intensity). Transitions from higher vibrational levels are *hot transitions*.

2.25

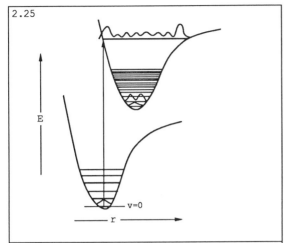

2.26

$$E_{exc} = E_{ES} - E_{GS}$$

E_{exc}...excitation energy
E_{ES}....energy of the excited state
E_{GS}....energy of the ground state

TRANSITION MOMENT ($\vec{M}$)

$$\vec{M} = \int \Psi^*_{v'}\Psi_v d\tau_n \int \Psi^*_{e'}\hat{\mu}_e\Psi_e d\tau_e \int \Psi^*_{s'}\Psi_s d\tau_s$$

Franck–		leading to
Condon	orbital	spin
factor	symmetry	rules

$\hat{\mu}$...operator of the dipole moment

Note: subscripts without (with) a prime
correspond to the ground (excited) state

2.27

OSCILLATOR STRENGTH

$$f = \frac{|\vec{M}|^2}{\Delta E} = \frac{(\int \Psi^*_{v'}\Psi_v d\tau_n)^2 \ (\int \Psi^*_{e'}\hat{\mu}_e\Psi_e d\tau_e)^2}{\Delta E}$$

SELECTION RULES OF ELECTRONIC TRANSITIONS

f is	large	small
absorption band in electronic spectrum is	strong	weak
halflife of emission of radiation is	short	long

2.25. The quantum mechanical interpretation of the Franck–Condon principle is that the intensity of a vibronic transition is proportional to the square of the overlap integral between the vibrational wave functions of the two states involved in the transition (Franck–Condon factor, see 2.26). Accordingly, the most intense vibronic absorption band corresponds to the transition into that vibronic state whose vibronic wave function has the most overlap with the vibronic wave function of the lower electronic state. With increasing vibrational quantum number (v), the wave functions have their largest amplitudes near the limits of the range of displacements (turning points). The vibronic transition to various other v' levels retains some probability, as shown by the line spectrum diagram.

2.26. For absorption of radiation, two conditions must be fulfilled. First, the frequency of radiation must satisfy the first equation (see also 1.4). Second, the *transition moment* ($\vec{M}$) must have a nonzero value. The higher the value of $\vec{M}$, the more efficient the absorption. It is composed of three integrals and is the basis of the *selection rules*, which determine whether a given transition is allowed or forbidden on the basis of symmetry or the spin of the wave functions of the initial and final states. If any of these integrals is zero, the corresponding transition is forbidden.

2.27. Another important characteristic is the intensity (probability) of the electronic transition. This probability is directly related to the theoretical *oscillator strength* (f), which is proportional to the square of the transition moment ($\vec{M}$) and is a measure of the intensity of a spectral band. If f is large (high probability of photon absorption), the resulting absorption band will have high intensity and, at the same time, the emission of the photon by an excited state will also be high, thus decreasing the lifetime of the excited state.

2.28   SUMMARY
      OF SELECTION RULES FOR POLYATOMIC MOLECULES

$$f = p_s \times p_0 \times p_m \times p_p f_A$$

Probability factor	Transition forbidden by	p_i-value
p_s	spin	$\sim 10^{-8};\,^a\ \sim 10^{-5\ b}$
p_0	orbital symmetry	$\sim 10^{-5\ c}$
p_m	momentum	$\sim 10^{-3} - 10^{-1\ d}$
p_p	parity	$\sim 10^{-1\ e}$

[a] For aromatic hydrocarbons. [b] For second row elements. [c] For n->π^* transitions. [d] For condensed ring systems. [e] For most transitions.

2.29

Non-degenerate orbital (state)	a,b (A,B)
Doubly degenerate	e (E)
Triply degenerate	T or F

Symmetry element	Sym.	Antisym.	Note
2π/n rotation about the main axis	A	B	
reflection in σ_h	'	"	superscript
reflection in σ_v	1	2	1st subscript
inversion	g	u	2nd subscript

2.30

2.28. *Selection rules* for electronic transitions in polyatomic molecules govern whether a transition is allowed or forbidden on the basis of symmetry or spin of the wave functions of the initial and final states. These rules express the probability of the transition rather than its strict allowedness or forbiddenness. As mentioned in 2.26, a transition occurs only if the transition moment $\vec{M} \neq 0$. The probability (intensity) of the transition is expressed by the oscillator strength (f) (see 2.5 and 2.27). The value of f for a transition can be viewed as a series of corrections (probability factors, p_i) to the oscillator strength (f_A) of a fully allowed $\pi \rightarrow \pi^*$ transition.[2]

2.29. Application of the *symmetry selection rules*, particularly those of *orbital symmetry* and *parity*, require a knowledge of symmetry operations and group theory. Although this approach will occasionally be used to describe and specify some electronic states, its background is beyond the scope of this book and only one example will be given. Each electronic state is associated with an *irreducible representation*, IRREP (symmetry type) on the basis of the symmetric or antisymmetric properties of the wave function with respect to the individual symmetry elements.

2.30. The principal six-fold axis (point group D_{6h}) in benzene is perpendicular to the plane of the molecule. Nondegenerate functions (orbitals) are labeled a or b, depending upon whether or not they change sign upon rotation by $(360/6)° = 60°$. Degenerate functions are labeled e. Subscripts 2 or 1 indicate that the sign of the function does not change as a result of mirroring at any σ_v plane passing through atoms. The indices g and u generally indicate the inversion of the wave function with respect to the center of symmetry. With no change, the function is symmetric (*gerade*—even); if it changes, the function is antisymmetric (*ungerade*—odd).

2.31

STATES

A x A = A	A x B = B
B x B = A	B x A = B

SUPERSCRIPTS

$X' \times X' = X'$	$X' \times X'' = X''$
$X'' \times X'' = X'$	$X'' \times X' = X'$

SUBSCRIPTS

$X_1 \times X_1 = X_1$	$X_1 \times X_2 = X_2$
$X_2 \times X_2 = X_1$	$X_2 \times X_1 = X_2$
$X_g \times X_g = X_g$	$X_g \times X_u = X_u$
$X_u \times X_u = X_g$	$X_u \times X_g = X_u$

EXAMPLES

POINT GROUP D_{6h} *POINT GROUP* $D_{6\infty}$

$A_{2u} \times A_{2u} = A_{1g}$ $\pi_g \times \pi_g = \Sigma^+_g + \Sigma^-_g + \Delta_g$

$A_{1g} \times B_{1u} = B_{1u}$

$B_{1u} \times B_{2u} = A_{2g}$ *POINT GROUP* D_{6h}

$E_{1g} \times E_{2g} = B_{1u} + B_{2u} + E_{1u}$

2.32

TRANSITION

$X_g \rightarrow X_g$ ... forbidden

$X_u \rightarrow X_u$ ... forbidden

$X_g \rightarrow X_u$ ... allowed

$X_u \rightarrow X_g$ ... allowed

2.33

2.31. The wave functions of electronic states are expressed in the form of *Slater determinants*, i.e., the sum of the products of the MOs occupied by electrons. The symmetry of state functions is, therefore, given by the product (called *direct product*) of the symmetries of the MOs. There are simple rules for the direct products of IRREPs of *nondegenerate* wave functions. If the representations are formed with *degenerate* levels (*E*, *T*..), the resulting *reducible representation* is expressed as a sum (called *direct sum*) of IRREPs. The example shown for point group D_{6h} will be used to describe the electronic states of benzene (see 2.33).

2.32. The IRREP, identified by the subscript *g*, is a symmetric function with respect to both *x* and *y* axes (benzene is planar). When the symmetry, with respect to any of these axes, changes, the function becomes antisymmetric, *u*, and its product with another function *g(u)* will be antisymmetric (symmetric). If both axes are considered, the *g* ↔ *g* (or *u* ↔ *u*) transition is forbidden (*parity selection rule*). However, the forbiddenness can be removed by vibrational deformation of the molecule.

2.33. The number of nodal planes in orbitals always increases with increasing energy. Accordingly, there are two degenerate HOMOs (Ψ_2, Ψ_3) and LUMOs (Ψ_4, Ψ_5) in the benzene molecule. They belong to the representations e_{1g} and e_{1u}, respectively (Ψ_1 to a_{2u}, Ψ_6 to b_{2g}; see also 2.30). The direct product of the reducible representations involved in the lowest electronic transitions give rise to three new functions: $E_{1g} \times E_{1u} = B_{1u} + B_{2u} + E_{1u}$. The ground state of benzene evidently has $^1A_{1g}$ symmetry. The transition $^1A_{1g} \rightarrow {}^1E_{1u}$ at 179 nm is allowed; the other two (at 207 and 264 nm, respectively) are forbidden. Because of vibrational deformations they are observed. The spin forbidden $^1A_{1g} \rightarrow {}^3B_{1u}$ transition has a maximum at ~310 nm.

2.34

SPIN SELECTION RULES

*The electronic transition is allowed by spin selection rules **only** if the multiplicities of the initial and final states are identical.*

TYPE OF TRANSITION	INTEGRAL VALUE	ALLOWEDNESS
$S \rightarrow S$	$\int \alpha\alpha \, d\tau_s = \int \beta\beta d\tau_s = 1$	FULLY ALLOWED
$T \rightarrow T$	$\int \alpha\alpha d\tau_s = 1$	FULLY ALLOWED
$S \rightarrow T$	$\int \alpha\beta \, d\tau_s = \int \beta\alpha d\tau_s = 0$	STRONGLY FORBIDDEN

2.35

TYPE OF TRANSITION	APPROXIMATE RANGE
SPIN AND SYMMETRY ALLOWED	$10^3 - 10^5$
SPIN ALLOWED/SYMMETRY FORBIDDEN	$10^0 - 10^3$
SPIN FORBIDDEN	$10^{-5} - 10^0$

2.36

ELECTRONIC TRANSITIONS

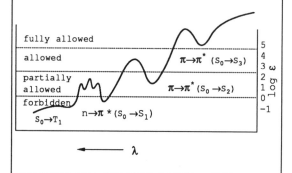

fully allowed

allowed — $\pi \rightarrow \pi^*$ $(S_0 \rightarrow S_3)$

partially allowed — $\pi \rightarrow \pi^*$ $(S_0 \rightarrow S_2)$

forbidden

$S_0 \rightarrow T_1$ — $n \rightarrow \pi^*$ $(S_0 \rightarrow S_1)$

$\log \varepsilon$: 5 4 3 2 1 0 -1

$\longleftarrow \lambda$

2.34. Previous discussions have centered only on the effects of the symmetry part of the electronic wave functions. Because spin functions are not affected by the dipole moment operator, it is possible to integrate them separately from molecular orbitals in the calculation of transition moment. Similarly, the result of the value of the integral for vibrational functions depends entirely on their overlap. Because the spin functions α and β are mutually orthogonal, this integral vanishes in the same way as a transition between states that differ in multiplicity, causing the transition moment to vanish. This conclusion is invalidated by spin-orbit coupling (see 3.10).

2.35. The allowedness or forbiddenness of an electronic transition is reflected in the intensity of the absorption band. The approximate values given in the table express the allowedness of the transitions permitted by one or more selection rules. However, they are valid only when other selection rules are not considered, e.g., the value of the spin forbidden transition does not include symmetry and vibrational terms.

2.36. Energies of electronic transitions in solution are specifically affected by solvent. By interaction with lone electron pairs, polar and protic solvents form either van der Waals complexes or, more commonly, hydrogen bonds of varying stability. This complex formation, along with the effect of dipole–dipole or dipole-induced dipole interactions, causes a stabilization of a polar molecule in both ground and excited states and can change the order of the excited states. As a consequence, the energy of the n $\rightarrow$ π^* transition is usually raised and the energy of the $\pi \rightarrow \pi^*$ transition is lowered. The result of the latter is a *bathochromic* (red) *shift* of the longest wavelength spectral transition. With the former, a *hypsochromic* (blue) *shift* occurs. The solvent effect is used to assign the type of electronic transition (see 2.46).[3]

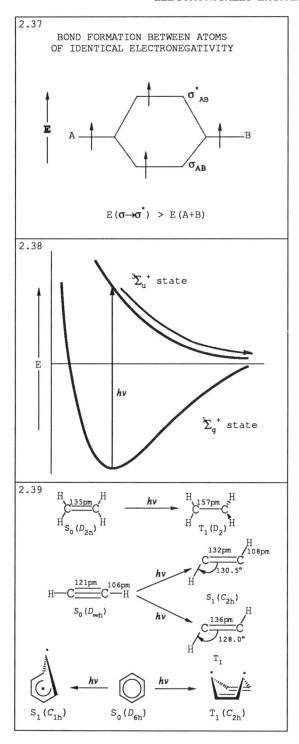

2.37

BOND FORMATION BETWEEN ATOMS
OF IDENTICAL ELECTRONEGATIVITY

$E(\sigma \rightarrow \sigma^*) > E(A+B)$

2.38

$^3\Sigma_u^+$ state

$h\nu$

$^1\Sigma_g^+$ state

2.39

2.37. Because of the short lifetimes and low attainable concentration of excited states, studies of their structure and properties are less common than those of the ground state molecules. Most studies utilize indirect methods. Some fundamental information can be obtained by using simple quantum chemical models. When an MO is formed from atomic orbitals of equal electronegativity, the HOMO is less bonding than the LUMO is antibonding. As a result, the energy of an excited isolated bond is higher than the energy of isolated atoms and the excited triplet state is an *unbound state* (see Fig. 2.38 for the $^3\Sigma_u^+$ state).

2.38. For this reason, a *bound state* originates, especially in excitations into π^* orbitals, because the energy difference *in both singlet and triplet state* is fully compensated by the energy of the remaining σ bond. This is the basis of the significance of π chromophores (see 2.48) in photochemistry. Dissociation is a consequence of the formation of an unbound state and is the fate of all states obtained by the triplet σ, σ^* excitation of single bonds.

2.39. Electronic transitions lead to changes in electron distribution and, consequently, to changes in the equilibrium internuclear distances (geometry), particularly of small molecules,[4] of the dipole moment, and the acid-base equilibrium. The antibonding character of the excited π bonds causes distortions of the molecular geometry, such as bond stretching, rotation to an orthogonal geometry (ethylene),[5] $sp \rightarrow sp^2$ rehybridization (acetylene),[6] out-of-plane bending of originally planar molecules (benzene),[7] etc. A knowledge of the geometries of excited molecules before they return to the ground state is essential for an understanding of the structures of photoproducts. Some rules are utilized for qualitative predictions.[8]

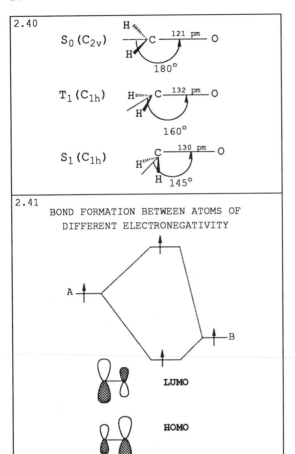

2.40

$S_0 (C_{2v})$

$T_1 (C_{1h})$

$S_1 (C_{1h})$

2.41

BOND FORMATION BETWEEN ATOMS OF
DIFFERENT ELECTRONEGATIVITY

A

B

LUMO

HOMO

2.42

MOLECULE	$D(S_0)$	$D(T_1)$	$D(S_1)$	Ref.
CH_2O	2.3	1.3	1.6	12
PhCH=O	2.8	–	–	13
Ph_2C=O	3.0	1.7	1.2	14
4-nitroaniline	6.0	–	14.0	15
4-aminobenzonitrile	–	–	12	16
4-amino-4-nitro BP[a]	6.0	–	18-23	15
pyridine	2.2	–	-1.0	17
pyrene excimer	–	–	2.5	18
exciplex N-DMC[b]	–	–	10.8	19

[a]BP=biphenyl; [b]N-DMC=naphthalene-dimethylcyclo-
hexene

2.40. For n → π* transitions and, in general, for electron transitions in which the charge is moved from one part of space to another, the conditions for the mutual isolation of the unpaired electrons are fulfilled automatically. Therefore, the molecular distortions, particularly changes in bond lengths, tend to be substantially smaller, as shown for formaldehyde.[9] The chemical behavior of the excited states corresponds to that of two separate radicals or zwitterions (see 4.17–4.20).

2.41. Polar bonds between atoms of different electronegativities are characterized by a HOMO with a shape reminiscent of the orbital of the more electronegative atom and a LUMO whose shape resembles the less electronegative atomic orbital. Therefore, excitation is accompanied by considerable changes in the charge distribution, direction of the dipole moment, and acidity. As a result, different positions in the benzene ring undergo attack by electrophilic and nucleophilic reagents (see 7.35–7.41).

2.42. Dipole moments of excited states can be determined by two methods. A direct method is the measurement of the linear polarization of the fluorescence of molecules in a strong electric field.[10] The degree of molecular orientation in such a field is a function of the dipole moment in the excited singlet state. Another method is based on the fact that absorption and emission spectral bands are shifted by the effect of solvent polarity resulting from a change in dipole moment upon excitation (see 2.36).[11] The values shown are only for illustration, since they are averages from a series of measurements. The values for the excimer (stabilized both by energy transfer and charge transfer—CT) and the exciplex (stabilized by CT only) are especially significant (see 3.26–3.30). The existence of polar structures in exciplexes is common.

2.43

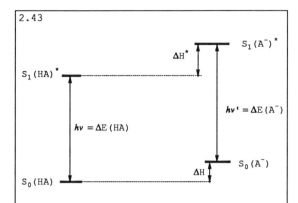

$$\Delta H^* = \Delta H + \Delta E(A^-) - \Delta E(HA)$$

$$\Delta H^* = 2.303RT \, \Delta pK$$

$$\Delta pK = 0.00209 \Delta \tilde{v}/cm^{-1} \quad (for \ T=298K)$$

2.44

MOLECULE	$pK_a(S_0)$	$pK_a(S_1)$
p-cresol	10.3	4.6
4-bromophenol	9.4	3.0
1-naphthol	9.2	2.0
2-naphthol (2-N)	9.5	3.2
5-cyano-2-N	8.75[a]	1.7[b]
6-cyano-2-N	8.40[a]	0.5[b]
7-cyano-2-N	8.75[a]	2.0[b]
8-cyano-2-N	8.35[a]	0.7[b]
naphthoic acid	3.7	10-12

[a]From absorption titration

[b]From fluorescence titration

2.45

REACTION TYPE	pK_a S_0	S_1	T_1
$-OH \rightleftharpoons -O^- + H^+$	9-10	2.5-3.4	7.8-8.6
$-NH_3^+ \rightleftharpoons -NH_2 + H^+$	4	(-1.5)-(-3)	3-3.5
$-COOH \rightleftharpoons -COO^- + H^+$	3.4-4.5	10-12	3.8-4.6
protonated naphthalene $\rightleftharpoons$ naphthalene + H^+	-4	12	–

2.43. The lifetimes of excited states are sufficiently long to establish an acid-base equilibrium.[20] pK_a values in excited states (pK_a^*) are determined by the Förster–Weller method, using a simple thermochemical cycle.[21] pK_a^* is derived from the known pK_a in the ground state and the excitation energy in the first excited state (the position of λ_{max} of the longest wavelength band) of the acid and conjugate base. A disadvantage of this method is the assumption that the entropy changes in the ground and the excited states are identical. In another method, the pK_a^* of, e.g., aromatic compounds, is measured directly by titration, i.e., the determination of the pH required to halve the luminescent intensity of the acid or double the luminescent intensity of the conjugate base. This method assumes direct proportionality between luminescent intensity and the concentration of the emitting molecules.

2.44. The excitation of organic molecules into the first singlet state (S_1) affects their acidity, depending on the functional groups present. Phenols and ammonium salts are much more acidic in the S_1 state than in the ground state [$\Delta pK_a(S_1)$ 5–6],[22] whereas aromatic acids and conjugate acids of some nitrogen-containing heterocycles exhibit a decrease in acidity by an $S_0 \rightarrow S_1$ transition.

2.45. The methods described in 2.43 can be employed for the determination of pK_a^* of organic compounds in the first excited triplet state [$pK_a(T_1)$], i.e., from the triplet-triplet absorption spectrum or from the location of the O–O' bands in the phosphorescence spectrum. While $pK_a(S_1)$ differs significantly from $pK_a(S_0)$, $pK_a(T_1)$ lies somewhere between the two, closer to $pK_a(S_0)$. The uncertainties in the pK_a values in acid-base reactions are primarily a result of the assumptions used in the different methods.

2.46

PROPERTIES OF $^1(n,\pi*)$ and $^1(\pi,\pi*)$ STATES

	$^1(n,\pi*)$	$^1(\pi,\pi*)$
INTENSITY OF ABSORPTION	weak	strong
SOLVENT EFFECT (polar solvent)	hypsochromic shift	bathochromic shift
POLARIZATION OF TRANSITION	perpendicular to molecular plane	parallel to molecular plane
LIFETIME	longer	shorter

2.47

PROPERTIES OF $^3(n,\pi*)$ and $^3(\pi,\pi*)$ STATE

	$^3(n,\pi*)$	$^3(\pi,\pi*)$
LIFETIME	short	long
VIBRATIONAL STRUCTURE	prominent	variable
S-T SPLITTING	low	high
INTENSITY OF S-T TRANSITION	high	low
HEAVY ATOM EFFECT	little or none	high
EPR SIGNAL	weak	present in frozen solution or matrix

2.48

CHROMOPHORE	E(S1) kJ/mol (kcal/mol)	LOG ε	TYPE	E(T1) kJ/mol (kcal/mol)	TYPE
C–C	>660 (>158)	3	$\sigma{\to}\sigma^*$	–	–
C–H	>660 (>158)	3	$\sigma{\to}\sigma^*$	–	–
C=C	660 (158)	4	$\pi{\to}\pi^*$	344 (82)	$\pi{\to}\pi^*$
C=C–C=C	544 (130)	4.3	$\pi{\to}\pi^*$	250 (60)	$\pi{\to}\pi^*$
benzene	460 (110)	2.3	$\pi{\to}\pi^*$	354 (85)	$\pi{\to}\pi^*$
naphthalene	386 (92)	2.3	$\pi{\to}\pi^*$	255 (61)	$\pi{\to}\pi^*$
anthracene	315 (75)	4	$\pi{\to}\pi^*$	180 (43)	$\pi{\to}\pi^*$
C=O	430 (103)	1.3	$n{\to}\pi^*$	302 (72)	$n{\to}\pi^*$
N=N	340 (81)	2	$n{\to}\pi^*$	–	–
N=O	180 (43)	2.3	$n{\to}\pi^*$	–	–
C=C–C=O	340 (81)	1.5	$n{\to}\pi^*$	291 (70)	$n{\to}\pi^*$
O=C–C=O	540 (129)	4.3	$\pi{\to}\pi^*$	230 (55)	$n{\to}\pi^*$

2.46. Since photochemical reactions are generally governed by both the lowest singlet and triplet states (see 1.7 and 15.4) their fundamental properties are summarized in this and the following table. $\pi \to \pi*$ absorption is strong only when the transition is symmetry allowed. On the other hand, $n \to \pi*$ transitions are characterized by the absence of absorption bands in the corresponding hydrocarbon. Also, the $n \to \pi*$ absorption band frequently disappears upon protonation. Polarization of an $n \to \pi*$ transition is a consequence of the fact that an electron in the lone pair orbital in the plane of the molecule is excited into an empty bonding $\pi*$ orbital, which is antisymmetric with respect to the plane of the molecule. The $\pi \to \pi*$ excitation does not change the antisymmetry relative to the plane of the molecule and, therefore, the $\pi \to \pi*$ transition is polarized in this plane.

2.47. Triplet lifetimes of n,$\pi*$ states are of the order of tenths or hundredths of milliseconds, while the lifetimes of triplet $\pi,\pi*$ states are frequently as long as several seconds. Singlet-triplet splittings for an n,$\pi*$ state are typically of the order of 60 kJ/mol (15 kcal/mol). This splitting for $\pi,\pi*$ states can be several hundred kJ/mol. Oscillator strength for the transition from the ground state to the n,$\pi*$ triplet state is larger (up to five orders of magnitude) than the oscillator strength for the transition to a triplet $\pi,\pi*$ state. EPR signals are measurable only in rigid systems, e.g., crystals and frozen solutions.

2.48. A summary of the basic spectroscopic and energetic properties of a selected series of chromophores is provided for later discussion of photochemical properties of chromophores. Excitation energies are given in kJ/mol and kcal/mol. The properties of excited states are one of the least understood areas of photochemistry and additional theoretical studies are needed.

REFERENCES

(1) Adapted with permission from Eckert, R. and Kuhn, H. *Z. Electrochem.* **1960**, *64*, 356.

(2) Platt, J. R. *J. Opt. Soc. Am.* **1953**, *43*, 252.

(3) From Turro, N. J. *Modern Molecular Photochemistry*, Benjamin/Cummings: Menlo Park, CA, 1978, p. 104.

(4) Innes, K. K. In *Excited States*, Lim, E. C., ed. Academic Press: New York, 1975, Vol. 2.

(5) Merer, A. J. and Mulliken, R. S. *Chem. Rev.* **1969**, *69*, 642.

(6) Winkelhofer, G., Janoschek, R., Fratev, F., and v.R. Schleyer, P. *Croat. Chem. Acta.* **1983**, *156*, 509.

(7) Arnold, D. R. *Photochemistry.* Academic Press: New York, 1974, p. 91.

(8) Imamura, A. and Hoffmann, R. *J. Am. Chem. Soc.* **1968**, *90*, 5379.

(9)(a) Kirchhoff, W. H., Lovas, F. J., and Johnson, D. R. *J. Phys. Chem. Ref. Data.* **1972**, *1*, 1011. (b) Brand, J. C. D. and Williamson, D. G. *Adv. Phys. Org. Chem.* **1963**, *1*, 403.

(10)(a) Czekalla, J. *Z. Electrochem.* **1960**, *64*, 1221. (b) For the excited state, dipole moments of some aromatic alkenes and alkynes, see Sinha, H. K., Thomson, P. C. P., and Yates, K. *Can. J. Chem.* **1990**, *68*, 1507.

(11) Lippert, E. *Z. Electrochem.* **1957**, *61*, 962.

(12) Buckingham, A. D., Ransay, D. A., and Tyrell, J. *Can. J. Phys.* **1970**, *48*, 1242.

(13) Labhart, H. *Experientia.* **1966**, *22*, 65.

(14) Hochstrasser, R. M. and Noe, L. J. *J. Mol. Spect.* **1971**, *38*, 175.

(15) Extracted from Barltrop, J. A. and Coyle, J. D. *Excited States in Photochemistry.* Wiley: New York, 1975, p. 52.

(16) Bischof, H., Baumann, W., Detzer, N., and Rotkiewicz, K. *Chem. Phys. Lett.* **1985**, *116*, 180.

(17) Hochstrasser, R. M. and Michaluk, J. W. *J. Chem. Phys.* **1971**, 55, 4668.

(18) Ghosh, N. R. and Basu, S. *J. Photochem.* **1974**, *3*, 247.

(19) Taylor, G. N. *Chem. Phys. Lett.* **1971**, *10*, 355.

(20) For reviews, see (a) Ireland, J. F. and Wyatt, R. A. H. *Adv. Phys. Org. Chem.* **1976**, *12*, 131. (b) Van der Donckt, E. *Prog. Reaction Kinetics.* **1970**, *5*, 273. (c) Becker, R. S. *Theory and Interpretation of Fluorescence and Phosphorescence*: Wiley–Interscience: London, 1969, p. 239.

(21)(a) Förster, Th. *Ber. Bunsenges. Phys. Chem.* **1950**, *54*, 531. (b) Weller, A. *Z. Phys. Chem.* **1958**, *15*, 438.

(22)(a) The enhanced photoacidities of phenols are the basis of several technological (e.g., photopolymerization, photodepolymerization) and mechanistic (e.g., pH jump experiments, biological probes) applications; see Tolbert L. M. and Haubrich, J. E. *J. Am. Chem. Soc.* **1990**, *112*, 8163, and references therein. Enhanced photoacidities of cyano-2-naphthols are discussed. (b) For the excited-state carbon acidity of 5*H*-dibenzo[*a*,*c*]cycloheptene derivatives, see Wan, P., Budac, D., Earle, M., and Shukla, D. *J. Am. Chem. Soc.* **1990**, *112*, 8048.

GENERAL READING IN ORGANIC SPECTROSCOPY

Fabian, J. and Hartmann, H. *Light Absorption of Organic Colorants.* Springer Verlag: New York, 1980.

Griffiths, J. *Colour and Constitution of Organic Molecules.* Academic Press: London, 1976.

Jaffé, H. H. and Orchin, M. *Theory and Applications of Ultraviolet Spectroscopy.* Wiley: New York, 1962.

Lamola, A. A. *Creation and Detection of Excited States.* Marcel Dekker: New York, 1971.

Murov, S. L. *Handbook of Photochemistry.* Marcel Dekker: New York, 1973.

Murrel, J. N. *The Theory of the Electronic Spectra of Organic Molecules.* Methuen: London, 1963.

Reichhardt, C. *Solvent Effects in Organic Chemistry.* 2nd ed. Verlag Chemie: Weinheim, 1988.

Sandorfy, C. *Electronic Spectra and Quantum Chemistry.* Prentice-Hall: Englewood Cliffs, N.J., 1964.

Scaiano, J. C., ed. *Handbook of Organic Photochemistry.* CRC: Washington, 1989; Vols. 1 and 2.

Simons, J. P. *Photochemistry and Spectroscopy.* Wiley: London, 1971.

Chapter 3
PHOTOPHYSICAL PROCESSES

PHOTOPHYSICAL PROCESSES

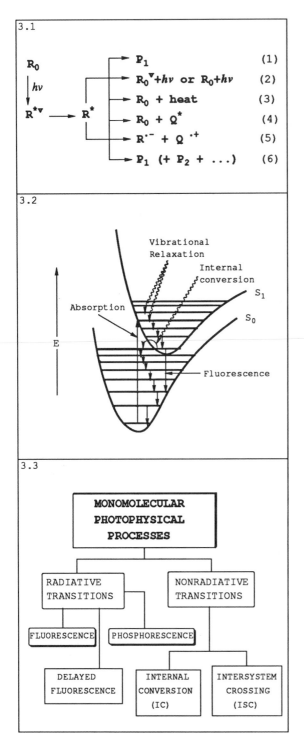

3.1 R_0

$\downarrow h\nu$

$R^{*v} \longrightarrow R^*$

$$\longrightarrow P_1 \qquad (1)$$
$$\longrightarrow R_0^v + h\nu \text{ or } R_0 + h\nu \qquad (2)$$
$$\longrightarrow R_0 + \text{heat} \qquad (3)$$
$$\longrightarrow R_0 + Q^* \qquad (4)$$
$$\longrightarrow R^{\cdot -} + Q^{\cdot +} \qquad (5)$$
$$\longrightarrow P_1 \; (+ \; P_2 + \dots) \qquad (6)$$

3.2

E

Vibrational
Relaxation

Internal
conversion

S_1

Absorption

S_0

Fluorescence

3.3

```
         MONOMOLECULAR
         PHOTOPHYSICAL
           PROCESSES
```

RADIATIVE TRANSITIONS — NONRADIATIVE TRANSITIONS

FLUORESCENCE — PHOSPHORESCENCE

DELAYED FLUORESCENCE — INTERNAL CONVERSION (IC) — INTERSYSTEM CROSSING (ISC)

3.1. Deactivation of an excited state occurs by a *photochemical* or *photophysical process*, either monomolecularly (*intra*molecularly) or biomolecularly (*inter*molecularly). In the monomolecular pathway, the vibronically excited molecule R^{*v} relaxes to R^* in the vibrational ground state level (*vibrational relaxation* or *cascade*). The lifetime of R^* is extremely short (10^{-12} s) and if the competing photoreaction producing P_1 [Eq. (1)] does not occur, R^* returns rapidly to the vibrationally relaxed ground state, R_0^v. This takes place due to the spontaneous emission of photon(s) (*radiative transition*) [Eq. (2)] or the nonradiative decay of electronic energy by its transformation into vibrational energy (*radiationless transition*) [Eq. (3)] with a resulting increase in the temperature of the system. In the bimolecular pathway, R^* can interact with and transmit its excitation energy to another molecule Q to form Q^* (*quencher* or *sensitizer*) in a process called *quenching* or *sensitization* [Eq. (4)], can lose or acquire an electron (*photoinduced single-electron-transfer, SET process*) [Eq. (5)], or react to yield the product(s) shown in Eq. (6).

3.2. Vibrational relaxation occurs from any vibronic state and represents a return to the thermal equilibrium of the vibrational ground state within the same electronic state, e.g., $S^v \rightsquigarrow S$ or $T^v \rightsquigarrow T$. This occurs in condensed systems by energy transfer to the environment caused by collision with solvent molecules.

3.3. A monomolecular photophysical decay process occurs in one of two ways: (1) within electronic states of equal multiplicity, e.g., radiative—$S_1 \rightarrow S_0$, *fluorescence* (*F*), or *delayed fluorescence*, or nonradiative—$S_x \rightsquigarrow S_{x-1}^v$ or $T_x \rightsquigarrow T_{x-1}$, *internal conversion* (*IC*) or (2) within electronic states of different multiplicity, e.g., radiative—$T_1 \rightarrow S_0$, *phosphorescence* (*P*), or nonradiative—$S_x \rightsquigarrow T_x$ or $T_x \rightsquigarrow S_x$, *intersystem crossing* (*ISC*).

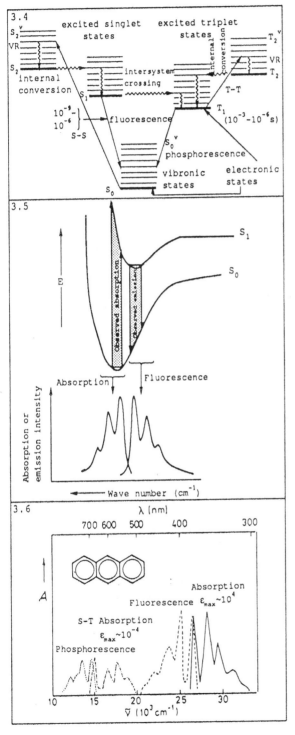

3.4. These processes, as well as *photoexcitation* (absorption, *A*) are well described by the *Jablonski (state) diagram*, in which states represented by horizontal lines are grouped into vertical columns according to their multiplicity. The individual processes are indicated by arrows (radiative processes = straight arrows, nonradiative processes = wavy arrows). The energy difference between the ground state (S_0), *singlet* (S_1, S_2), and *triplet* (T_1, T_2) states is progressively lower at higher vibronic states. Each process is characterized by a rate constant (k_i); the sum of the reciprocal values of the deactivation rates of a vibronic state determines its lifetime (τ_i).

3.5. *Fluorescence* is a spontaneous ("prompt," $\tau_f \sim 10^{-8}$ s) spin-allowed emission of radiation (*luminescence*) within vibronic states of the same multiplicity, usually from the thermally relaxed S_1 (see the Kasha rule in 3.7) to S_0^v and S_0 states. The vibrational structure of the fluorescence band is a mirror image of the longest wavelength band in the absorption spectrum, shifted to longer wavelengths. The difference between the position of the band maxima in absorption and fluorescence spectra is called *Stokes shift*. At a high temperature, the band intensity decreases and the vibrational structure vanishes. For *delayed fluorescence*, see 3.23.

3.6. *Phosphorescence* is the spin-forbidden emission of radiation between vibronic states of different multiplicity, generally from the T_1 to the S_0^v state. The vibrational structure of a phosphorescence spectrum is a mirror image of the phosphorescence excitation spectrum ($S_0 \rightarrow T_1^v$). Since the T_1 state always lies below the S_1 state, this band occurs at longer wavelengths than that of fluorescence. However, it is generally not observed in gas or fluid systems; the use of rigid glasses is standard in phosphorescence studies. The lifetimes are relatively long ($\tau_p \sim 10^{-4} - 10$ s).

3.7

KASHA RULE

Fluorescence almost invariably occurs from the S_1 state; phosphorescence from the T_1 state, independent of which state is initially created. Emissions from higher states are rare.

VAVILOV RULE

Quantum yields of either fluorescence or phosphorescence are not dependent on the excitation energy of the initially-formed excited state.

ERMOLAEV RULE

The sum of the quantum yields of fluorescence and phosphorescence approaches 1.

3.7. Some important rules concerning the emission of radiation should be mentioned. These are a consequence of the fact that the energy difference between the ground and lowest excited singlet or triplet state is usually much larger than the energy difference between two excited states of the same multiplicity. It is not surprising that the rules are not obeyed when the energy difference, e.g., $S_2 - S_1$ or $T_2 - T_1$, is comparable to the deactivation energy for reversion to the ground state. The observed $S_2 \rightarrow S_0$ emission of azulenes[1] and thioketones[2] and the $T_2 \rightarrow T_1$ emission of anthracenes[3] represent such exceptions. The rules of Vavilov and Ermolaev are merely consequences of Kasha's rule. Ermolaev's rule is violated in the case of interaction with another molecule that can proceed either by a quenching mechanism or by a subsequent photochemical reaction. The nonradiative transitions $S_1 \rightsquigarrow S_0$ and $T_1 \rightsquigarrow T_0$ are slow, essentially because of the large energy differences involved.

3.8

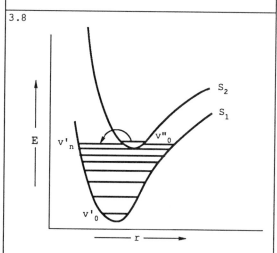

Mechanism of IC through potential curves for S_1 and S_2 states; horizontal lines are the vibrational states; r is the internuclear distance.

3.8. *Internal conversion (IC)* is an isoenergetic nonradiative transition between two electronic states of equal multiplicity, usually $S_x \rightsquigarrow S_{x-1}^v$ or $T_x \rightsquigarrow T_{x-1}^v$. The S_{x-1}^v or T_{x-1}^v state usually decays rapidly by vibrational relaxation (VR) into a thermally equilibrated S_{x-1} or T_{x-1} state. Therefore, VR is often implicated in IC and, consequently, IC is understood as an $S_x \rightsquigarrow S_{x-1}$ or $T_x \rightsquigarrow T_{x-1}$ process. IC competes with fluorescence and its rate ($k_f \sim 10^6 - 10^9$ s^{-1}) decreases with increasing energy difference (gap) between the lowest vibrational levels of the electronic states involved in the process; the smaller the separation, the faster the process. Since IC for $S_x \rightsquigarrow S_1$ proceeds much faster than radiative decay ($k_{IC} > k_f$), generally no fluorescence is observed from higher vibronic states (*energy gap law*). In contrast, k_f is greater than k_{IC} for the $S_1 \rightarrow S_0$ transition and fluorescence is usually observed.

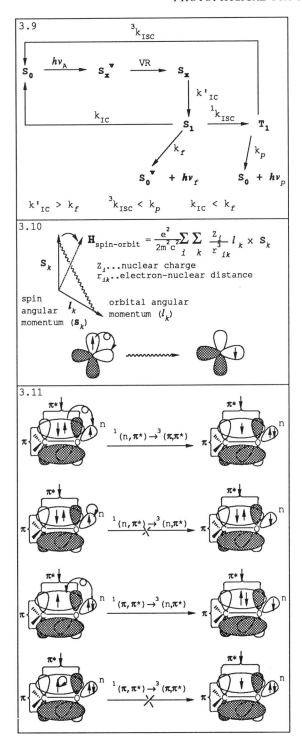

3.9. *Intersystem crossing (ISC)* is an isoenergetic nonradiative transition between two vibronic states of different multiplicity. ISC often results in a higher vibronic state, which rapidly deactivates to its lowest vibrational level, i.e., $S_x^v \rightsquigarrow T_{x-1}$ or $T_x^v \rightsquigarrow S_{x-1}$. The ISC processes, $S_1 \rightsquigarrow T_1$ and $T_1 \rightsquigarrow S_0$, are particularly interesting. ISC competes with phosphorescence and its rate is also governed by the *energy gap law*. The energy gap order for ISC is $T_1 \rightsquigarrow S_0 > S_1 \rightsquigarrow T_1$; the corresponding rate constants are in the opposite order. Although ISC is a spin-forbidden process, it occurs as a consequence of *spin-orbit coupling*.

3.10. Spin-orbit coupling is a process by which an electron flips its spin and leads to an $S \rightsquigarrow T$ or $T \rightsquigarrow S$ transition by ISC. It results from mixing an electron's spin-magnetic moment and the orbital electron angular moment, as shown by the approximate Hamiltonian operator. The interaction of the orbital moment and the spin moment permits the spin moment to be pulled around by the orbital moment, as if there was a spring between the two moments. The process is exceedingly effective if the electron moves from one orbital to another.[4] Spin-orbit coupling is enhanced by either paramagnetic compounds or heavy atoms (see 3.13).

3.11. Spin-orbit coupling is particularly effective between those states in which the change of an electron spin is accompanied by a large change in the orbital function (location of the electron). A $^1(n,\pi^*) \rightsquigarrow {}^3(\pi,\pi^*)$ transition is accompanied by an electron transfer from a bonding π MO into a nonbonding (n) MO. The $^1(\pi,\pi^*) \rightsquigarrow {}^3(n,\pi^*)$ transition is accompanied by the transfer of an electron from an n MO into a bonding π MO (effective spin-orbit coupling). In contrast, $^1(n,\pi^*) \rightsquigarrow {}^3(n,\pi^*)$ and $^1(\pi,\pi^*) \rightsquigarrow {}^3(\pi,\pi^*)$ transitions involve a change in electron spin without changing orbital types (relatively ineffective spin-orbit coupling).

3.12

EL SAYED SELECTION RULES

TRANSITIONS BETWEEN STATES

$^1(n,\pi^*) \Longleftrightarrow ^3(\pi,\pi^*)$	allowed
$^3(n,\pi^*) \Longleftrightarrow ^1(\pi,\pi^*)$	allowed
$^1(n,\pi^*) \Longleftrightarrow ^3(n,\pi^*)$	forbidden
$^1(\pi,\pi^*) \Longleftrightarrow ^3(\pi,\pi^*)$	forbidden

3.13

INCREASE IN PROBABILITY OF SINGLET-TRIPLET TRANSITIONS

1) Interaction with paramagnetic compounds (O_2, NO, *etc.*)

2) Heavy atom effect (I, Br, Xe, Hg, *etc.*)

KINDS OF INTERACTIONS

a) INTERNAL - paramagnetic group or heavy atom is part of the excited molecule

b) EXTERNAL - paramagnetic compound or heavy atom (or heavy atom-containing compound) is present in the irriadiated mixture

3.14

ATOM	N	ξ	ATOM	N	ξ
C	6	0.4[a]	Kr	36	63.0
O	8	1.7[a]	Rb	37	4.2
F	9	2.9[a]	I	53	63.0
S	16	4.2	Xe	54	117.2
Cl	17	7.1	Hg	80	75.3
K	19	0.8	Pb	82	87.9
Br	35	29.8			

[a] extrapolated value

3.12. The above relations are known as the *El-Sayed selection rules*, which state that in nonradiative transitions from the lowest singlet state to the triplet state manifold (intersystem crossing), a change of orbital type enhances the rate of the process. The forbiddenness and allowedness, however, must be understood in a relative sense. Forbidden processes still occur, although with a probability $\sim 10^{-3}$ less than the allowed process. As shown previously, all transitions accompanied by a spin flip are substantially weaker than transitions between states of equal multiplicity.

3.13. The probability of a spin-forbidden singlet-triplet ($S \rightsquigarrow T$) transition (rate of intersystem crossing) increases in the presence of paramagnetic molecules or heavy atoms (atoms of high atomic weight) that are either part of or external to the excited molecule. If external, the ground state molecule is complexed with another molecule to form a metastable or contact complex, i.e., one whose stabilization energy is smaller than its vibrational energy at a given temperature. Interaction with a higher multiplicity state, e.g., O_2 or NO may permit transfer of spin (see 3.36) from a triplet to a singlet. Therefore, photoreactions are usually carried out in an inert atmosphere. When a sample is saturated with a paramagnetic gas under a pressure of 10–18 MPa, the intensity of an $S \rightarrow T$ transition in an electronic spectrum is clearly enchanced.

3.14. The heavy atom effect is caused by the interaction of the wave function of an excited molecule partially delocalized on the heavy atom. The interaction is related to the magnitude of the principal quantum number N of the heavy atom by a factor of 10^4. The spin-orbital coupling constant (ξ) for several selected elements is given in kJ/mol.[5]

3.15

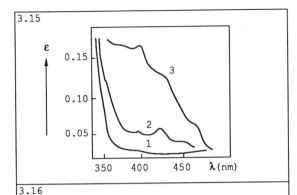

3.15. The S → T absorption spectra of chloronaphthalene (1), chloronaphthalene and hexane saturated with xenon at high pressure (2), and chloronaphthalene in ethyl iodide solution (3) are shown. Both (2) and (3) are examples of the external heavy atom effect, resulting in not only a higher intensity of S → T transitions, but also an increase in the efficiency of ISC and a decrease in the lifetime of the excited triplet states. This effect is frequently used to determine whether a reaction proceeds in an excited singlet or triplet state or to increase the quantum yields of triplet reactions.[6]

3.16

S_1-T_1 SPLITTING IN $\pi,\pi*$ and $n,\pi*$ STATES

COMPOUND	CONFIGURATION	$\Delta E(S_1-T_1)$ kJ/mol (kcal/mol)
ethylene	$\pi,\pi*$	290 (69)
butadiene	$\pi,\pi*$	271 (65)
benzene	$\pi,\pi*$	223 (53)
naphthalene	$\pi,\pi*$	160 (38)
anthracene	$\pi,\pi*$	126 (30)
formaldehyde	$n,\pi*$	36 (8.6)
acrolein	$n,\pi*$	19 (4.5)
benzaldehyde	$n,\pi*$	22 (5.3)
acetone	$n,\pi*$	29 (6.9)
benzophenone	$n,\pi*$	27 (6.5)

3.16. The ground state of most organic molecules is a singlet state, S_0, in which the MOs are doubly occupied. Their energetically lowest excited state is the triplet state, T_1. The triplet state is nearly always lower in energy than the corresponding singlet state of the same electronic configuration. The energy difference between the ground vibrational state of both the first singlet and triplet state is known as *singlet-triplet splitting* [$\Delta E(S_1 - T_1)$]. As shown in the table, this splitting has large values for $\pi,\pi*$ and low values for $n,\pi*$ states, and is one of the factors that determines the rate of ISC.

3.17

$$K_{23}=\iint (\phi_2\phi_3)_1\ 1/r_{12}(\phi_2\phi_3)_2 dr_1 dr_2$$

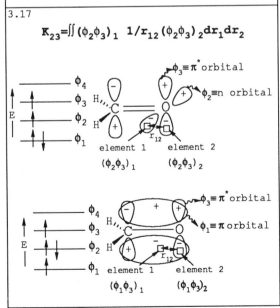

3.17. Singlet-triplet splitting is a measure of the difference in the distribution of electrons between the two states. According to the MO theory, the single-triplet splitting for singly excited singlet and triplet configurations is twice the exchange integral (K_{ij}). Its magnitude is obtained as the sum of elementary contributions, each of which is the product of the MO product ϕ_2, ϕ_3 in the space elements 1 and 2, divided by the distance between the elements (r_{12}). Thus, the magnitude of K_{ij} is a function of the overlap of the SOMOs that participate in the corresponding excited state.

3.18

SINGLET TRIPLET SPLITTING VALUES

OCCUPIED ORBITAL \ UNOCCUPIED ORBITAL	π	σ
π	L	S
n	S	M
σ	S	L

S...small; M...medium; L...large

3.19

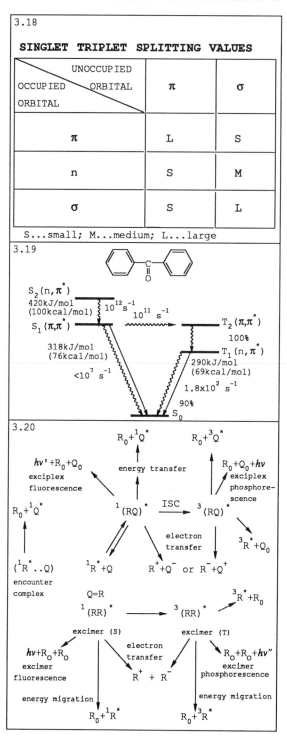

$S_2(n,\pi^*)$
420kJ/mol
(100kcal/mol)
$10^{12}\,s^{-1}$
$10^{11}\,s^{-1}$

$S_1(\pi,\pi^*)$

318kJ/mol
(76kcal/mol)

$<10^7\,s^{-1}$

$T_2(\pi,\pi^*)$
100%

$T_1(n,\pi^*)$
290kJ/mol
(69kcal/mol)
$1.8\times10^2\,s^{-1}$
90%

S_0

3.20

$R_0+{}^1Q^*$ $R_0+{}^3Q^*$

$h\nu'+R_0+Q_0$
exciplex fluorescence

energy transfer

$R_0+Q_0+h\nu$
exciplex phosphorescence

$R_0+{}^1Q^*$

${}^1(RQ)^*$ — ISC → ${}^3(RQ)^*$

electron transfer

${}^3R^*+Q_0$

$({}^1R^*..Q)$
encounter complex

${}^1R^*+Q$ R^++Q^- or R^-+Q^+

Q=R

${}^1(RR)^*$ → ${}^3(RR)^*$

${}^3R^*+R_0$

excimer (S)   excimer (T)

$h\nu+R_0+R_0$
excimer fluorescence

electron transfer

$R^+ + R^-$

$R_0+R_0+h\nu''$
excimer phosphorescence

energy migration   energy migration

$R_0+{}^1R^*$ $R_0+{}^3R^*$

3.18. Consequently, the magnitude of the singlet-triplet splitting will be large if the orbitals involved in the electronic transition are located approximately in the same area of space (more overlap). This condition is best shown by complementary pairs of bonding and antibonding orbitals produced from two identical wave functions, e.g., σ,σ^*, π,π^*. Small splitting is typical of SOMOs located in different parts of space (less overlap).

3.19. A state diagram, equivalent to the Jablonski diagram, for a particular organic compound can be designed from experimental, especially spectroscopic, data. The relative energies of the lowest excited states are given, along with the probability of radiative and radiationless transitions (implied by their rate constants). ISC from singlet to triplet is extremely efficient, both because of the miniscule energy difference between the S_1 and T_2 states and because the transition obeys the El-Sayed selection rule.

3.20. The *bimolecular (intermolecular) deactivation* of an excited state is commonly described as *quenching*; the component that accelerates this process is the *quencher*. There are basically two major routes to quenching. *Photochemical quenching*, in which the quencher transforms the excitation energy into chemical energy and a product is formed; this process is the subject of later chapters (see 5-13–5.15). *Photophysical quenching* can be divided into (a) self-quenching or concentration quenching (see 3.24) and (b) impurity quenching, which is further classified as quenching by electron transfer (see 3.30–3.32), electronic energy transfer (see 3.33), and heavy atom quenching. In some of these processes, bimolecular entities (associations of two identical or different species) are formed, with or without electron transfer. The new excited state [a unimolecular entity or a complex (encounter)] can undergo photochemical or photophysical monomolecular deactivation, as previously discussed.

3.21 WIGNER SPIN CONSERVATION RULE

In the reaction $A + B \rightarrow C + D$, where the total spins are S_A, S_B, S_C, and S_D, the total spin of the transition complex can have only the following magnitudes:

$$|S_A+S_B|, |S_A+S_B-1|, |S_A+S_B-2|, \ldots |S_A-S_B|; \text{ or}$$
$$|S_C+S_D|, |S_C+S_D-1|, |S_C+S_D-2|, \ldots |S_C-S_D|$$

ALLOWED PROCESSES: $^1D^* + {^1A} \rightarrow {^1D} + {^1A^*}$; $^3D^*+{^1A} \rightarrow {^1D} + {^3A^*}$; $^3A^*+{^3A^*} \rightarrow {^1A} +{^{5,3,1}A}$; etc

3.22

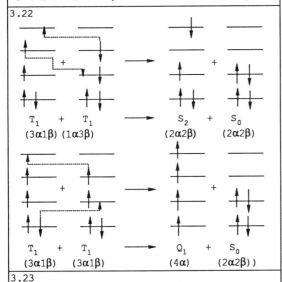

$T_1 + T_1 \longrightarrow S_2 + S_0$
$(3\alpha1\beta)(1\alpha3\beta)$ $(2\alpha2\beta)$ $(2\alpha2\beta)$

$T_1 + T_1 \longrightarrow Q_1 + S_0$
$(3\alpha1\beta)$ $(3\alpha1\beta)$ (4α) $(2\alpha2\beta)$)

3.23

P-TYPE DELAYED FLUORESCENCE

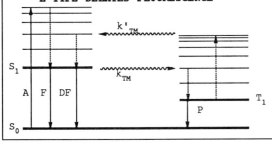

D^{**} - excimer-like entity; $^1D^*$ - excimer

E-TYPE DELAYED FLUORESCENCE

3.21. Quenching processes are assumed to be essentially collisional and, therefore, should obey the *Wigner spin conservation rule*, which states that upon transfer of electronic energy between an excited molecule and another molecule in its ground or excited state, the overall spin of the system should not change. Accordingly, products C and D, formed by the collision of reactants A and B, must correlate with each other and with the transition complex. Thus, the total spin of the transition complex can have only one of the values of the sequences shown. The process will be spin allowed if the sequences have at least one number in common.

3.22. According to the Wigner rule, the difference in the number of spin electrons with spins α and β must remain constant during the energy transfer, and the interaction of two molecules in a triplet state can lead to two singlet states, a singlet and a triplet state, or a singlet and a quintet state. Since quintet states are doubly excited relative to the ground state, they generally have very high energies and this transition is energetically unfavorable. While this interaction has not yet been observed, it cannot be excluded theoretically. The interaction can occur in two ways, either at a long distance in the repulsion part of the energy surface or at a short distance with the formation of a metastable or encounter complex.

3.23. *Delayed fluorescence* is a spin-allowed emission process that occurs at a much slower rate than "prompt" fluorescence (by a factor of $\sim 10^3$). It can arise from two different mechanisms: triplet-triplet annihilation (*P-type* delayed fluorescence) and reversible ISC due to thermal activation (*E-type*). In contrast to "prompt" and E-type delayed fluorescence, the P-type follows second-order kinetics and the emission intensity is proportional to the second power of concentration.

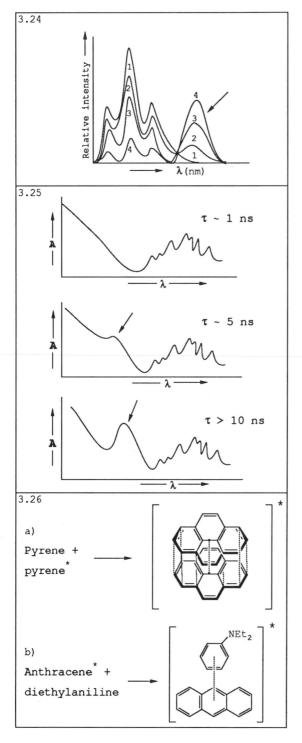

3.24. Fluorescence is often quenched by an increase in the solute concentration (*self-quenching* or *concentration quenching*). Simultaneously, a new structureless emission band, whose intensity is proportional to the concentration of the solute, appears at longer wavelengths. This new emission band is caused by the fluorescence of a bimolecular entity of definite stoichiometry (usually 1:1) that results from the collision of a molecule in the ground state with another molecule in an excited state. The fluorescence spectra of mixtures of anthracene and dimethylaniline at four different concentrations are shown.[7]

3.25. The two emission bands can be distinguished by *time-resolved spectroscopy*. The formation of a new band is the result of a bimolecular reaction that follows second order kinetics and is slower than the emission itself. It is, therefore, possible to observe pure emission of the excited molecule immediately after irradiation. The emission from the new bimolecular entity shown appears in a few nanoseconds.

3.26. The bimolecular entity mentioned above is a stable species resulting from the collision of a molecule in an excited state with a molecule in the ground state. If it is only stable in the excited state, and if, after transition to the ground state, it dissociates into its components, it is an *excimer* (*excited dimer*, *a*), assuming the two interacting components are identical. If they are different, an *exciplex* (*excited complex*, *b*) is formed. The structures shown are hypothetical since the actual orientation of the components has not yet been determined unequivocally by either experiment or theory. It is assumed that excimers and exciplexes are stabilized by transfer of energy, with exciplexes stabilized by charge transfer as well. In the formation of some exciplexes, a change in dipole moment was observed (see 2.42).[7,8]

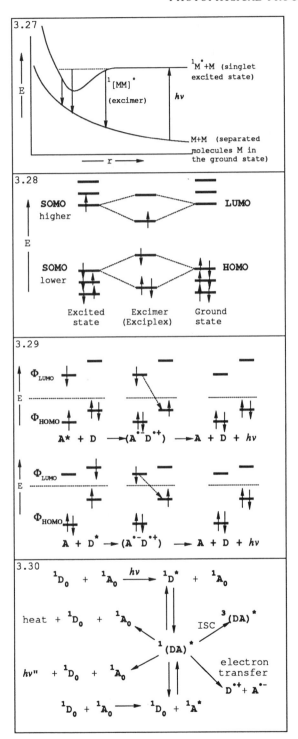

3.27. The formation of an excimer or exciplex is affected by the existence of a minimum on the energy curve of an excited state in the region where the ground state is purely repulsive. It was shown that the formation of these complexes is responsible for certain photochemical processes, particularly photocycloadditions.[9] The products of these reactions can be stable, even in the ground state. This phenomenon is characteristic of singlet states.

3.28. According to the frontier orbital theory, the formation of an excimer (exciplex) results from the interaction between the lower SOMO and higher SOMO of the excited molecule with the HOMO and LUMO of the ground state, respectively. Since the corresponding orbitals are usually close in energy, both interactions are strong and, consequently, the formation of the complex is often very fast.[10]

3.29. The two molecules involved in exciplex formation can have different *ionization potentials* (IP) and *electron affinities* (EA) and, therefore, different electron donor (D)—acceptor (A) properties. Since a molecule always possesses a lower IP and a higher EA in its excited than in its ground state, electron transfer either from an excited donor D* to an acceptor in the ground state A, or from D to A*, is possible and an exciplex $^1(A^{\cdot-}D^{\cdot+})^*$ is formed.[11] The fluorescence of exciplexes exhibits a bathochromic shift, diminishes in intensity with increasing polarity of the solvent, and vanishes in highly polar solvents.[7]

3.30. Exciplexes (excimers), $^1(DA)^*$, are intermediates in the quenching process. As mentioned in 3.27, they can be deactivated by a photochemical reaction, either directly or after conversion into $^3(DA)^*$ by ISC.[12] Other routes available to exciplexes are radiative or radiationless deactivation, energy transfer, electron transfer, and back-electron transfer quenching.[13]

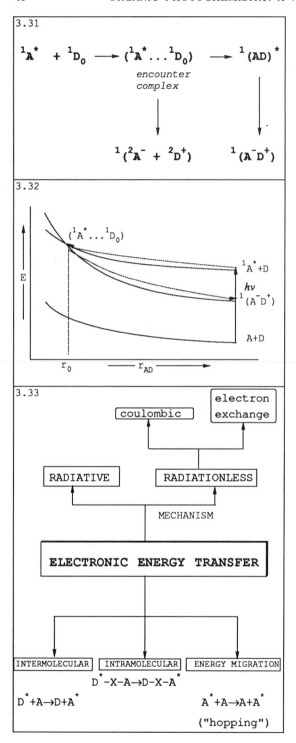

3.31. The suppression of the intensity of excimer fluorescence by polar solvents (see 3.29) can sometimes be explained by a competing electron transfer within the encounter complex, followed by dissociation into solvated radical ion pairs in a doublet state. This *single-electron-transfer* (*SET*) quenching mechanism plays a significant role in a number of photochemical reactions, and in recent years has been the center of considerable attention.[9a,14]

3.32. There is experimental evidence that electron transfer and subsequent dissociation sometimes proceed *via* an encounter complex prior to the formation of the exciplex.[15] By collision of acceptor $^1A*$ with donor D, an encounter complex is formed at the *effective interaction distance* (r_0) (~ 700 pm). An electron transfer takes place within this complex with the formation of an ion pair in the excited singlet state $^1(A^-D^+)$, which can then undergo radiative or radiationless transition to the ground state or, more commonly, subsequent reactions of the individual ions occur. This is called a *charge transfer process*.

3.33. Electronic excitation *energy transfer* is a photophysical process in which the excitation energy is transferred from an excited donor to an acceptor in its ground state with the simultaneous formation of the donor in the ground state and the acceptor in the excited state. Some of these processes occur through exciplexes (excimers). During the energy transfer, one observes either (1) suppression of the luminescence or a photochemical reaction of the donor (the process is called *quenching*) or (2) initiation of luminescence or a photochemical reaction of the acceptor (the process is called *photosensitization*). The donor of the excitation energy is the (*photo*)-*sensitizer*. Various kinds of energy transfers of different mechanisms are known.[16]

3.34

$$^1R^* + {}^3O_2 \longrightarrow {}^3R^* + {}^3O_2 \quad (1)$$

$$^3R^* + {}^3O_2 \longrightarrow {}^1R + {}^3O_2 \quad (2)$$

$$^3R^* + {}^3O_2 \longrightarrow {}^1R + {}^1O_2 \quad (3)$$

ENERGY TRANSFER	EXAMPLE
singlet-singlet	$^1D^* + {}^1A_0 \rightarrow {}^1D_0 + {}^1A^*$
triplet-singlet	$^3D^* + {}^1A_0 \rightarrow {}^1D_0 + {}^1A^*$
triplet-triplet	$^3D^* + {}^1A_0 \rightarrow {}^1D_0 + {}^3A^*$

3.35

S-S ENERGY TRANSFER:

$$^1D^* \longrightarrow {}^1D + h\nu_D$$

$$h\nu_D + {}^1A \longrightarrow {}^1A^*$$

FLUORESCENCE FROM A:

$$^1A^* \longrightarrow {}^1A + h\nu_A$$

T-S ENERGY TRANSFER:

$$^3D^* \longrightarrow {}^1D + h\nu_P$$

$$h\nu_P + {}^1A \longrightarrow {}^1A^*$$

SPECTRAL OVERLAP INTEGRAL:

$$\int_0^\infty F_D(\tilde{\nu}) \, \varepsilon_A(\tilde{\nu}) d\tilde{\nu}$$

$F_D(\tilde{\nu})$ ...the emission spectrum of the excited donor;

$\varepsilon_A(\tilde{\nu})$ ...the absorption spectrum of the acceptor

3.34. Quenching and photosensitization are processes controlled by various factors that impose restrictions on the multiplicity of the donor and the acceptor involved in the process (e.g., the spin-conservation rule). They are differentiated by the multiplicity of the starting state of the donor and the multiplicity of the final state of the acceptor. Quenching processes of importance in preparative organic photochemistry include the quenching by *dioxygen* (molecular oxygen). Dioxygen is a common and usually undesirable quencher. In photochemical reactions dioxygen must be removed, e.g., by bubbling an inert gas (nitrogen or, preferably, argon) through the solution to be irradiated, by repetitive "freeze-pump-thaw" cycles, or by sonification (see 14.30). Process (3) is an example of *triplet-triplet annihilation*, which takes place by the dye-sensitized generation of singlet oxygen (see 13.6).

3.35. In *radiative electronic energy transfer* a photon emitted by the donor molecule is reabsorbed by a molecule of the acceptor. Consequently, the probability of this energy transfer is a function of the *spectral overlap integral*, the light absorbing power of the acceptor, its concentration and absorption path length (see Beer-Lambert Law, 2.4), and the quantum efficiency of emission of the donor. In this process radiative energy transfer takes place between identical molecules. It causes self-quenching or the *inner filter effect*, i.e., a decrease in the intensity of fluorescence of the solute, particularly in that part of the fluorescence spectrum overlapping the absorption spectrum. The luminescence spectra of very dilute solutions should, therefore, be recorded. Radiative energy transfer is a long-range phenomenon that occurs over distances exceeding molecular dimensions (>5000 pm).

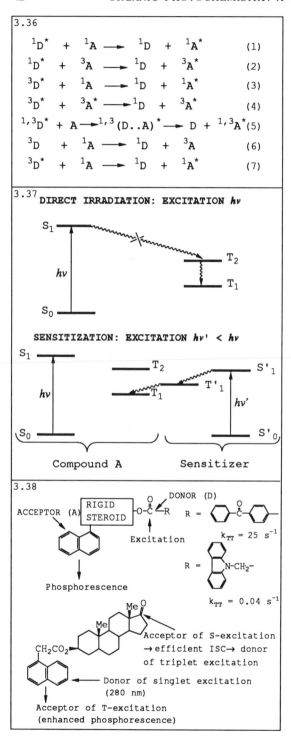

3.36

$$^1D^* + \ ^1A \longrightarrow \ ^1D + \ ^1A^* \qquad (1)$$

$$^1D^* + \ ^3A \longrightarrow \ ^1D + \ ^3A^* \qquad (2)$$

$$^3D^* + \ ^1A \longrightarrow \ ^1D + \ ^1A^* \qquad (3)$$

$$^3D^* + \ ^3A^* \longrightarrow \ ^1D + \ ^3A^* \qquad (4)$$

$$^{1,3}D^* + A \longrightarrow ^{1,3}(D..A)^* \longrightarrow D + \ ^{1,3}A^* \ (5)$$

$$^3D + \ ^1A \longrightarrow \ ^1D + \ ^3A \qquad (6)$$

$$^3D^* + \ ^1A \longrightarrow \ ^1D + \ ^1A^* \qquad (7)$$

3.37 DIRECT IRRADIATION: EXCITATION hv

SENSITIZATION: EXCITATION hv' < hv

Compound A   Sensitizer

3.38

ACCEPTOR (A)   RIGID STEROID   DONOR (D)

Excitation

$k_{TT} = 25 \ s^{-1}$

$R =$ ... N-CH₂-

$k_{TT} = 0.04 \ s^{-1}$

Phosphorescence

Acceptor of S-excitation → efficient ISC → donor of triplet excitation

Donor of singlet excitation (280 nm)

Acceptor of T-excitation (enhanced phosphorescence)

3.36. *Nonradiative energy transfer* generally occurs over short distances by: (1) *Coulombic (dipole-dipole) interaction*, which is theoretically allowed only if there is no change in the spin multiplicity of the reactants during the energy transfer process [Eqs. (1) and (2)]. However, energy transfers have been observed with a spin change of the donor (never the acceptor) [Eqs. (3) and (4)]. This is apparently due to the long lifetimes of ^3D, particularly in a frozen solution or a matrix. Or (2) *electron exchange excitation transfer*, either S–S or T–T, illustrated by Eq. (5). Since (D . . . A)* can be either an excimer or an encounter complex, energy transfer must obey the Wigner rule. Accordingly, in Eq. (6) the process is spin-allowed by this mechanism, whereas in Eq. (7) it is spin-forbidden.

3.37. Photosensitization plays an important role in photochemical reactions, especially those which involve an excited state of a spin-forbidden transition. This is particularly important for T–T energy transfer, since the lowest triplet states (T_1) of compounds with large S–T splittings, e.g., alkenes, cannot be attained by direct irradiation. These lead instead to products of S_1 reactions since the singlet-triplet ISC is very inefficient. With a sensitizer of appropriate triplet energy (the energy gap between $^3D^*$ and $^3A^*$ must be > 12.5 kJ/mol), the triplet state of the acceptor is produced by radiation incapable of populating its S_1 state.

3.38. The probability of T–T energy transfer is enhanced by an increased lifetime of the donor triplet. T_1–T_1 energy transfer is used in photosensitization when a close approach of D and A is frequent. When D and A are in the same molecule, an *intramolecular energy transfer* occurs, whose rate is a function of both the distance and structure.[17] Intramolecular transfer of both singlet and triplet electronic excitation between an arene and a $>C=O$ moiety has been reported recently.[18]

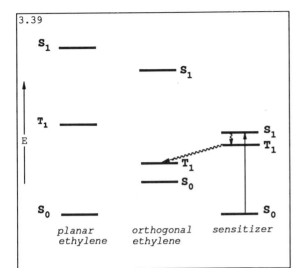

3.39. In a photosensitized reaction a complex is formed between the substrate and sensitizer. The time required for production of this complex depends on the rate of diffusion. The energy transfer occurs adiabatically, directly into the stable configuration of the molecule in its triplet state. The necessary excitation energy of the sensitized reaction can be even smaller than the energy of the vertical singlet-triplet transition of the substrate. The sensitization of ethylene by acetone or acetophenone is an example. The relaxed triplet, e.g., orthogonal ethylene, is lower in energy and was originally called a *phantom triplet* (see 5.10). Sensitized energy transfer to produce a phantom triplet is a typical phenomenon for alkenes and polyenes, in which excitation produces substantial changes in the geometry of the molecule and is accompanied by a large stabilization energy.

3.40.

MOLECULE	E_T	
	kJ/mol	kcal/mol
mercury	472	113
pyridine	355	85
benzene	351	84
anisole	339	81
acetone	326	78
benzonitrile	322	77
benzaldehyde	301	72
benzophenone	288	69
biphenyl	276	66
trans-stilbene	209	50
anthracene	180	43

3.40. A good sensitizer should possess a small difference in the energies of the S_1 and T_2 states so that ISC is sufficiently efficient. Another requirement is compliance with the El-Sayed selection rules (see 3.11 and 3.12). For these reasons carbonyl compounds are excellent sensitizers for the formation of triplet π,π^* states. Certain benzenoid hydrocarbons are also good sensitizers in those cases where the triplet state is sufficiently populated by ISC between close-lying S_1 and T_2 states.

3.41.

$$\Phi_i = \frac{\text{number of events}}{\text{number of photons absorbed}}$$

QUANTUM YIELD	EVENT	PROCESS
Φ_f	fluorescence	$S_1 \rightarrow S_0^v$
Φ_p	phosphorescence	$T_1 \rightarrow S_0^v$
Φ_{IC}	internal conversion	$\begin{cases} S_x \rightsquigarrow S_{x-1}^v \\ T_x \rightsquigarrow T_{x-1}^v \end{cases}$
Φ_{ISC}	intersystem crossing	$\begin{cases} S_1 \rightsquigarrow T_1^v \\ T_1 \rightsquigarrow S_0^v \end{cases}$

3.41. The overall efficiency of an excited state process is usually described by its *quantum yield* (Φ), i.e., the number of events divided by the number of photons absorbed by the system. The event may be either photophysical or photochemical. According to the Stark–Einstein law (see 1.2), Φ ranges from 0 to 1. However, Φ's of the order of 10–100 have been reported for photoinitiated chain reactions.

3.42

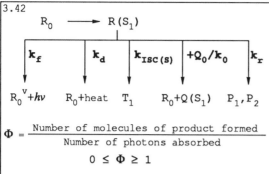

$$\Phi = \frac{\text{Number of molecules of product formed}}{\text{Number of photons absorbed}}$$

$$0 \leq \Phi \geq 1$$

3.43

PROCESS	REACTION	RATE	k_i (s^{-1})
EXCITATION	$S_0 + h\nu \to S_1$	I_a	
FLUORESCENCE	$S_1 \to S_0 + h\nu$	$k_f[S_1]$	$10^6 - 10^9$
INTERNAL CONVERSION	$S_1 \to S_0$	$k_{IC}[S_1]$	$10^7 - 10^{12}$
INTERSYSTEM CROSSING	$S_1 \to T_1$	$k_{ST}[S_1]$	$10^8 - 10^{11}$
	$T_1 \to S_0$	$k_{TS}[T_1]$	$10^{-2} - 10^3$
PHOSPHORESCENCE	$T_1 \to S_0 + h\nu''$	$k_p[T_1]$	$10^{-2} - 10^3$
FLUORESCENCE QUENCHING	$S_1 + Q \to S_0 + Q$	$k_q^f[S_1][Q]$	
PHOSPHORESCENCE QUENCHING	$T_1 + Q \to S_0 + Q$	$k_q^P[T_1][Q]$	

3.44

$$\frac{d[S_1]}{dt} = I_a - k_d[S_1]$$

$$I_a = k_d[S_1]$$

$$\Phi_i = \frac{k_i[S_1]}{I_a} = \frac{k_i}{k_d} = \frac{k_i}{\sum_i k_1} = k_i \tau_f$$

$$k_d = k_f + k_{IC} + k_{ST} + k_{TS} + \cdots$$

$$\tau_s = \frac{1}{k_d} = \frac{1}{k_i}; \quad \tau_0 = \frac{1}{k_f}$$

I_a = intensity (rate) of absorption
k_d = total rate constant for disappearance of S_1
k_i = rate constant of the individual event
τ_f = average lifetime of S_1
τ_0 = radiative (natural) lifetime
τ_s = see 3.45

3.42. The monomolecular and bimolecular photophysical processes mentioned in 3.1 and discussed in detail throughout this chapter compete with chemical changes of the excited molecule. A chemical change occurs only if the reaction rate (k_r) is comparable to the rates of all dissipative processes. The efficiency of a photochemical reaction is given by its quantum yield Φ, defined as the number of molecules (mols) of product formed, divided by the number of photons (mols of photons) absorbed by the system. If more than one photoproduct is formed, P_1, P_2, . . . , P_i, *quantum yields*, ϕ_1, ϕ_2, . . . , ϕ_i, can be assigned for each product. A photoreaction with a low Φ can have a high chemical yield but requires long irradiation times. For most photochemical reactions, the observed Φ is generally less than 1.

3.43. The basis of photochemical kinetics is the assumption that monomolecular deactivation processes follow the first-order rate law under steady-state illumination. The k_i's are first-order rate constants of the individual events, I_a is the intensity of absorption, and Q represents any molecule other than the reactant R in the system. The following example is limited to the kinetics of disappearance of the S_1 state.

3.44. Using the above assumption, the total rate constant of disappearance (k_d) of the singlet state (S_1) of the reactant is equal to the sum of the rate constants of the individual events. Accordingly, the average lifetime (τ) of the excited state is the reciprocal of the total rate constant. In the absence of radiationless transitions (IC and ISC) the lifetime of the S_1 state, τ_0, is the reciprocal of the first-order rate constant of fluorescence and is called the *radiative lifetime*. Thus, the quantum yield of any event is given by the ratio of its rates, $k_i[S_1]$, to the sum of the rates of all events.

3.45 **STERN–VOLMER EQUATION**

$$\Phi_f^0 = \frac{k_f}{k_d} = k_f\tau_s \qquad \Phi_f = \frac{k_f}{k_q+k_q[Q]}$$

$$\boxed{\frac{\Phi_f^0}{\Phi_f} = 1 + k_q\tau_s[Q]}$$

Φ_f^0 ...fluorescence quantum yield

in the absence of quencher

Φ_f ...fluorescence quantum yield

in the presence of quencher

k_q ...quenching rate constant

τ_s ...excited state half-life in

the absence of quencher

[Q] ...concentration of the quencher

3.46 **STERN–VOLMER PLOT**

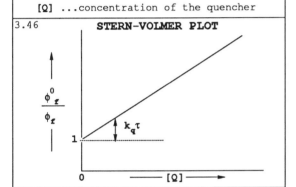

3.45. Along with the rate of disappearance of S_1, expressed by the rate constant of disappearance, the presence of a quencher Q requires addition of the term $k_q[Q]$ to the rate expression. The *Stern–Volmer equation*, which states that the emission quantum yield is a linear function of the concentration of the quencher, can be derived from the fluorescence quantum yields: Φ_f^0 in the absence of a quencher and Φ_f in its presence.[19]

3.46. The magnitude of the quenching constant k_q can be obtained experimentally from the Stern–Volmer plot and τ is also available independently from time-resolved spectroscopy. For many systems the value of the quenching constant is close to that of the diffusion constant. These are rapid quenching events in which the rate-determining step is diffusion in the solvent and the quenching constant depends on the solvent viscosity. When the rate of deactivation by the quencher, either alone or in a transition complex, is slow, the quenching constant is independent of solvent viscosity. Nonlinear Stern–Volmer plots always indicate that the quenching mechanism is complicated and, in many instances, provide indirect proof of the existence of the competing photochemical reactions that are responsible for the complication.

3.47

PHOTOPHYSICAL QUANTITY	SYMBOL	RELATION TO RATE CONSTANT	LIFETIME
fluorescence quantum yield	Φ_f	$\dfrac{k_f}{k_{IC}+k_{ST}+k_f}$	$k_f\tau_f$
phosphorescence quantum yield	Φ_p	$\dfrac{k_p}{k_{TS}+k_p} \times \dfrac{k_{ST}}{k_{IC}+k_f+k_{ST}}$	$k_p\tau_p \times k_{ST}\tau_f$
first singlet state lifetime	τ_f	$\dfrac{1}{k_{IC}+k_{ST}+k_f}$	
radiative lifetime	τ_f^0	$\dfrac{1}{k_f}$	
first triplet state lifetime	τ_p	$\dfrac{1}{k_{TS}+k_p}$	
radiative lifetime	τ_p^0	$\dfrac{1}{k_p}$	

3.47. The kinetic equations for quantum yields, lifetimes, and first-order rate constants for any of the events that take place in a photochemical process can be derived. The quantum yields of fluorescence and phosphorescence, as well as the lifetimes of the first excited singlet (S_1) and triplet (T_1) states are experimentally available. Their relations to the corresponding rate constants in the absence of a chemical reaction are given in the table.

REFERENCES

(1)(a) Murata, S., Iwanaga, C., Toda, T., and Kokubun, H. *Chem. Phys. Lett.*, **1972**, *13*, 101. (b) Eaton, D. F., Evans, T. R., and Leermakers, P. A. *Mol. Photochem.* **1969**, *1*, 347.

(2) McGimpsey, W. G. In *Handbook of Organic Photochemistry.* Scaiano, J. C. ed. CRC Press: Boca Raton, Florida, Vol. I, p. 419.

(3)(a) Gillispie, G. D. and Lim, E. C. *Chem. Phys. Lett.* **1979**, *63*, 355. (b) For the heavy-atom effects on the excited singlet-state electron-transfer reactions, see Kikuchi, K., Hoshi, M., Niwa, T., Takahashi, Y., and Miyashi, T. *J. Phys. Chem.* **1991**, *95*, 38.

(4) Adapted from Salem, L. *Electrons in Chemical Reactions: First Principles.* Wiley: New York, 1982, p. 208.

(5)(a) Koziar, J. C. and Cowan, D. O. *Acc. Chem. Res.* **1978**, *11*, 334. (b) Adapted from Turro, N. J. *Modern Molecular Photochemistry.* Benjamin/Cummings: Menlo Park, California, 1978, p. 118. (c) Richards, W. W., Trivedi, H. P., and Cooper, D. L. In *Spin-Orbit Coupling in Molecules.* Clarendon: Oxford, 1971, p. 17.

(6) See Ref. 2(b), p. 125.

(7) Adapted from Weller, A. *Pure Appl. Chem.* **1968**, *16*, 115.

(8)(a) For a review, see Stevens, B. *Adv. Photochem.* **1971**, *8*, 161. (b) Mattes, S. L. and Farid, S. *Science.* **1984**, *226*, 117. (c) Julliard, M. and Chanon, M. *Chem Rev.* **1983**, *83*, 425. (d) Julliard, M. and Chanon, M. *Chem. Scripta.* **1984**, *24*, 11.

(9)(a) Figure adapted from Mattay, J. *Angew. Chem.*, *Int. Ed. Engl.* **1987**, *26*, 825. (b) Mattes, S. L. and Farid, S. *Acc. Chem. Res.* **1982**, *15*, 80. (c) Caldwell, R. A. *Acc. Chem. Res.* **1980**, *13*, 45. (d) Lewis, F. D. *Acc. Chem. Res.* **1979**, *12*, 152.

(10) Fleming, I. *Frontier Orbitals and Organic Chemical Reactions.* Wiley: New York, 1976, p. 209.

(11) Leonhardt, H. and Weller, A. *Ber. Bundesges. Phys. Chem.* **1963**, *67*, 791.

(12)(a) For a review of molecular triplet excimers, see Lim, E. C. *Acc. Chem. Res.* **1987**, *20*, 8. (b) For a review of intramolecular excimer formation, see DeSchryver, F. C., Collart, P., Vandendriessche, J., Goedeweeck, R., Swinnen, A. M., and Van der Auweraer, M. *Acc. Chem. Res.* **1987**, *20*, 159.

(13) For the flash-spectroscopy detection of radical ions, see (a) Gersdorf, J., Mattay, J., and Görner, H. *J. Amn. Chem. Soc.* **1987**, *109*, 1203. (b) Mataga, N., Karen, A., Tadashi, O., Nishitani, S., Kurata, N.,

Sakata, Y., and Misumi, S. *J. Phys. Chem.* **1984**, *88*, 5138.

(14)(a) Fox, M. A. *Adv. Photochem.* **1986**, *13*, 237. (b) Kavarnos, G. J. and Turro, N. J. *Chem. Rev.*, **1986**, *86*, 401. (c) Mariano, P. S. and Stavinoha, J. L. In *Synthetic Organic Photochemistry.* Horspool, W. M., ed., Plenum: New York: 1984. (d) Davidson, R. S. *Adv. Phys. Org. Chem.* **1983**, *19*, 1. (e) Weller, A. *Pure Appl. Chem.* **1982**, *54*, 1885.

(15) Rehm, D. and Weller, A. *Ber. Bundesges. Phys. Chem.* **1969**, *73*, 834.

(16)(a) Turro, N. J. *Pure Appl. Chem.* **1977**, *49*, 405. For reviews, see (b) Wilkinson, F. *Adv. Photochem.* **1964**, *3*, 241. (c) Lamola, A. A. In *Energy Transfer and Organic Photochemistry.* Leermakers, P. A. and Weissberger, A., eds. Vol. 14 of *Technique of Organic Chemistry.* Interscience: New York, 1969, p. 17.

(17) Keller, R. A. and Dolby, L. J. *J. Am. Chem. Soc.* **1969**, *91*, 1293.

(18) Tong, Z., Yang, G., and Wu, S. *Acta Chim. Sin. (Engl. Ed.).* **1989**, *5*, 450 [*C. A.* **1991**, *114*, 24299c].

(19)(a) Wagner, P. J. In *Creation and Detection of the Excited State.* Lamola, A. A., ed. Marcel Dekker: New York, 1971, Vol. 1, Part A, p. 173. (a) Dalton, J. C. and Turro, N. J. *Mol. Photochem.* **1970**, *2*, 133.

GENERAL READING IN PHOTOPHYSICAL PROCESSES

Becker, R. S. *Theory and Interpretation of Fluorescence and Phosphorescence.* Wiley–Interscience: New York, 1969.

Birks, J. B. *Photophysics of Aromatic Molecules.* Wiley: New York, 1970.

Bowen, E. J., ed. *Luminescence.* Van Nostrand: London, 1968.

Dewar, M. J. S. *The Molecular Orbital Theory of Organic Chemistry.* McGraw-Hill: New York, 1969.

Dewar, M. J. S. and Dougherty, R. C. *The PMO Theory of Organic Chemistry.* Plenum: New York, 1975.

Foster, R. *Organic Charge-Transfer Complexes.* Academic Press: New York, 1969.

Lim, E. C., ed., *Excited States.* Academic Press: New York, Vols. 1–6.

McGlynn, S. P., Azumi, T., and Kinoshita, M. *Molecular Spectroscopy of the Triplet State.* Prentice-Hall: Englewood Cliffs, New Jersey, 1969.

Parker, C. A. *Photoluminescence in Solution.* Elsevier: Amsterdam, 1968.

Chapter 4
THEORETICAL CONCEPTS IN ORGANIC PHOTOCHEMISTRY

THEORETICAL CONCEPTS IN ORGANIC PHOTOCHEMISTRY

"Love is a good thing but a

golden bracelet is forever."

Anita Loos[1]

"Theory is a good thing but a

good experiment is forever."

Peter L. Kapitza[2]

PERICYCLIC REACTIONS

All intra- or intermolecular transformations in which chemical bonds are formed or broken proceed in a concerted manner *via* a cyclic transition state.

FORMATION OF BONDS BY ORBITAL INTERACTIONS

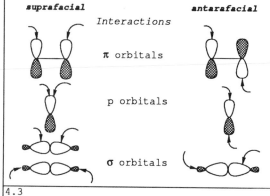

suprafacial   **antarafacial**

Interactions

π orbitals

p orbitals

σ orbitals

GENERAL RULE FOR PHOTOCHEMICAL PERICYCLIC REACTIONS

An excited state pericyclic reaction is symmetry allowed when:

1. The excited state originates by the transition of an electron from the **HOMO** *to the* **LUMO**.

2. The total number of electrons involved can be divided by 4 and the total number of antarafacial interactions is **even** *or when the total number of electrons involved is expressed by the formula (4n+2) and the total number of antarafacial interactions is* **odd**.

4.1. The use of these quotations[1,2] to introduce a chapter dealing with the theoretical aspects of photochemical reactions may seem contradictory. However, the beauty of a good theory is that it inspires good experiments that last forever. While experimental organic photochemistry has witnessed an explosive development in the last few decades, no comprehensive theory of reactivity of electronically excited states existed until recently. Photochemical reactions generally do not occur on a single energetic hypersurface. In quantum chemical terms it is inevitable that the effect of correlation energy must be considered. For this reason, still more reliable theoretical studies are needed.

4.2. Historically, the most important breakthrough in understanding the mechanism of photochemical reactions was the formulation of the Woodward–Hoffmann rules of conservation of orbital symmetry.[3] These apply to concerted reactions involving a cyclic activated complex (pericyclic reactions), e.g., electrocyclic, cycloaddition, sigmatropic, and cheletropic reactions. According to these rules, the allowedness or forbiddenness of these reactions depends primarily on the number of π electrons in the cyclic-activated complex and on the way in which the orbitals overlap.

4.3. The correct prediction of the stereospecificity of products from pericyclic reactions is the basis of the success of the Woodward–Hoffmann rules. One of the consequences of these rules is that thermal [2+2] cycloadditions (except for those with heterocumulenes) and [1+3] sigmatropic reactions in the ground electronic state are shown to be practically impossible. The necessary antarafacial interactions are sterically unfavorable and these reactions generally do not occur.

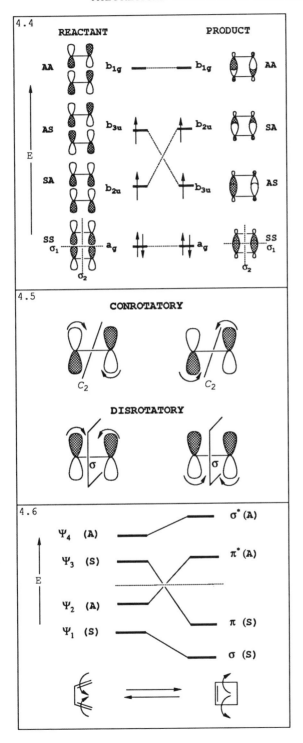

4.4

REACTANT PRODUCT

CONROTATORY

DISROTATORY

4.5

4.6

4.4. In a concerted reaction bond-breaking/forming processes occur simultaneously. The pathway of a concerted reaction can be illustrated by a correspondence in energy and symmetry between the MOs of the reactant and product. The relevant orbitals (*symmetry-adapted orbitals*) are *symmetric* (S) or *antisymmetric* (A) with respect to the symmetry element (mirror plane of symmetry, σ, or two-fold axis of rotation, C_2), which is retained at all points along the reaction coordinate. Consider the cycloaddition (dimerization) of two molecules of ethylene. The reaction involves four π electrons located on the C atoms in the plane of the "complex" and proceeds by the cleavage of two π bonds and the formation of two new σ bonds. The molecular complex maintains two planes of symmetry, σ_1 and σ_2. By generating the lowest excited state of the reactant and considering the electrons moving along the connection lines, the lowest excited state of the product is generally formed. Such a reaction is symmetry allowed.

4.5. Another example is an electrocyclic reaction that proceeds either by a *conrotatory* (antarafacial overlap of orbitals) or a *disrotatory* (suprafacial overlap) mode with a C_2-axis or σ-plane symmetry element, respectively.

4.6. The mechanism of the 1,3-butadiene $\leftrightharpoons$ cyclobutene interconversion can be illustrated by an *orbital correlation diagram*, in which the bonding and antibonding orbitals of reactant and product of the same symmetry are connected by solid lines. If the symmetry element remains unchanged during the conversion, the reaction occurs with *conservation of orbital symmetry* and is *allowed*. The correlation diagram provides useful information about the region between the reactant and product, i.e., the transition state.

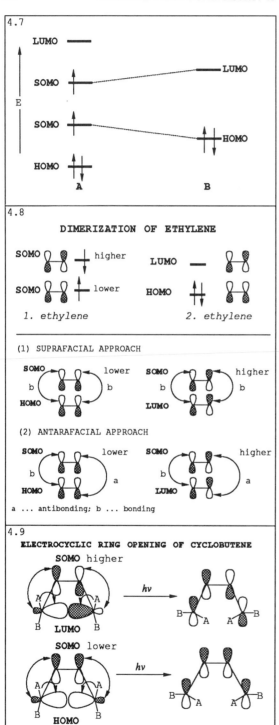

4.7

LUMO

SOMO

E

SOMO

HOMO

A   B

LUMO

HOMO

4.8

DIMERIZATION OF ETHYLENE

SOMO ⟶ higher   LUMO

SOMO ⟶ lower   HOMO

1. ethylene   *2. ethylene*

(1) SUPRAFACIAL APPROACH

SOMO — lower   SOMO — higher
b   b   b   b
HOMO   LUMO

(2) ANTARAFACIAL APPROACH

SOMO — lower   SOMO — higher
b   a   b   a
HOMO   LUMO

a ... antibonding; b ... bonding

4.9

ELECTROCYCLIC RING OPENING OF CYCLOBUTENE

SOMO higher

A   A
B LUMO B

$\xrightarrow{h\nu}$

B   A   A   B

SOMO lower

A   A
B HOMO B

$\xrightarrow{h\nu}$

B   A   A   B

4.7. An extension of the Woodward–Hoffmann approach, based on perturbation theory (*frontier–orbital theory*), was introduced by Fukui.[4] According to Fukui, the stabilization energy is composed of electrostatic and/or covalent contributions. Organic photochemical reactions are mainly governed by a second-order perturbation term, called a covalent term (*orbital-controlled reactions*). The most important contributions result from the interaction of either the HOMO with the lower SOMO or the LUMO with the higher SOMO. These contributions are particularly important if the participating orbitals are close in energy. Orbital interactions are stabilizing when the total overlap of the corresponding wave functions is not zero.

4.8. Along the concerted path of a supra-suprafacial interaction it is possible to orient the frontier orbitals in such a way that both interactions are bonding. The Fukui theory leads to the same prediction as the Woodward–Hoffmann rules. In a supra-antarafacial interaction, the orbitals can only be oriented so that one interaction is bonding and the other antibonding (shown as *b* and *a*). An advantage of the Fukui theory is the ability to make predictions for those reactions that do not preserve any symmetry element. However, it is also applicable only to states that can be described by a single configuration and to reactions that are *concerted*.

4.9. According to the Fukui approach, e.g., the ring opening of cyclobutene with the formation of butadiene can be viewed as the cycloaddition of a σ-bond to a π bond with suprafacial interactions (shown by double arrow curves) on both the σ and π bond. Ring opening proceeds by a disrotatory mode and the reaction is classified as a $[_\pi 2_s + _\sigma 2_s]$ cycloaddition reaction, allowed by the Woodward–Hoffmann rules for the excited state.

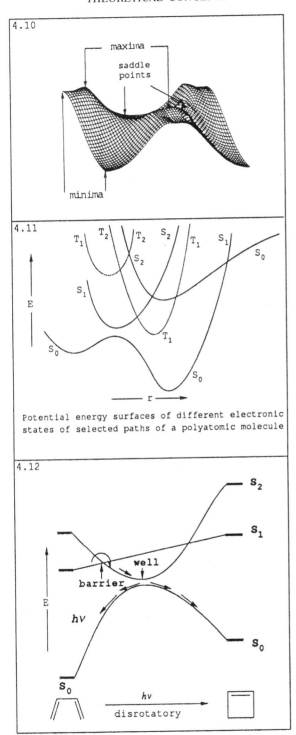

Potential energy surfaces of different electronic states of selected paths of a polyatomic molecule

4.10. In the framework of the Born–Oppenheimer approximation, ground and excited state energies for every arrangement of atomic nuclei can be assigned by a solution of the Schrödinger equation for the motion of electrons in the field of stationary nuclei. Repetitive calculations for different nuclear configurations result in a set of molecular energies that form a potential energy surface. This surface for a system of N nuclei possesses $3N$-5 dimensions. Its characteristic features are: (1) *minima* (stable or metastable configurations); (2) first-order *saddle points* (the saddle points of lowest energy correspond to the activated complex); and (3) *maxima* and higher-order saddle points have no value in the description of a chemical reaction.[5]

4.11. The potential energy surface of a diatomic molecule can be illustrated for different electronic states on the same plot. Polyatomic molecules usually select a one-dimensional path for the change of the potential energy, with respect to the change of nuclear coordinates, and the resulting curve is similar to that of a diatomic molecule.

4.12. According to Woodward–Hoffmann, a pericyclic reaction is restricted by an energy barrier between reactants and products. Van der Lugt and Oosterhoff have shown that the driving force for an allowed pericyclic reaction may be the existence of a minimum (well) on the excited state energy surface. Their calculations[6] indicate that, although the reactant starts on the singly excited state surface, it crosses near the intersection of the doubly excited state surface and the doubly excited state becomes the lowest excited state along the reaction pathway. At the geometry configuration corresponding to a "*nonspectroscopic minimum*," the molecules "jump" to the ground state surface to complete the overall nonadiabatic process.

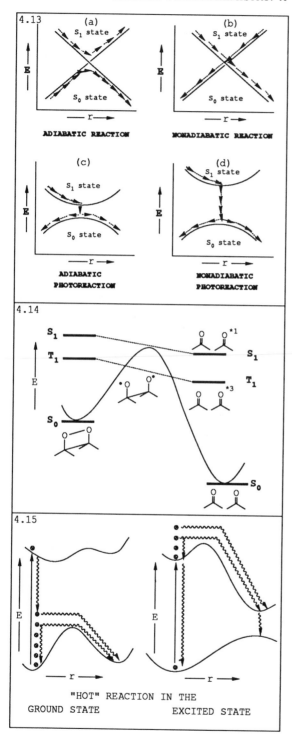

4.13. Reactions are treated by Woodward–Hoffmann as they proceed on either the ground or the excited state potential energy surface [*adiabatic reaction (a)*]. Reactions may, however, proceed on different energy surfaces [*diabatic* or *nonadiabatic reactions (b–d)*], either with *crossing* of the surfaces, if such crossing is allowed (*b*), or by weak (*c*) or strong (*d*) *avoided crossing*,[7] if the crossing is forbidden (*noncrossing rule*).[8] A photoreaction either occurs on a single energy surface or is accompanied by a nonradiative jump to a surface of lower energy. The probability of such a transition is strongly dependent on the energy difference (see 3.16). Therefore, *diabatic transitions* are the most probable in the regions of crossing or avoided crossing.[9,10]

4.14. A nonadiabatic transition takes place not only from an excited to a ground state surface but also the reverse. Thermal decomposition of highly energetic molecules may produce activated complexes of higher energy than that of the excited state of the product. If crossing or avoided crossing of this excited state occurs in the region of the activated complex, a nonadiabatic transition into the excited state can occur, followed by a radiative transition (*chemiluminescence*) into the ground electronic state. The chemiluminescence mechanism of a 1,2-dioxetane calculated on a semiempirical MO level (orbital crossing) is shown.[11]

4.15. *Hot ground state reactions* are special types of nonadiabatic reactions in photochemistry. A molecule is converted by a nonradiative transition into a high vibrational excited ground state in the region of a local minimum in the potential energy curve. Further progress of the reaction requires the molecule to overcome a barrier on the surface of the ground state, which complicates the mechanism. In practice, it is often difficult to distinguish between a true photoreaction and a hot ground state reaction leading to the same products.

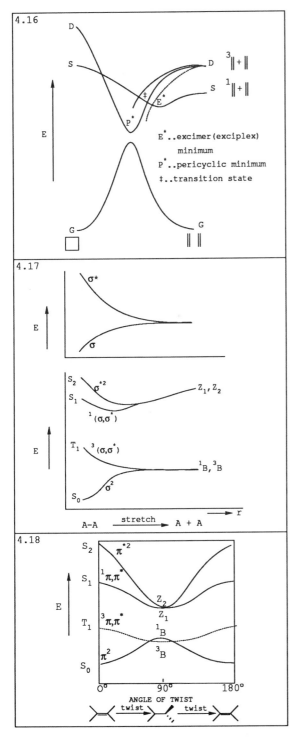

E*..excimer (exciplex) minimum

P*..pericyclic minimum

‡..transition state

A–A $\xrightarrow{\text{stretch}}$ A + A

ANGLE OF TWIST

twist   twist

4.16. The Woodward–Hoffmann rules apply to excited states only if the lowest excited state is well described by a single configuration and if the reaction is concerted. They cannot truly reflect the mechanism of the reaction as do energy profile diagrams. The characteristic features of the energy profile of a pericyclic reaction are the *minima* and *maxima* on the ground state and singlet state surfaces and the *pericyclic* and *exciplex (excimer) minima*. The region where the surfaces approach one another provides information about the electronic states and the geometries of the molecules.[12]

4.17. Many photoreactions, especially those involving triplet states, are stepwise processes accompanied by the formation of biradical intermediates (see 2.18). There are two fundamental pathways to biradical formation: by stretching the σ bond or by twisting the π bond. Stretching the σ bond within homonuclear systems at some distance between the atoms results in nearly degenerate σ and σ^* orbitals that are transformed into nonbonding p orbitals. In this geometry, the S_0 (σ^2) state forms a singlet biradical, ^1B; the T_1 (σ, σ^*) state leads to a triplet biradical, ^3B (both are degenerate with half-filled orbitals). The S_1 (σ,σ^*) and S_2 (σ^{*2}) states form degenerate zwitterionic states, Z_1 and Z_2, both of which have spin-paired electrons in one of the orbitals.[13a]

4.18. By twisting a π bond in a homonuclear system, the π orbitals of the reactant (0° of twisting) and of the product (180° of twisting) cross at a perpendicular geometry (90°). Consequently, the orbitals are degenerate in energy and are subsequently transformed into two p nonbonding orbitals. In the process of twisting, the electronic energies of T_1, S_1, and S_2 decrease significantly and that of S_0 increases sharply. The states of the system in a perpendicular geometry correspond to two biradicals and two zwitterions.[13b]

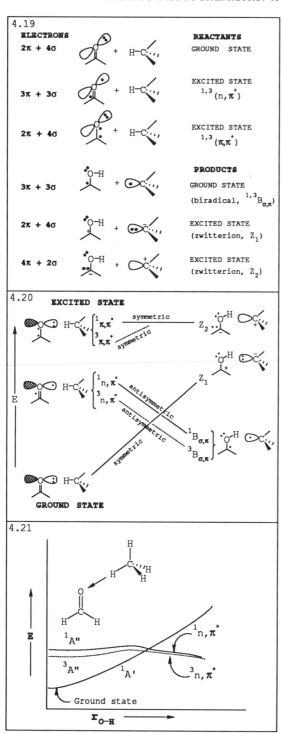

4.19. The concept of singlet and triplet biradicals and zwitterions as primary intermediates in photochemical processes has been employed in the construction of state correlation diagrams for nonconcerted photoreactions (Salem diagrams).[14] An example is the application to the photochemical H abstraction by an excited carbonyl group. The electrons, n, σ, π, and π^* of the reactant and the product involved in the process must first be classified.

4.20. Hydrogen abstraction is assumed to occur in the plane of the carbonyl group. Accordingly, all orbitals lying in this plane (reactant orbitals, σ_{CH}, n_O, σ^*_{CH} and product orbitals, σ_{OH}, σ^*_{OH}, p_C from C—H) are symmetric (S) with respect to the plane. All orbitals lying above or below this plane (reactant, π_{CO}, π^*_{CO}; product, p_O, p_C of the C=O) are antisymmetric (A). Taking into account the noncrossing rule (4.12) and the fact that S_0, S_1, and S_2 must correlate with ^1B, Z_1, and Z_2, respectively, and that T_1 must correlate with ^3B, the correlation diagram can be constructed by connecting states of the same symmetry. Crossing is allowed since the states have different symmetries. Both S_1 and T_1 states correlate directly with the degenerate ground state of the product radical pair, $^1(n,\pi^*)$ with ^1B and $^3(n,\pi^*)$ with ^3B. Therefore, coplanar hydrogen abstraction to form the ketyl radical from either S_1 or T_1 n,π^* states is symmetry allowed. It is important to note that the excited state leads directly downward to the product.

4.21. *Ab initio* calculations slightly modify this result.[15] The calculated surfaces for n,π^* excited singlet and triplet states in the formaldehyde-methane reaction are almost flat rather than directed downward. Another difference not predicted by the correlation diagram is the small activation energies on the calculated surfaces for n,π^* singlet [32 kJ/mol (7.6 kcal/mol)] and triplet [73 kJ/mol (17 kcal/mol)] which have also been observed experimentally.

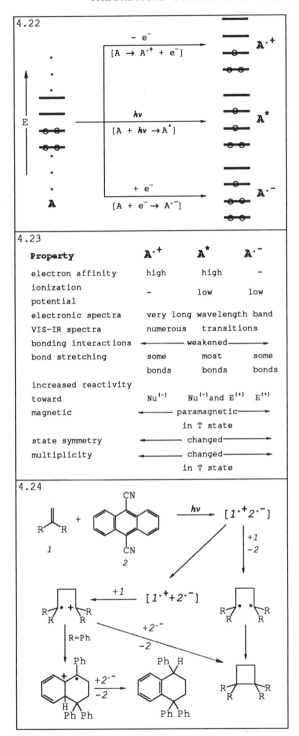

4.22. A diverse group of reactions is based on the formation of radical ions, especially of radical cations, in the photochemical process. A comparison of the orbital energy diagrams of radical ions with that of the excited state demonstrates the similarity in the physical and chemical properties of these diverse systems. They are open-shell systems (see 2.14) that have one (radical ion) or two (electronically excited states) singly occupied MOs, e.g., both the radical anion and the excited state have an odd electron in the LUMO.

4.23. The table reflects the similarity in properties of radical ions and excited states as compared to the singlet ground state. It is important to note that: (1) according to the frontier-orbital theory the existence of a very long wavelength band in the electronic spectrum is a necessary condition for easy monomolecular decomposition; (2) the bond stretching (change of bond order) is generally different for a radical cation and radical anion.

4.24. A versatile method for the formation of radical cation-radical anion pairs in solution is based on *photoinduced- (photosensitized-) electron transfer.*[16] It utilizes the fact that the oxidative ability of an electron acceptor is substantially increased by photoexcitation. The radical cations formed from alkenes or dienes are known to undergo [2+2] or [4+2] cycloaddition reactions by different mechanisms. The originally formed contact ion pair $[1^{\cdot+}2^{\cdot-}]$ either undergoes a further reaction by a back-electron transfer with the formation of a 1,4-biradical or the contact ion pair is separated by solvation to form $[1^{\cdot+} + 2^{\cdot-}]$. The latter then produces a radical cation that yields either a [2+2] dimer or, when R=Ph, a [4+2] dimer.[17]

4.23

Property	$A^{\cdot+}$	A^{*}	$A^{\cdot-}$
electron affinity	high	high	–
ionization potential	–	low	low
electronic spectra	very long wavelength band		
VIS-IR spectra	numerous transitions		
bonding interactions	← weakened →		
bond stretching	some bonds	most bonds	some bonds
increased reactivity toward	$Nu^{(-)}$	$Nu^{(-)}$ and $E^{(+)}$	$E^{(+)}$
magnetic	← paramagnetic in T state →		
state symmetry	← changed →		
multiplicity	← changed in T state →		

REFERENCES

(1) Adapted from Anita Loos, *Gentlemen Prefer Blondes*. Boni & Liveright: New York, 1925, p. 100: ". . . kissing your hand may make you feel very, very good but a diamond bracelet lasts forever."

(2) Kapitza, P. L. "Experiment, Theory, Practice". In *Boston Studies in the Philosophy of Science*. Cohen, R. S. and Wartofsky, M. W., eds. D. Reidel: Dordrecht, Holland, 1980, Vol. 46, p. 160.

(3)(a) Michl, J. and Bonacic-Koutecky, V. *Electronic Aspects of Organic Photochemistry*. Wiley–Interscience: New York, 1990. (b) Woodward, R. B. and Hoffmann, R. *The Conservation of Orbital Symmetry*. Verlag Chemie: Weinheim, 1970. (c) Gilchrist, T. L. and Storr, R. C. *Organic Reactions and Orbital Symmetry*, 2nd ed. Cambridge University Press: Cambridge, 1979. (d) *Orbital Symmetry Papers*. ACS Reprint Collection: Washington, 1974. (e) Dewar, M. J. S. *Angew. Chem., Int. Ed. Engl.* **1971,** *10,* 761. (f) Zimmerman, H. E. *Acc. Chem. Res.* **1971,** *4,* 272. (g) Salem, L. *Electrons in Chemical Reactions*. Wiley: New York, 1982.

(4)(a) Fukui, K. *Topics Curr. Chem.* **1970,** *15,* 1. (b) Fleming, I. *Frontier Orbitals and Organic Chemical Reactions*. Wiley: New York, 1976. (c) Fukui, K. *Acc. Chem. Res.* **1971,** *4,* 57. (d) For a predictor of reactivity in photodimerizations and photocycloadditions, see Caldwell, R. A. *J. Am. Chem. Soc.* **1980,** *102,* 4004.

(5)(a) Turro, N. J. *Modern Molecular Photochemistry*. Benjamin/Cummings: Menlo Park, California, 1978, p. 12. (b) Bonacic-Koutecky, V., Koutecky, J., and Michl, J. *Angew. Chem., Int. Ed. Engl.* **1987,** *26,* 170. (c) Michl, J. *Photochem. Photobiol.* **1977,** *25,* 141. (d) Michl, J. *Mol. Photochem.* **1972,** *4,* 243. (e) Michl, J. *Topics Curr. Chem.* **1974,** *46,* 1. (f) Müller, K. *Angew. Chem., Int. Ed. Engl.* **1980,** *19,* 1. (g) Gerhartz, W., Poshusta, R. D., and Michl, J. *J. Am. Chem. Soc.* **1976,** *98,* 6427.

(6)(a) Adapted from Van der Lugt, W. T. A. M. and Oosterhoff, L. J. *J. Am. Chem. Soc.* **1969,** *91,* 6042. (b) For *ab initio* calculations, see Grimbert, D., Segal, G., and Devaquet, A. *J. Am. Chem. Soc.* **1975,** *97,* 6629. (c) For an MC-SCF study of electrocyclization of *cis*-butadiene and [2+2] dimerization of ethylene, see Bernardi, F., De, S., Olivucci, M., and Robb, M. A. *J. Am. Chem. Soc.* **1990,** *112,* 1737. (d) For a theoretical study of the ring opening reactions of dioxetene, oxetene, dithiete, and thiete, see Yu, H., Chan, W.-T., and Goddard, J. D. *J. Am. Chem. Soc.* **1990,** *112,* 7529.

(7)(a) See ref. 3(f), p. 141. (b) See ref. 5(a), p. 154. (c) Klessinger, M. *Elektronenstruktur organischer Moleküle*. Verlag Chemie: Weinheim, 1982. (d) Klessinger, M. and Michl, J. *Lichtabsorption und Photochemie organischer Moleküle*. Verlag Chemie: Weinheim, 1989, p. 177.

(8) Coulson, C. A. *Valence*, 2nd ed. Oxford University Press: Oxford, 1961, p. 65.

(9)(a) See ref. 5(a), p. 73. (b) See ref. 7 (d) p. 172.

(10)(a) Michl, J. In *Chemical Reactivity and Reaction Path*. Klopman, G., ed. Wiley: New York, 1974. (b) Pearson, R. G. *Symmetry Rules for Chemical Reactions*. Wiley: New York, 1976. (c) Bigot, B., Devaquet, A., and Turro, N. J. *J. Am. Chem. Soc.* **1981,** *103,* 6.

(11) Figure adapted from ref. 5(a), p. 601.

(12)(a) Figure adapted from ref. 4(c). (b) See ref. 5(a), p. 215. (c) See ref. 5(b). (d) See ref. 5(d), p. 257. (e) Salem, L. and Rowland, C. *Angew. Chem., Int. Ed. Engl.* **1972,** *11,* 92. (f) For a theoretical study of spin-orbit coupling in biradicals, see Carlacci, L., Doubleday, C., Jr., Furlani, T. R., King, H. F., and McIver, J. W., Jr. *J. Am. Chem. Soc.* **1987,** *109,* 5323. (g) For a review of a valence-bond model for [2+2] cycloaddition reactions, see Bernardi, F., Olivucci, M., and Robb. M. A. *Acc. Chem. Res.* **1990,** *23,* 405.

(13)(a) Adapted from ref. 5(a), p. 215. (b) Adapted from ref. 5(a), p. 217. (c) See ref. 5(e), p. 220. (d) For a unified view of biradicals and biradicaloids, see Michl, J. and Bonacic-Koutecky, V. *Tetrahedron.* **1988,** *44,* 7559.

(14)(a) Dauben, W. G., Salem, L., and Turro, N. J. *Acc. Chem. Res.* **1975,** *8,* 41. (b) Salem, L. *J. Am. Chem. Soc.* **1974,** *96,* 3486. (c) See ref. 3(f), p. 135. (d) Salem, L. *Isr. J. Chem.* **1975,** *14,* 89. (e) Salem, L. *Science.* **1976,** *191,* 822.

(15) Salem, L., Leforesmer, C., Segal, G., and Wetmore, R. *J. Am. Chem. Soc.* **1975,** *97,* 479.

(16)(a) Mattay, J. *Angew. Chem., Int. Ed. Engl.* **1987,** *26,* 825. (b) Mattes, S. L. and Farid, S. *Org. Photochem.* **1983,** *6,* 233. (c) Mariano, P. S. In *Synthetic Organic Photochemistry*. Horspool, W. M., ed. Plenum: New York, 1984, p. 145. (d) For differences in the reactivity of cyano aromatic radical anions generated by photoinduced electron transfer, see Kellet, M. A., Whitten, D. G., Gould, I. R., and Bergmark, W. R. *J. Am. Chem. Soc.* **1991,** *113,* 358. (e) For exhaustive reviews of the utilization of photoinduced electron transfer in various areas, see *Topics Curr. Chem.* **1990, 156.**

(17) Mattes, S. L. and Farid, S. *J. Am. Chem. Soc.* **1986,** *108,* 7356.

Chapter 5
PHOTOCHEMISTRY OF ALKENES

PHOTOCHEMISTRY OF ALKENES

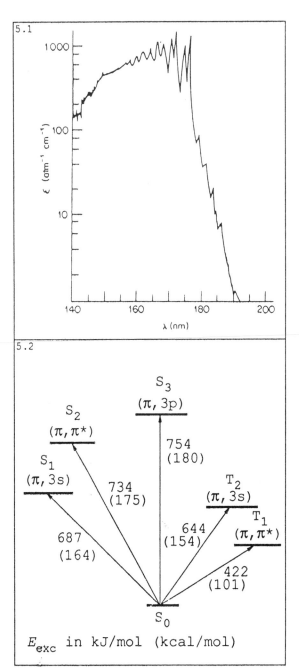

5.1. Electronic absorption spectra of simple alkenes consist of an intense broad band from 140 to 190 nm (for ethylene) with an absorption threshold of 200 (ethylene) to 240 nm (tetramethylethylene). The diffuse bands in the spectrum of ethylene are attributed[1] to a $\pi \rightarrow \pi^*$ transition; the absorption at 174 nm is the first singlet Rydberg transition, $\pi \rightarrow 3s$. Recent theoretical and experimental studies have shown, however, that the electronic spectrum of ethylene is much more complicated. The fine structure on the intense band is the result of at least three additional electronic transitions that may have fundamental significance in the study of the mechanisms of singlet photoreactions of alkenes. However, it is virtually impossible to correlate alkene photoreactivity with either known[2,3] or calculated[4,5] excited states.[6]

5.2. The excited states of alkenes are complex because the different electronic states have different electronic configurations. Although the nature of most electronic states of alkenes is not known in detail, some are well understood.[7] Alkenes have two low-lying excited singlet states: the $^1(\pi,3s)$ Rydberg state[8] and the $^1(\pi,\pi^*)$ valence state. Calculations indicate that there is apparently an additional state in the vicinity of these two states, i.e., the $(\pi, 3p)$ Rydberg state. The relatively high energy of the valence singlet π,π^* state can be understood in view of strong singlet-triplet (S–T) splitting (see 3.18). This splitting almost vanishes for electronic transitions between the π and 3s orbitals. The lowest triplet state is practically pure $^3(\pi,\pi^*)$ and T_2 is essentially a pure Rydberg $^3(\pi,3s)$ state. The excitation energies of the lowest excited states of ethylene are shown.

5.3

ELECTRONIC STATES OF ETHYLENE

State	Symmetry	Wavelength (nm)	Energy (eV)
S_0	$^1A_{1g}$	----	----
T_1	$^3B_{2g}$	284	4.36
T_2	$^3B_{2u}$	186	6.66
S_1	$^1B_{3u}$	174	7.11
S_2	$^1B_{1u}$	163	7.60
S_3	$^1B_{1g}$	159	7.80

5.4

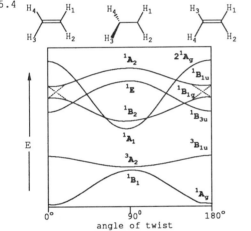

$^1B_1$=^{1}Biradical; $^3A_2$=^{3}Biradical; $^1A_1$= (+)Zwitterion; $^1B_2$=(-) Zwitterion; ^1A_g=Ground state $^1(\pi^2)$; $^3B_{1u}$= $^3(\pi,\pi*)$; $^1B_{3u}$=Rydberg $^1(\pi,3s)$; $^1B_{1g}$=Rydberg $^1(\pi,3p_y)$; $^1B_{1u}$=Valence $^1(\pi,\pi*)$; 2^1A_{1g}=Valence $^1(\pi,\pi*)^2$.

5.5

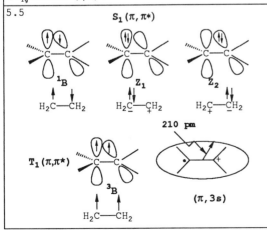

5.3. Most alkenes have several low-lying singlet states of comparable energy, while their triplet states are well separated. The S–T splitting is generally large for alkenes and dienes [ethylene = 292 kJ/mol (70 kcal/mol)]. As a result, ISC is slow and inefficient. Therefore, direct irradiation of alkenes induces singlet excited state reactions, and sensitization is required for triplet state reactions.[9]

5.4. The potential energy diagram for several of the lowest electronic states of ethylene shows the dependence of their energy on the angle of rotation about the C—C bond.[10] Excitation followed by a rapid relaxation to orthogonal geometry and the release of vibrational energy is due to the high energy of the π^* MO compared to the π MO. Thus, there is effectively no π bond in the π,π^* excited state of the alkene, and the rotation about the σ bond produces a geometry of the lowest energy and minimum electronic interaction (orthogonal geometry). In addition, minima in this and other surfaces occur at orthogonal geometry as a result of an avoided crossing of excited states of the same symmetry. Consequently, the lowest excited singlet state of orthogonal ethylene is the $^1(\pi,\pi^*)$ state. Many similar potential energy diagrams of ethylene on different levels of sophistication have been reported.[5,11]

5.5. Orthogonal ethylene has degenerate frontier orbitals and is characterized by the diagram in 2.18 indicating that twisted ethylene in the S_1 state has a polar-biradical (biradicaloid) structure [a linear combination of a biradical structure and two polar (zwitterionic) structures], whereas the orthogonal T_1 and S_0 states are pure biradicals.[12] The Rydberg $(\pi,3s)$ state can be visualized as a radical cation orbited by an electron (not an ionized state; see also 4.18).[13,14]

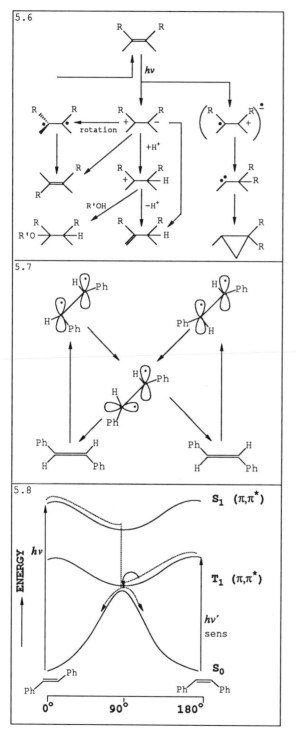

5.6 The photoreactions of simple alkenes, upon direct excitation or in sensitized processes, are quite complex because excited states of different electronic configurations compete.[13(b), 15] The participation of the $^1(\pi,\pi^*)$ state results mainly in *cis-trans* isomerization, [1,3] hydrogen shift or alkoxylation (hydration), and [2+2] cycloaddition (dimerization) reactions *via* an exciplex (excimer). The Rydberg $^1(\pi,3s)$ state participates in [1,2] hydrogen or alkyl shifts with the formation of a carbene intermediate and in photosolvolysis reactions. The $^3(\pi,\pi^*)$ is involved in photosensitized *cis-trans* isomerizations and [2+2] cycloadditions via 1,4-biradicals.

5.7. The most facile photochemical process of alkenes, as well as azomethines and aromatic azocompounds (see 11.8 and 11.9), is *cis-trans* isomerization. Using the simple geometric model of alkenes in an electronically excited state (see 5.4 and 5.5) the individual steps in this isomerization can be easily visualized.

5.8. There are two possible pathways for *cis-trans* isomerization.[16] (1) After absorption of a photon the alkene *via* one of the singlet states, by twisting, reaches the biradicaloid minimum (orthogonal geometry) on the hypersurface of S_1. Since singlet-singlet (S → S) transitions are allowed, it can efficiently return to the hypersurface of the S_0 state, forming either geometric isomer.[17] (2) After sensitization, the alkene in the biradical geometry at the minimum on the T_1-hypersurface will reach the S_0 hypersurface through a very inefficient ISC process. Although triplet-singlet (T → S) transitions are forbidden, the higher efficiency of ISC results from the fact that twisted T_1 and S_0 states are nearly degenerate (very small energy difference). Direct or triplet-sensitized *cis-trans* isomerization is an important procedure for the synthesis of the thermodynamically less stable *cis* isomers.[18, 19]

5.9

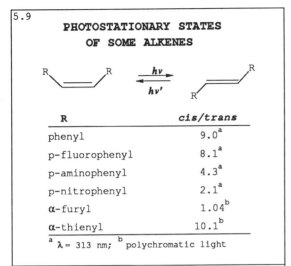

**PHOTOSTATIONARY STATES
OF SOME ALKENES**

R	cis/trans
phenyl	9.0[a]
p-fluorophenyl	8.1[a]
p-aminophenyl	4.3[a]
p-nitrophenyl	2.1[a]
α-furyl	1.04[b]
α-thienyl	10.1[b]

[a] λ = 313 nm; [b] polychromatic light

5.10

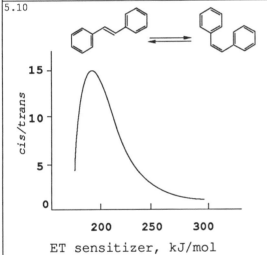

ET sensitizer, kJ/mol

5.11

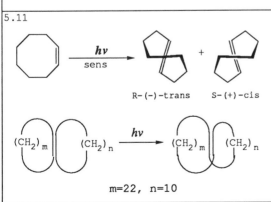

R-(-)-trans   S-(+)-cis

m=22, n=10

5.9. For most alkenes $(\epsilon_{max})_{trans}$ is greater than $(\epsilon_{max})_{cis}$ of the longest wavelength band, e.g., 16,300 and 2,280 Lmol^{-1} for stilbene at 313 nm. Therefore, in an equimolar mixture more molecules of *trans* (t) than *cis* (c) isomer will reach the excited singlet state, followed by rapid relaxation to S_0 (see 5.8). Since the rates of c → t and t → c isomerization become equal, no change in the composition of the reaction mixture occurs upon further irradiation (*photostationary state*). This rate is also a function of $\Phi_{t \to c}/\Phi_{c \to t}$ (1.5 for stilbene at 313 nm) and temperature. The composition of the photostationary state is a function of concentration[20] and the wavelength of the incident radiation[21] (the rates of isomerization depend on ϵ_λ, which varies with λ).[22,23]

5.10. The composition of the photostationary state in triplet-sensitized *cis-trans* isomerizations depends on the E_T of the sensitizer and, generally, is different than that from direct irradiation.[23,27] The isomerization of stilbene[24-26] can be induced even by sensitizers whose triplet energy lies below that of both geometric isomers. This was originally explained by assuming the existence of a *phantom triplet*,[27(b),28] later shown to be the triplet state of the twisted alkene, whose E_T lies below that of both geometric isomers. The advantage of this method is its applicability to alkenes, which absorb at short wavelengths (conventional techniques require substrates that absorb at λ > 230 nm).

5.11. Direct or sensitized irradiation of *cis*-cycloalkenes yields the often highly strained trans isomers. Some can be isolated ($\geq C_8$)[29]; others are stable in the form of complexes (C_7)[30] or intermediates (C_6, C_7) which undergo further transformations in the S_0 state,[31] e.g., trapping by a molecule of cycloalkene to form a dimer. Photoisomerization of bicyclic alkenes yields [a.b]*betweenanenes* in which a tetrasubstituted double bond is shared by a pair of trans cycloalkyl groups.[32]

5.12

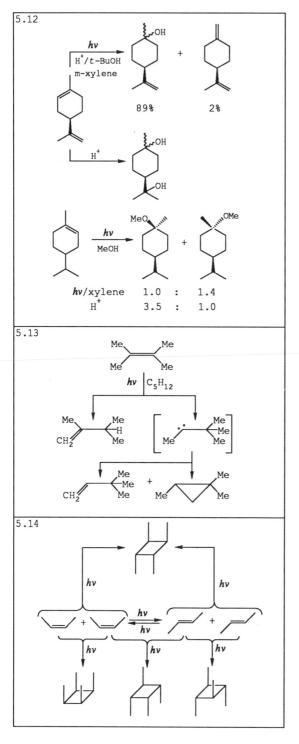

5.13

5.14

5.12. Direct or sensitized irradiation of cyclic alkenes (C_6, C_7) in protic solvents, e.g., methanol, water, or acetic acid, follows another pathway. The xylene-sensitized photolysis of a cycloalkene in protic solvents causes protonation of the highly strained trans cycloalkene with the formation of a cycloalkane carbocation, stabilized by a subsequent thermal reaction (rearrangement of a double bond, hydration, or alkoxylation).[15(e),(f);31;33] This mechanism is supported by experiments with CH_3OD, in which the expected deuterium substitution pattern was observed.[34] The hydration reaction gives products whose structures are different than those of ground state hydration products.[35] The acid-catalyzed addition of methanol to a cycloalkene, unlike the photoaddition of methanol, produces the thermodynamically more stable trans isomer. The photoaddition of methanol is a kinetically controlled reaction.[35]

5.13. The Rydberg $^1(\pi,3s)$ state of trialkyl and tetraalkyl ethylenes is lower in energy than the $^1(\pi,\pi^*)$ state because the energy of the latter is less affected by alkyl substituents than the Rydberg state. The major products of this photochemical singlet reaction are the result of Rydberg $\pi \rightarrow 3s$ excitation, which leads to a carbene intermediate.[36]

5.14. Reactions competing with *cis-trans* isomerization at increased concentrations of the alkene(s) are photocycloaddition reactions. An alkene in its S_1 or T_1 state reacts with either the same (*photodimerization*) or a different (*photocycloaddition*) alkene in the ground state to produce a cyclobutane derivative. In both these processes two π bonds vanish with the formation of two new σ bonds. Since 2π electrons of each alkene are involved, they are called [$_\pi 2 + _\pi 2$], or more simply, [2+2] cycloaddition reactions.[37]

5.15. A [2+2] photocycloaddition (dimerization) proceeds from the S_1 state via an *exciplex minimum 1* (*excimer*, see 3.26). Return to the S_0 hypersurface from S_1 occurs through a pericyclic intermediate (P*) at the pericyclic minimum 2. Since the bonds in P* are in a cyclic array, the stereochemistry of the starting alkene is preserved (stereospecific) and it is a concerted reaction. The decay of P* on the S_0 hypersurface may form either the cycloadduct (dimer) or the starting alkene. P* formed in the T_1 state has the geometry of a 1,4-biradical with free rotation about the σ bonds. This leads to the loss of stereocontrol; however, the reaction is still regiocontrolled through the most stable 1,4-biradical.[38]

5.16. The stereochemistry of photocycloaddition (dimerization) from the S_1 state along a concerted path obeys the Woodward–Hoffmann selection rules: *suprafacial* (*antarafacial*) interaction leads to retention (inversion) of configuration at the terminal atoms (see also 4.2–4.4 and 4.8–4.9). The allowed suprafacial interaction in the exciplex* is favored by electronic factors and results in *cis* or *syn* stereochemistry of the products.[39] The same factors favor the head-to-head regiochemistry in the exciplex, although head-to-tail products have been reported in cycloadditions in the absence of apparent stereochemical restrictions.[40] Several models have been developed for the prediction of stereo- and regiochemistry of these reactions.[38(d),41–43]

5.17. Although the dimerization of simple alkenes by direct irradiation is technically difficult ($\lambda_{abs} \le 200$ nm), increased alkyl substitution of ethylene causes a bathochromic shift of the absorption band, so that direct irradiation ($\lambda = 214$ nm) of neat *cis*- and *trans*-2-butenes gives the corresponding dimers.[44] Photosensitized dimerization occurs easily with cycloalkenes, particularly small ring alkenes, because they cannot undergo *cis-trans* isomerization.[45]

5.15

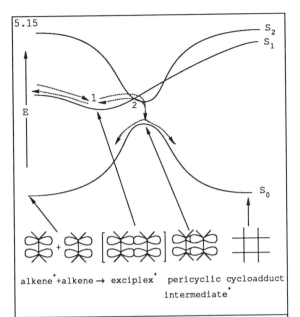

alkene*+alkene → exciplex* pericyclic cycloadduct
                            intermediate*

5.16

**'SELECTION RULES
OF CONCERTED PHOTOCYCLOADDITIONS**

Number of electrons	Photocycloaddition	
	allowed	forbidden
$4q$ ($q=1,2,3...$)	supra–supra antara–antara	supra–antara antara–supra
$4q+2$	supra–antara antara–supra	supra–supra antara–antara

5.17

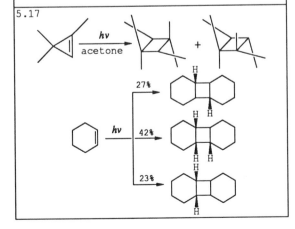

5.18

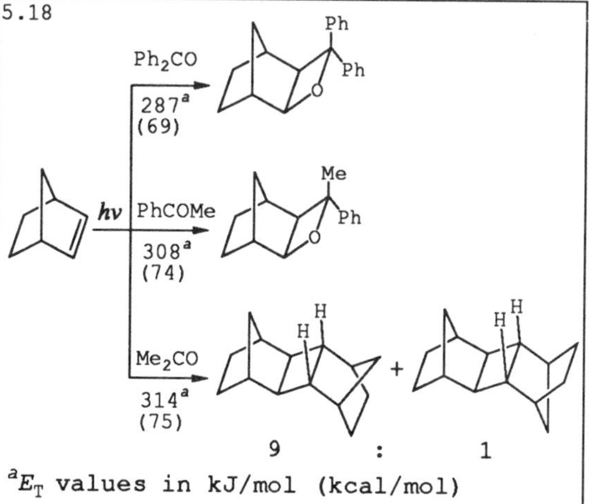

^aE_T values in kJ/mol (kcal/mol)

5.19

5.20

X = CH$_2$, C=O, C=CH$_2$

5.18. Triplet-sensitized photodimerization is applicable to simple alkenes that absorb at very short wavelengths. Carbonyl compounds are widely used as triplet sensitizers if their E_T is greater than that of the alkenes. The E_T of an alkene can be so estimated (see acetone *vs.* acetophenone). The dimerization of alkenes, especially small ring cycloalkenes, is best sensitized by aliphatic ketones.[46] Oxetanes are formed from aromatic ketones.[47] The photodimerization of norbornene sensitized by Cu triflate gives a different ratio of dimers than that of the acetone-sensitized reaction (see 15.44).

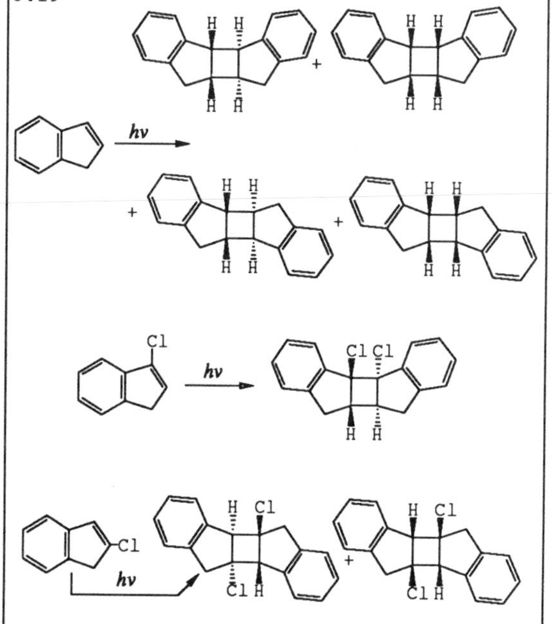

5.19. Four dimers that differ in regio- and stereochemistry are formed by the photodimerization of indene (IN). The ratio of dimers depends on the irradiation conditions. The triplet-sensitized dimerization, as a stepwise bond formation process, generally produces the thermodynamically more stable exo products from the more stable 1,4-biradical intermediate (head-to-head). The singlet reaction of IN leads to the endo head-to-head dimer as the major product (probably from an excimer with overlapping π orbitals).[48] This dimer is only 3% of the product from the triplet-sensitized reaction. However, the major product in both dimerizations is the less sterically hindered *exo* head-to-head dimer (83% and 69%).[48] It is unclear whether the regioselectivity in the photodimerization of chloroindene is the result of a steric or a heavy-atom effect.[49]

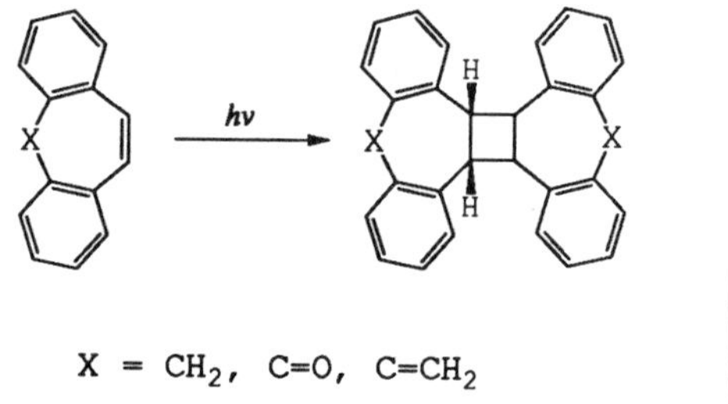

5.20. The double bond in dibenzo[a,e]cycloheptatriene (X=CH$_2$) is arranged as in *cis*-stilbene. This compound, as well as its oxo (X=CO) and methylene (X=C=CH$_2$) derivatives, dimerizes readily upon direct irradiation. In agreement with HMO calculations, no dimerization or cycloaddition products of exocyclic double bonds have been observed. In all cases photodimers and cycloadducts have an *anti* configuration.[50]

5.21

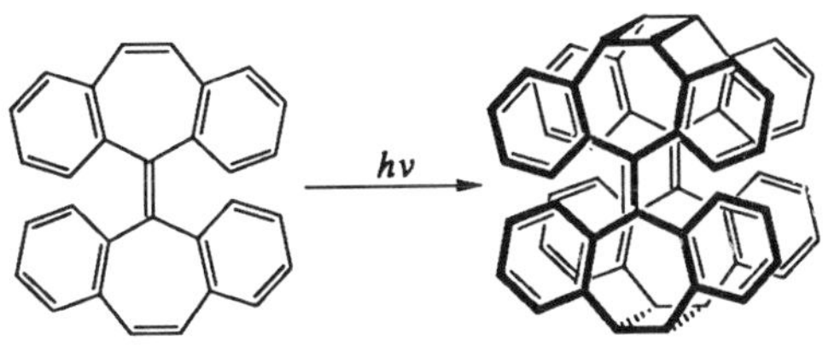

5.21. The photodimerization of tetrabenzoheptafulvalene occurs by a double dimerization of the endocyclic double bonds, with the formation of a remarkable cage compound in which the exocyclic double bonds of the heptafulvalene system are not involved.[51]

5.22

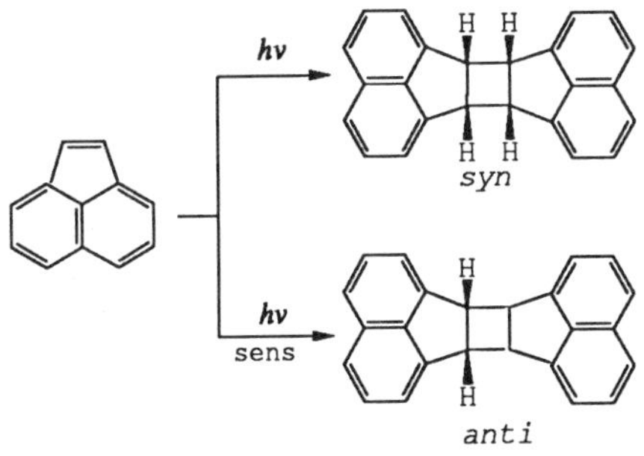

Solvent	mol%	Yield of dimer, g		
		syn	anti	s+a
cyclohexane	100	5.4	1.1	6.5
1-chlorobutane	100	5.0	2.1	7.1
1-bromobutane	10	2.5	3.4	5.9
	50	3.5	7.5	11.0
	100	4.1	10.0	14.1
ethyl iodide	5	2.0	8.4	10.4
	10	2.8	11.0	13.8
ethylene dibromide	10	1.8	4.3	6.1

5.22. The photodimerization of acenaphthylene produces a mixture of *syn* and *anti* dimers (heptacyclenes).[52] Direct irradiation yields the *syn* dimer as the major product in a stereospecific reaction. The singlet excited complex (exiplex) of two acenaphthylene molecules apparently prefers mutual overlap of the naphthalene rings of acenaphthylene. The major product of the sensitized dimerization is the *anti* dimer.[53] Irradiation of acenaphthylene in heavy-atom (H–A) solvents (BrCH$_2$CH$_2$Br, EtI, etc.) shifts the product distribution in favor of the triplet-derived *anti* dimer.[54] This is an example of the *heavy-atom effect* (see 3.14 and 3.15).[55] An increase in the concentration of the H–A solvent enhances the proportion of the *anti* dimer and the total yield of dimers. In a heavy-atom environment, nonradiative S–T transitions (ISC) in the π,π^* state are accelerated, but not in the n,π^* state. Utilization of the heavy-atom effect in photochemical syntheses is discussed in 15.15.[56] In light-atom solvents ISC is relatively inefficient and dimerization occurs primarily from the S$_1$ state to give predominantly the *syn* dimer.

5.23

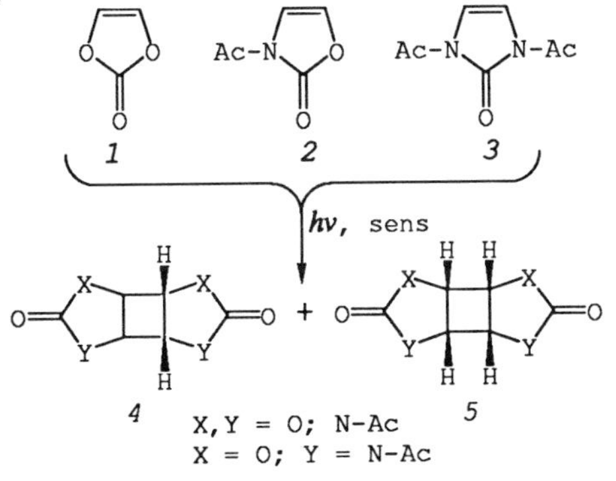

5.23. Heterocyclic *ene* systems, such as vinylene carbonate (*1*),[57] 3-acetyl-4-oxazolin-2-one (*2*),[58] and 1,3-diacetylimidazolin-2-one (*3*),[59] undergo sensitized dimerization to produce the corresponding *anti* dimer *4* and *syn* dimer *5*, with (*1*). If benzophenone is used as the sensitizer, the corresponding oxetane is formed, instead of the dimer. [2+2] photocycloaddition reactions of these compounds have been reported and some have been utilized in the synthesis of natural products (see, e.g., 5.27).

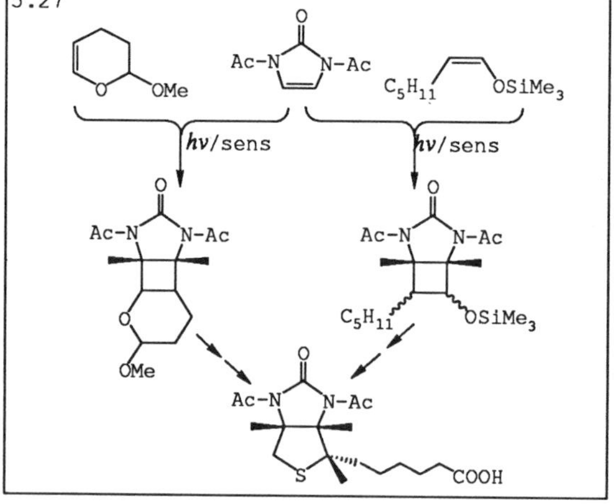

5.24. In addition to prerequisites for a dimerization reaction, a successful [2+2] photocycloaddition with different alkenes requires that the two reactants have sufficiently different absorption properties so that only one component is excited. The second component, in large excess or as a solvent, acts as a trap and, thereby, increases the yield of photocycloaddition and prevents dimerization. Under these conditions, charge transfer bands may appear at longer wavelengths and can be used for cycloaddition (see 7.21 and 7.22). When the photocycloaddition occurs from the S_1 state, it is a $[\pi 2_s + \pi 2_s]$ concerted process with retention of stereochemistry. No *cis-trans* isomerization of any component in Eq. (1) was observed.[60] Singlet cycloaddition reactions are often highly regioselective.

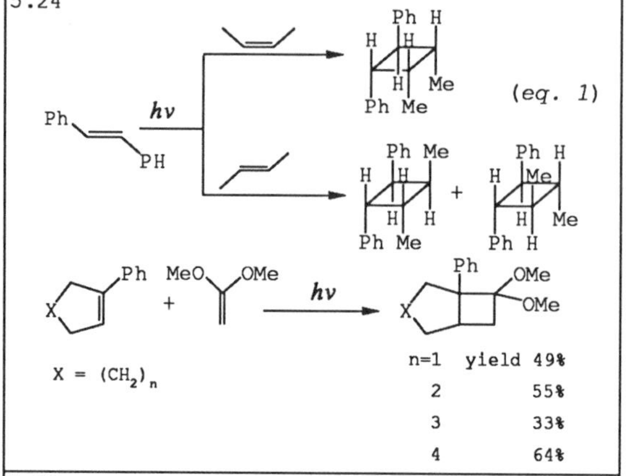

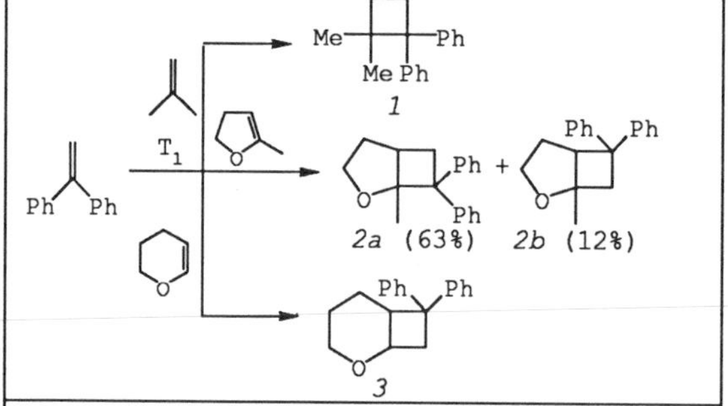

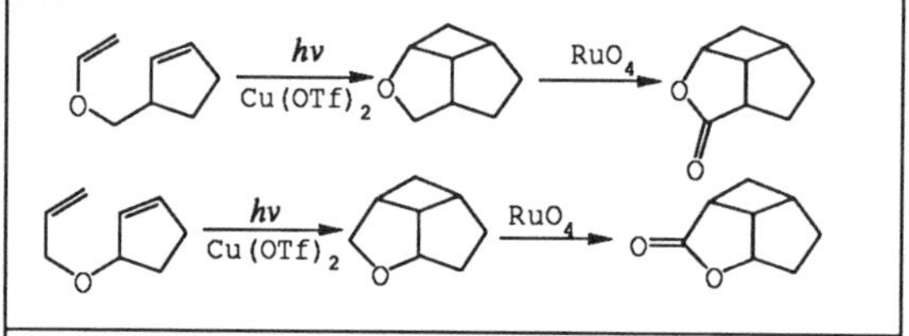

5.25. Triplet-sensitized [2+2] cycloaddition generally proceeds through the most stable 1,4-biradical (free rotation and loss of stereochemical memory) and is, therefore, regioselective with preferred formation of head-to-head products (*1,2a*).[62] However, some cycloaddition reactions do not involve the most stable 1,4-biradical (*2b,3*).[62(b),63] A method based on HMO coefficients has been developed for the prediction of regioselectivity.[40]

5.26. Intramolecular Cu(I)-catalyzed [2+2] photocycloaddition of homoallyl, vinyl, or diallyl ethers provides an effective route for the synthesis of polycyclic tetrahydrofurans whose further oxidation (RuO$_4$) yields *cis*-fused cyclobutanated butyrolactones.[64]

5.27. It is advantageous if the C=C bonds in both components of a sensitized [2+2] cycloaddition are part of a small ring (C$_3$—C$_6$) since the T$_1$ state will not dissipate its energy by *cis-trans* isomerization. The key step in the synthesis of (±)-biotin is the [2+2] cycloaddition of an enol ether to 1,3-diacetylimidazolin-2-one. The cycloaddition of the cyclic enol ether is more favorable.[65]

5.28

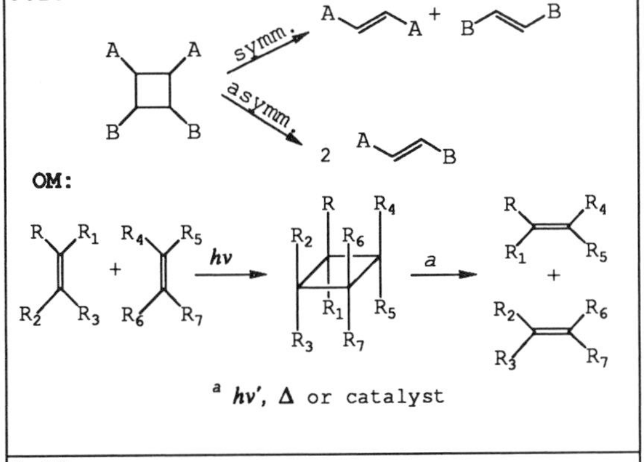

OM:

^{a} hv', Δ or catalyst

5.29

HO

oxid

5.30

R=H; yield 69-72%

5.28. Derivatives of cyclobutane undergo ring opening to the original alkenes with short wavelength irradiation, thermally or in the presence of a catalyst. This scission can proceed in two ways: symmetrically, producing the original alkenes, or asymmetrically, producing two different alkenes. The cleavage is part of the multistep *olefin metathesis* (OM), a method which involves the exchange of alkylidene groups between two olefins. Multistep OM is a valuable synthetic method since: (1) cyclobutanes are generally accessible by [2+2] photocycloaddition; (2) by altering the structure of the starting alkenes, a variety of skeletal changes are possible by performing the [2+2] cycloaddition intra- or intermolecularly, with open-chain or cyclic alkenes, etc.[66] The retrocleavage of a cyclobutane ring is also achieved enzymatically and is an important protecting mechanism in biological systems (see 16.29).

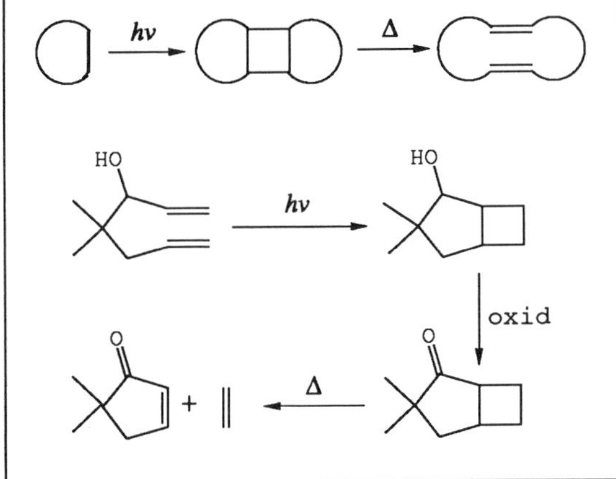

5.29. The stepwise *OM* method has little synthetic value for simple alkenes. A few important examples involve the synthesis of macrocyclic dienes.[66(a)] Its synthetic application is largely utilized in inter- and intramolecular [2+2] photocycloaddition reactions of ketones and enones to alkenes with subsequent fragmentation of the resulting cyclobutane derivative. With dienes, the first step is usually an *intramolecular* [2+2] photocycloaddition.[67] Additional examples will be given in appropriate chapters.

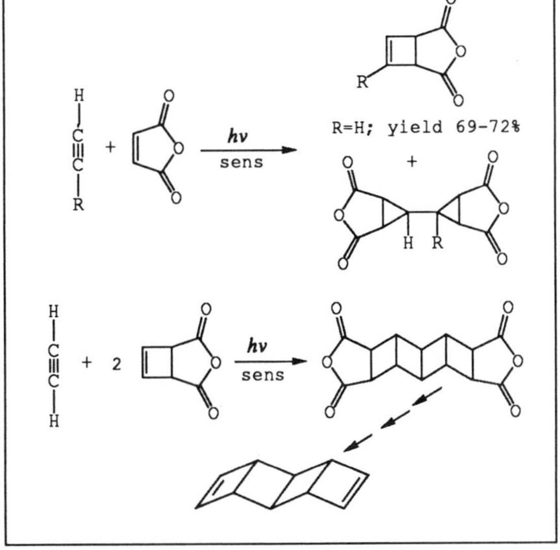

5.30. Alkynes undergo acetone- or acetophenone-sensitized [2+2] photocycloaddition to maleic anhydride with the formation of cyclobutene derivatives that can undergo further cycloaddition, as shown by the synthesis of pterodactyladiene.[68,69] If R=H or CH$_3$, a mixture of stereoisomeric bicyclopropanes is also formed, apparently via a temperature- and concentration-dependent carbene mechanism.[70,71]

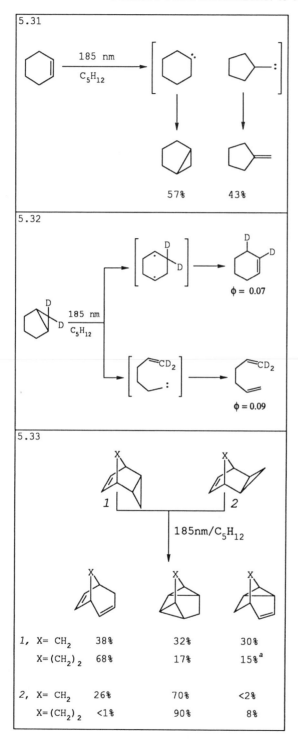

5.31

185 nm
C₅H₁₂

57%   43%

5.32

185 nm
C₅H₁₂

φ = 0.07

φ = 0.09

5.33

1   2

185nm/C₅H₁₂

1, X= CH₂	38%	32%	30%
X=(CH₂)₂	68%	17%	15%ᵃ
2, X= CH₂	26%	70%	<2%
X=(CH₂)₂	<1%	90%	8%

5.31. Organic photoreactions in solution are usually accomplished with UV radiation of λ >250 nm. However, this radiation is insufficient to promote most simple unsubstituted chromophores to their lowest singlet state. To excite such molecules far-UV radiation (185–214 nm) is required. The energy (598 kJ/mol, 143 kcal/mol) corresponding to these wavelengths is more than sufficient to cleave any bond in a typical organic molecule.[72] Surprisingly, irradiation with far-UV light in solution usually results in highly selective and efficient photochemical reactions with high product quantum yields.[73]

5.32. The products of the previous reaction result from a competition between a 1,2-alkyl migration (methylenecyclopentane) and a 1,2-hydrogen shift (bicyclo[3.1.0]hexane) in the Rydberg state. Further irradiation of the latter yields a product that results from one-bond fragmentation followed by a 1,2-hydrogen shift (cyclohexene), together with a product from two-bond fragmentations (alkenyl carbene intermediate), followed by a 1,2-hydrogen migration (1,5-hexadiene). Similar transformations have been observed with other bicyclo[n.1.0]alkanes.[74]

5.33. Competition between the one-bond and two-bond fragmentation with far-UV irradiation is the general behavior of alkylcyclopropanes in their excited singlet state. However, allylcyclopropanes with two chromophores present in the molecule give new products that are usually not observed with isolated chromophores, as shown in 5.31 and 5.32. Because of through-space interaction in allylcyclopropanes, the products are the result of either a formal [2π + 2σ] intramolecular cycloaddition (predominantly, the exo isomer) or an initial cyclopropane bond cleavage with the formation of the diene and subsequent transformation into the tricyclic compound (predominantly, the *endo* isomer).[75]

5.34

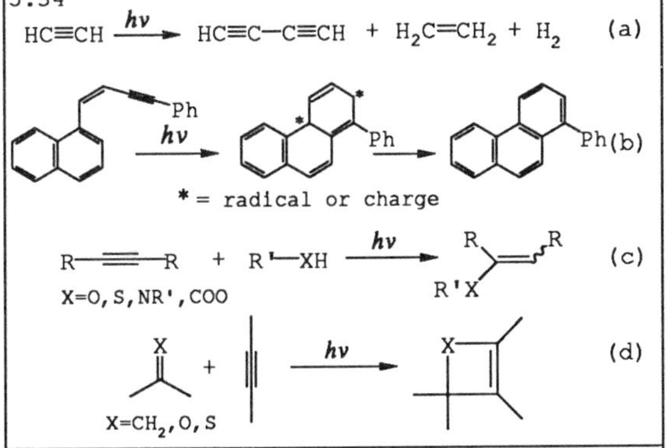

$HC\equiv CH \xrightarrow{h\nu} HC\equiv C-C\equiv CH + H_2C=CH_2 + H_2$ (a)

(b)

* = radical or charge

(c)

X=O, S, NR', COO

(d)

X=CH$_2$, O, S

5.35

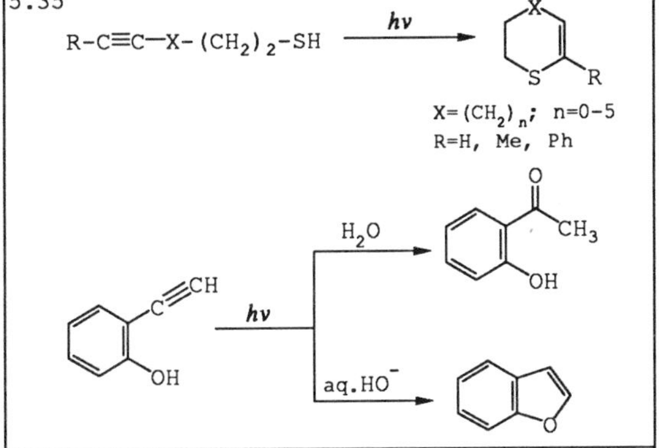

X=(CH$_2$)$_n$; n=0-5
R=H, Me, Ph

5.36

X=CH$_2$, O

X=CH$_2$; yield 82%
X=O; yield 64%

yield 88%

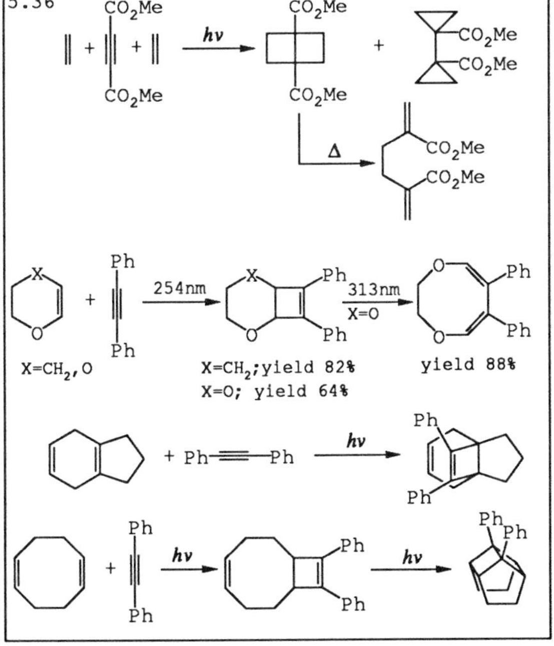

5.34. The electronic configuration of both the lowest singlet (S_1) and triplet (T_1) states of simple alkynes (acetylenes) is π,π^*. These states possess a nonlinear transoid geometry (see 2.39).[76] Unlike alkenes that exhibit diverse photochemical reactions because of a number of low-lying electronically excited states of comparable energy (see 5.1 through 5.3) and that are more readily available (including cycloalkenes), the only reactive states of simple alkynes of interest to organic photochemists are the S_1 and T_1 π,π^* excited states. Although alkynes undergo fragmentation (a) and rearrangement (b) reactions, the more important photochemical reactions of simple alkynes are inter- and intra-molecular (c) addition reactions, and cycloaddition reactions (d).[77]

5.35. Irradiation of an alkyne in the presence of protic solvents,[78-80] such as acetic acid, alcohols, thiols, or secondary amines, produces enolacetates, enol ethers, or enamines. The intramolecular arrangement gives rise to a cyclic product and the reaction pathway is sometimes dependent on the reaction conditions, as shown by the hydration of *o*-hydroxyphenylacetylene.[81]

5.36. Photocycloaddition reactions of alkynes occur with one or two[82] molecules of alkenes, aromatic compounds, carbonyl, and thiocarbonyl compounds, etc. Some of these are mentioned in 5.30, 7.25 through 7.27, 9.43, and 15.54. Bicyclic products having a cyclobutene moiety are formed from cycloalkenes.[83] Some of these products exhibit a further ring expansion photoreaction at a different wavelength.[84] The photocycloaddition of alkynes to bicyclo[4.n.0]dienes leads to propellane derivatives.[85] Although the initial product of the photoreaction of diphenylacetylene with 1,5-cyclooctadiene is the cyclobutene derivative, subsequent photocycloaddition of the cyclobutene double bond with the remaining isolated double bond results in a cage compound in high yield.[86]

REFERENCES

(1)(a) Spectrum adapted from Zelikoff, M. and Watanabe, K. *J. Opt. Soc. Amer.* **1953**, *43*, 756.   (b) See also Merer, A. J. and Mulliken, R. S. *Chem. Rev.* **1968**, *69*, 639.

(2) Robin, M. B. In *Higher Excited States of Polyatomic Molecules*. Academic Press: New York, 1985, Vol. III, pp. 213–236, and references therein.

(3) Lee, Y.-P. and Pimentel, G. C. *J. Chem. Phys.* **1981**, *75*, 4241.

(4) Bouman, T. D. and Hansen, A. E. *Chem. Phys. Lett.*, **1981**, *117*, 461.

(5) Palmer, M. H., Beveridge, A. J. Walker, I. C., and Abuain, T. *Chem. Phys.* **1986**, *102*, 63.

(6) For an excellent review of electronic states of simple alkenes, see Collin, G. J. *Adv. Photochem.* **1988**, *14*, 135.

(7) Robin, M. B. In *Higher Excited States of Polyatomic Molecules*. Academic Press: New York, 1975, Vol. II.

(8) For a review of Rydberg states, see Mulliken, R. S. *Acc. Chem. Res.* **1976**, *9*, 7.

(9)(a) The data in the table has been extracted from Ref. 6, p. 141.   (b) For relaxed and spectroscopic energies of triplet alkenes, see Ni, T., Caldwell, R. A., and Melton, L. A. *J. Am. Chem. Soc.* **1989**, *111*, 457.

(10)(a) Buenker, R. J., Bonačić-Koutecký, V., and Pogliani, L. *J. Chem. Phys.* **1980**, *73*, 1836, and references therein.   (b) Bonačić-Koutecký, V., Pogliani, L., Persico, M., and Koutecký, J. *Tetrahedron*. **1982**, *38*, 741, and references therein.

(11)(a) Buenker, R. J., Peyerimhoff, S. D., and Kramer, W. E. *J. Chem. Phys.* **1971**, *55*, 814.   (b) Petrogonlo, C., Buenker, R. J., and Peyerimhoff, S. D. *J. Chem. Phys.* **1982**, *76*, 3655.

(12)(a) Wulfman, C. E. and Kumei, S. E. *Science*. **1971**, *172*, 1061.   (b) For a review, see Michl, J. *Mol. Photochem.* **1972**, *4*, 243.

(13)(a) Robin, M. B. In *Higher Excited States of Polyatomic Molecules*. Academic Press: New York, 1975, Vol. II, and references therein.   (b) Kropp, P. J. *Org. Photochem.* **1978**, *4*, 1.

(14) For a symmetry-based procedure for molecular geometry changes of alkenes following electronic excitation, see Bachler, V. *J. Am. Chem. Soc.* **1988**, *110*, 5977.

(15) For selected reviews of photochemistry of alkenes, see   (a) Caldwell, R. A. and Creed, D. *Acc.*

Chem. Res. **1980**, *13*, 45.   (b) Coyle, J. D. *Chem. Soc. Rev.* **1974**, *3*, 329.   (c) Fonken, G. J. *Org. Photochem.* **1967**, *1*, 197.   (d) Kaupp, G. In *Methoden der Organischen Chemie*/(Houben-Weyl). G. Thieme: Stuttgart, 1975, Vol. 4/5a: Photochemie, p. 278.   (e) Kropp, P. J. *Mol. Photochem.* **1978–1979**, *9*, 39.   (f) Kropp, P. J. *Pure Appl Chem.* **1970**, *24*, 585.   (g) Swenton, J. S. *J. Chem. Educ.* **1969**, *46*, 7.   (h) Dorofeev, Y. I. and Skurat, V. E. *Russ. Chem. Rev.* **1982**, *51*, 527.   (i) For a recent survey of the photochemistry of simple alkenes, see Steinmetz, M. G. *Org. Photochem.* **1987**, *8*, 67.   (j) For a review of the photochemistry of allyl-, vinyl-, and alkylidene-cyclopropane, see Leigh, W. J. and Srinivasan, R. *Acc. Chem. Res.* **1987**, *20*, 107.

(16) Bruckmann, P. and Salem, L. *J. Am. Chem. Soc.* **1976**, *98*, 5073.

(17) For the dynamics of the singlet *cis-trans* photoisomerization, see Persico, M. and Bonacic-Koutecky, V. *J. Chem. Phys.* **1982**, *76*, 6018.

(18)(a) For multiple Z/E-photoisomerizations of macrocycles with configurational change up to six C=Cs on a single excitation, see Sundahl, M., Wennerström, O., Sandros, K., and Norinder, U. *Tetradhedron Lett.* **1986**, *27*, 106.   (b) For regioselective *cis-trans* photoisomerization of bichromophoric styrylstilbenes, see Ito, Y., Uozu, T., Dote, T., Ueda, M., and Matsuura, T. *J. Am. Chem. Soc.* **1988**, *110*, 189, and references therein.

(19)(a) For the Schenck mechanism of sensitized *cis-trans* isomerization, see Schenck, G. O. and Steinmetz, R. *Bull. Soc. Chim. Belges.* **1962**, *71*, 781. See also Caldwell, R. A. *J. Am. Chem. Soc.* **1970**, *92*, 1439, and references therein.   (b) For the halogen atom catalysis of *cis-trans* isomerization, see Ref. 23.   (c) For tetraalkyltin catalysis, see Hemmerich, H.-P., Warwel, S., and Asinger, F. *Chem. Ber.* **1973**, *106*, 505.

(20) Zimmerman, A. A., Orlando, C. M., Gianni, M. H., and Weiss, K. *J. Org. Chem.* **1969**, *34*, 73.

(21) For the effect of excitation wavelength of the photolysis of *trans*-1-phenyl-2-butene, see Comtet, M. *J. Am. Chem. Soc.* **1969**, *91*, 7761.

(22) For the *cis-trans* isomerization of stilbenes, see   (a) Hammond, G. S. and Saltiel, J. *J. Am. Chem. Soc.* **1962**, *84*, 4983.   (b) Hammond, G. S., Saltiel, J., Lamola, A. A., Turro, N. J., Bradshaw, J. S., Cowan, D. O., Counsell, R. C., Vogt, V., and Dalton, C. *J. Am. Chem. Soc.* **1964**, *86*, 3197.   (c) Hammond, G. S. and Turro, N. J. *Science.* **1963**, *142*, 1541.   (d) For the photochemical generation, isomerization, and oxygenation of stilbene cation radicals, see Lewis, F. D.,

Bedell, A. M., Dykstra, R. E., Elbert, J. E., Gould, I. R., and Farid, S. *J. Am. Chem. Soc.* **1990**, *112*, 8055.

(23) For reviews on *cis-trans* isomerization, see (a) Meier, H. In *Methoden der Organischen Chemie/* (Houben–Weyl). G. Thieme: Stuttgart, 1975, Vol. 4/ 5a: Photochemie, p. 189. (b) Saltiel, J., D'Agostino, J., Megarity, E. D., Metts, L., Neuberger, I. R., Wrighton, M., and Zafiriou, O. C. *Org. Photochem.* **1973**, *3*, 1. (c) Saltiel, J., Chang, D. W. L., Megarity, E. D., Rousseau, A. D., Shannon, P. T., Thomas, B., and Uriarte, A. K. *Pure Appl. Chem.* **1975**, *41*, 559. (d) Saltiel, J. and Charlton, J. L. In *Rearrangements in Ground and Excited States*. De Mayo, P., ed. Academic Press: New York, 1980, Vol. 3, p. 25.

(24) See Ref. 23(a), p. 191.

(25) For the effect of substituents on photostationary state of stilbene, see (a) Schulte-Frohlinde, D., Blume, H., and Gusten, H. *J. Phys. Chem.* **1962**, *66*, 2486. (b) Gusten, H. and Klasinc, L. *Tetrahedron Lett.* **1968**, 3097.

(26) Gusten, H. and Schulte-Frohlinde, D. *Chem. Ber.* **1971**, *104*, 402.

(27) For a review of triplet-sensitized *cis-trans* isomerization of stilbenes, see Stephenson, L. M. and Hammond, G. S. *Angew. Chem., Int. Ed. Engl.* **1969**, *8*, 261.

(28) Yamauchi, S. and Azumi, T. *J. Am. Chem. Soc.* **1973**, *95*, 2709, and references therein.

(29)(a) Swenton, J. S. *J. Org. Chem.* **1969**, *34*, 3217. (b) Deyrup, J. A. and Betkouski, M. *J. Org. Chem.* **1972**, *37*, 3561. (c) Inoue, Y., Takamuku, S., and Sakurai, H. *Synthesis*, **1977**, 111. (d) For asymmetrically photosensitized *cis-trans* isomerization of cyclooctene, see Inoue, Y., Yokoyama, T., Yamasaki, N., and Tai, A. *J. Am. Chem. Soc.* **1989**, *111*, 6480.

(30)(a) Evers, J. T. M. and Mackor, A. *Rec. Trav. Chim. Pays-Bas*. **1979**, *98*, 423. (b) Spee, T. and Mackor, A. *J. Am. Chem. Soc.* **1981**, *103*, 6901.

(31) For a review of these processes, see Kropp, P. J. *Org. Photochem.* **1979**, *4*, 1.

(32)(a) For a review of the chemistry of betweenanenes, see Marshall, J. A. *Acc. Chem. Res.* **1980**, *13*, 213. (b) [11.11]Betweenanene, see Nickon, A. and Zurer, P. S. J. *Tetrahedron Lett.* **1980**, 3527.

(33) Marshall, J. A. *Acc. Chem. Res.* **1969**, *2*, 33.

(34) Kropp, P. J. and Drauss, H. J. *J. Am. Chem. Soc.* **1967**, *89*, 5199.

(35) Kropp, P. J. *J. Org. Chem.* **1970**, *35*, 2435.

(36)(a) Kropp, P. J., Reardon, E. J., Jr., Gaibel, Z. L. F., Williard, K. F., and Harraway, J. H., Jr. *J. Am. Chem. Soc.* **1973**, *95*, 7058. (b) Fields, T. R. and Kropp, P. J. *J. Am. Chem. Soc.* **1974**, *96*, 7559. (c) Inoue, Y., Mukai, T., and Hakushi, T. *Chem. Lett.* **1983**, 1665. (d) Collin, G. J. and Deslauriers, H. *Can. J. Chem.* **1985**, *63*, 1424.

(37)(a) For a leading reference to earlier reviews, see Kaupp, G. In *Methoden der Organischen Chemie/* (Houben–Weyl). G. Thieme: Stuttgart, 1975, Vol. 4/ 5a: Photochemie, p. 278 (dimerizations) and p. 360 (cycloadditions). For recent reviews of [2+2] photocycloaddition reactions of alkenes with excited state alkenes, dienes, and chromophores incorporating these groups, see (b) Bauslaugh, P. G. *Synthesis.* **1970**, 287. (c) Baldwin, S. W. *Org. Photochem.* **1981**, *5*, 123. (d) Dilling, W. L. *Photochem. Photobiol.* **1977**, *25*, 605. (e) Kossanyi, J. *Pure Appl. Chem.* **1979**, *51*, 181. (f) Lenz, G. *Rev. Chem. Intermed.* **1981**, *4*, 369. (g) Weedon, A. C. In *Synthetic Organic Photochemistry*. Horspool, W., ed. Plenum: New York, 1984, p. 61. (h) Wender, P. A. In *Photochemistry in Organic Synthesis*. Coyle, J. D., ed. The Royal Society of Chemistry: Burlington House, 1986, p. 163. (i) Schreiber, S. L. *Science.* **1985**, *227*, 857.

(38)(a) Gerhartz, W., Poshusta, R. D., and Michl, J. *J. Am. Chem. Soc.* **1976**, *98*, 6427. (b) Michl, J. *Pure. Appl. Chem.* **1975**, *41*, 507. (c) Michl, J. *Photochem. Photobiol.* **1977**, *25*, 141. (d) *Figure adapted from* Bonačić-Koutecký, V., Koutecký, J., and Michl, J. *Angew. Chem., Int. Ed. Engl.* **1987**, *26*, 170.

(39)(a) Woodward, R. B. and Hoffman, R. *The Conservation of Orbital Symmetry*. Verlag Chemie International: Deerfield, Florida, 1970. (b) Gilchrist, T. L. and Storr, R. C. *Organic Reactions and Orbital Symmetry*. 2nd ed. Cambridge University Press: London, 1979.

(40) Herndon, W. C. *Top. Curr. Chem.* **1974**, *46*, 141.

(41) For examples of frontier-orbital analysis of photocycloaddition regioselectivity, see (a) Epiotis, N. D. *Angew. Chem., Int. Ed. Engl.* **1974**, *13*, 751. (b) Herndon, W. C. *Chem. Rev.* **1972**, *72*, 157. (c) Houk, K. N. *Acc. Chem. Res.* **1975**, *8*, 361. (d) Lewis, F. D., Hoyle, C. E., and Johnson, D. E. *J. Am. Chem. Soc.* **1975**, *97*, 3267.

(42)(a) Halevi, E. A. *Helv. Chim. Acta.* **1975**, *58*, 2136. (b) Katriel, J. and Halevi, E. A. *Theoret. Chim. Acta.* **1975**, *40*, 1. (c) Halevi, E. A. *Nouv. J. Chim.* **1977**, *1*, 229.

(43)(a) Caldwell, R. A. *J. Am. Chem. Soc.* **1980**, *102*, 4004. (b) Caldwell, R. A., Mizuno, K., Hansen, P. E., Vo. L. P., Frentrup, M., and Ho, C. D. *J. Am.*

Chem. Soc. **1981**, *103*, 7263. (c) Caldwell, R. A. and Creed, D. *Acc. Chem. Res.* **1980**, *13*, 45. (d) For applications, see McCoullough, J. J., MacInnis, W. K., Lock, C. J. L, and Faggiani, R. *J. Am. Chem. Soc.* **1982**, *104*, 4664. (d) For another reactivity pattern than that predicted by Caldwell's method, see Smothers, W. K., Meyer, M. C., and Saltiel, J. *J. Am. Chem. Soc.* **1983**, *105*, 545. (e) For a review of the valence-bond model of [2+2] cycloaddition reactions, see Bernardi, F., Olivucci, M., and Robb, M. A. *Acc. Chem. Res.* **1990**, *23*, 405.

(44)(a) Yamazaki, H., Cvetanovic, R. J., and Irwin, R. S. *J. Am. Chem. Soc.* **1976**, *98*, 2198. (b) For photodimerization of 2,3-dimethyl-2-butene, see Arnold, D. R. and Abraitys, V. Y. *J. Chem. Soc. Chem. Commun.* **1967**, 1053.

(45) Stechl, H. H. *Chem. Ber.* **1964**, *94*, 2681.

(46)(a) Arnold, D. R., Trecker, D. J., and Whipple, E. B. *J. Am. Chem. Soc.* **1965**, *87*, 2596. (b) Trecker, D. J. and Foote, R. S. *Org. Photochem. Synth.* **1971**, *1*, 81. (c) Cupas, C., Schleyer, P. v.R., and Trecker, D. J. *J. Am. Chem. Soc.* **1965**, *87*, 917.

(47)(a) Scharf, H. D. and Korte, F. *Tetrahedron Lett.* **1963**, 821. (b) Arnold, D. R., Glick, A. H., and Abraitys, V. Y. *Org. Photochem. Synth.* **1971**, *1*, 51.

(48) Metzner, W. and Wendisch, D. *Liebigs Ann. Chem.* **1969**, *730*, 111.

(49) Griffin, G. W. and Heep, U. *J. Org. Chem.* **1970**, *35*, 4222.

(50)(a) Kopecký, J. and Shields, J. E. *Tetrahedron Lett.* **1968**, 2821. (b) Kopecký, J. and Shields, J. E. *Coll. Czech. Chem. Commun.* **1971**, *36*, 3517. (c) For photocycloaddition of the X=NCOR type systems, see Kricka, L. J., Lambert, M. C., and Ledwith, A. *J. Chem. Soc. Perkin 1.* **1974**, 52. (d) For [2+2] cycloaddition reactions of model *cis*-stilbenes, see Penn, J. H., Gan, L.-X., Eaton, T. A., Chan, E. Y., and Lin, Z. *J. Org. Chem.* **1988**, *53*, 1519.

(51) Schönberg, A., Sodtke, U., and Praefcke, K. *Tetrahedron Lett.* **1968**, 3669.

(52)(a) Dzievonski, K. and Rapalski, K. *Chem. Ber.* **1912**, *45*, 249. (b) Dzievonski, K. and Paschalski, C. *Chem. Ber.* **1913**, *46*, 1986. (c) Griffin, G. W. and Veber, D. F. *J. Am. Chem. Soc.* **1969**, *82*, 6417. (d) Hartmann, I.-M., Hartmann, W., and Schenck, G. O. *Chem. Ber.* **1967**, *100*, 3146.

(53) Schenck, G. O. and Wolgast, R. *Naturwiss.* **1962**, *49*, 36.

(54) Cowan, D. O. and Drisco, R. L. *J. Am. Chem. Soc.* **1967**, *89*, 3068.

(55) For a review, see Koziar, J. C. and Cowan, D. O. *Acc. Chem. Res.* **1978**, *11*, 34.

(56) For photodimerization of aceanthrylene, see Plummer, B. F. and Singleton, S. F. *Tetrahedron Lett.* **1987**, *28*, 4801, and references to the photochemistry of acenaphthylene therein.

(57)(a) Hartmann, W. and Steinmetz, R. *Chem. Ber.* **1967**, *100*, 217. For its dichloroderivatives, see (b) Scharf, H.-D. *Angew Chem.*, **1974**, *86*, 567. (c) Scharf, H.-D. and Seidler, H. *Chem. Ber.* **1971**, *104*, 2995.

(58)(a) Scholz, K. H., Hinz, J., Heine, H. G., and Hartmann, W. *Liebigs Ann. Chem.* **1981**, 248. (b) Sekretar, S., Kopecky, J., and Martvon, A. *Coll. Czech. Chem. Commun.* **1982**, *47*, 1848.

(59) Steffan, G. and Schenck, G. O. *Chem. Ber.* **1967**, *100*, 3961.

(60) Chapman, O. L., Lura, D. R., Owens, R. M., Plank, E. D., Shim, S. C., Arnold, D. R., and Gillis, L. B. *Can. J. Chem.* **1972**, *50*, 1984.

(61) Tada, M., Shinozaki, H., and Sato, T. *Tetrahedron Lett.* **1970**, 3879.

(62)(a) Cantrell, T. S. *J. Chem. Soc.* (D) **1970**, 1633. (b) Rosenberg, H. M. and Serve, M. P. *J. Org. Chem.* **1971**, *36*, 3015.

(63)(a) Serve, M. P., Rosenbert, H. M., and Rondeau, R. *Can. J. Chem.* **1969**, *47*, 4295. (b) For the mechanism of the photocycloaddition of cyclopentene to N-benzoylindole, see Disanayaka, B. W. and Weedon, A. C. *Can. J. Chem.* **1990**, *68*, 1685.

(64)(a) Ghosh, S., Raychaudhuri, S. R., and Salomon, R. G. *J. Org. Chem.* **1987**, *52*, 83. (b) See also Evers, J. T. M. and Mackor, A. *Tetrahedron Lett.* **1978**, 821. (c) For a review of homogeneous metal catalysis in organic photochemistry, see Salamon, R. G. *Tetrahedron Lett.* **1983**, *39*, 485.

(65) Whitney, R. A. *Can. J. Chem.* **1981**, *59*, 2650; **1983**, *61*, 1158.

(66)(a) Mehta, G. *J. Chem. Ed.* **1982**, *59*, 313, and references therein. (b) Oppolzer, W. *Acc. Chem. Res.* **1982**, *15*, 135.

(67) Salomon, R. G., Coughlin, D. J., and Easler, E. M. *J. Am. Chem. Soc.* **1979**, *101*, 3961.

(68)(a) Martin, H. D., Mayer, B., Puetter, M., and Hoeschstetter, H. *Angew. Chem.* **1981**, *93*, 695. (b) For the photochemical synthesis of pterodactyladiene derivatives, see Martin, H. D. and Hekman, M. *Tetrahedron Lett.* **1978**, 1183.

(69) For the preparative procedure of acetylene pho-tocycloaddition to maleic anhydride, see Bloomfield, J. J. and Owsley, D. C. *Org. Photochem. Synth.* **1976,** *2,* 36.

(70) Hartmann, W. *Chem. Ber.* **1969,** *102,* 3974.

(71) For the photochemical cycloaddition reactions of cyanoacetylene and dicyanoacetylene, see Ferris, J. P. and Guillemin, J. C. *J. Org. Chem.* **1990,** *55,* 5601.

(72) For a review, see (a) Steinmetz, M. G. *Org. Photochem.* **1987,** *8,* 67. (b) Leigh, W. J. and Sri-nivasan, R. *Acc. Chem. Res.* **1987,** *20,* 107.

(73) Inoue, Y., Mukai, T., and Hakushi, T. *Chem. Lett.* **1982,** 1045, and references therein.

(74)(a) Srinivasan, R. and Ors, J. A. *J. Am. Chem. Soc.* **1978,** *100,* 7081. (b) Srinivasan, R., Ors, J. A., and Baum, T. H. *J. Org. Chem.* **1981,** *46,* 1950.

(75) Srinivasan, R., Ors, J. A., Brown, K. H., Baum, T. H., White, L. S., and Rossie, A. R. *J. Am. Chem. Soc.,* **1980,** *102,* 5297.

(76)(a) Nakayama, T. and Watanabe, K. *J. Chem. Phys.* **1964,** *40,* 558. (b) Bowman, E. R. and Miller, W. D. *J. Chem. Phys.* **1965,** *42,* 681.

(77) Coyle, J. D. *Org. Photochem.* **1985,** *7,* 1.

(78) Gallagher, M. J. and Noerdin, H. *Aust. J. Chem.* **1985,** *38,* 997.

(79)(a) Roberts, T. D. *Chem. Commun.* **1971,** 362. (b) Fujita, K., Yamamoto, K., and Shono, T. *Tetrahedron Lett.,* **1973,** 3865.

(80) Kawanisi, M. and Matsunaga, K. *Chem. Commun.* **1972,** 313.

(81)(a) Ferris, J. P. and Antonucci, F. R. *Chem. Commun.* **1972,** 126. (b) For the pH dependence of these reactions, see Isaks, M., Yates, K., and Kalandero-poulos, P. *J. Am. Chem. Soc.* **1984,** *106,* 2728.

(82) Owsley, D. C. and Bloomfield, J. J. *J. Am. Chem. Soc.* **1971,** *93,* 782.

(83)(a) Kaupp, G. and Stark, M. *Chem. Ber.* **1978,** *111,* 3608. (b) Rosenberg, H. M. and Servé, P. *J. Org. Chem.* **1968,** *33,* 1653.

(84) Kaupp, G. and Stark, M. *Angew. Chem.* **1977,** *89,* 555.

(85) Kaupp, G. and Stark, M. *Chem. Ber.* **1981,** *114,* 2217.

(86) Kubota, T. and Sakurai, H. *J. Org. Chem.* **1973,** *38,* 1762.

Chapter 6
PHOTOCHEMISTRY OF DIENES AND POLYENES

PHOTOCHEMISTRY OF DIENES AND POLYENES

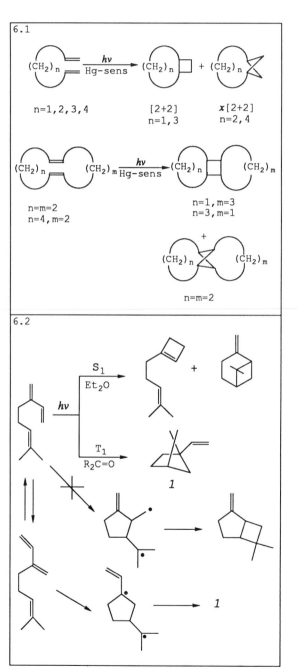

6.1. The photochemistry of dienes with isolated double bonds resembles that of simple alkenes, i.e., [2+2] cycloaddition reactions, but in an intramolecular fashion to give bicyclic products. The products of sensitized intramolecular cycloadditions depend on the number of CH_2 groups between the double bonds (sensitization with Hg for simple dienes, carbonyl sensitizers for dienes substituted by Ph, $C\equiv N$, CO_2R that lower the E_T). If the number of intervening CH_2 groups is odd, e.g., 1,4-pentadiene or 1,6-heptadiene, the major product results from a "normal" [2+2] cycloaddition. If the number is even, e.g., 1,5-hexadiene, the major product is formed via a "cross", i.e., x[2+2] cycloaddition to form two crossed σ bonds.[1] This feature is even more prominent in cyclic dienes, e.g., $n = m = 2$ (1,5-cyclooctadiene)[2] in which the x[2+2] cycloadduct is the only product. If $n = 1$ and $m = 3$ or *vice versa* (1,4-cyclooctadiene),[3] the normal [2+2] cycloadduct is formed. However, 1,4-pentadienes ($n = 1$) undergo predominantly a unique rearrangement (see 6.8).

6.2. Intramolecular cycloadditions of dienes proceed differently in the S_1 and T_1 states. In the S_1 state of myrcene, electrocyclic ring closure occurs (a 1,3-diene reaction), in addition to the normal [2+2] cycloaddition.[4] In the T_1 state, myrcene yields a cross cycloaddition product similar to that of 1,5-hexadiene.[5] The photoreactions of 1,5-dienes from the T_1 state follow the empirical *rule of five*,[1(a),6] which states that if triplet cyclization can lead to rings of different size, the one formed by 1,5 addition is preferred kinetically. If there are several possibilities for 1,5 addition leading to different biradicals, the most stable one is formed.

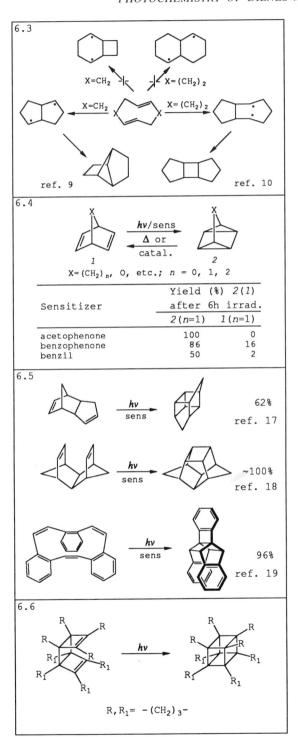

6.3

ref. 9   ref. 10

6.4

$$X = (CH_2)_n, \ O, \ etc.; \ n = 0, 1, 2$$

Sensitizer	Yield (%) 2(1) after 6h irrad.	
	2(n=1)	1(n=1)
acetophenone	100	0
benzophenone	86	16
benzil	50	2

6.5

$\xrightarrow{hv/\text{sens}}$ 62%   ref. 17

$\xrightarrow{hv/\text{sens}}$ ~100%   ref. 18

$\xrightarrow{hv/\text{sens}}$ 96%   ref. 19

6.6

$\xrightarrow{hv}$

$$R, R_1 = -(CH_2)_3-$$

6.3. The kinetic preference for the formation of five-membered rings is analogous to the tendency of a 5-hexenyl radical to produce a cyclopentyl methyl radical rather than a cyclohexyl radical.[7] A theoretical approach[8] indicates that the energy barrier for the $x[2+2]$ reaction path is lower (higher) than that of the $[2+2]$ path for dienes, which contain an even (odd) number of CH_2 groups. The validity of the rule of five has been confirmed in a large number of compounds: open-chain, cyclic, with or without heteroatoms, etc. In the example shown, the products are formed quantitatively according to the rule.[9-11]

6.4. While intramolecular cycloaddition of linear nonconjugated dienes leads to bicyclic products, bicyclic dienes frequently produce unique cage systems, if there is an interaction between both π systems.[12] One of the most studied examples is the transformation of norbornadiene to quadricyclane $[x = CH_2; n = 1]$[13,14] in which the photochemical equilibrium between the two compounds is a function of the sensitizer (see also 16.22).[15]

6.5. The thermodynamic requirement of decreasing free energy in the transformation of reactants to products leads to a definite limitation in the synthesis of molecules having high energy content, e.g., highly strained molecules by thermal processes. Photochemical intramolecular cycloadditions of nonconjugated cyclic dienes are particularly suited for the facile preparation of cage compounds, often in high yields.[16] The figure shows an intramolecular cycloaddition in the dimer of cyclopentadiene, which serves as a model for the synthesis of the hydrocarbon cubane from the dimer of 2-bromocyclopentadienone.[20]

6.6. An interesting example is the light-induced ring closure that occurs with the bridged tricyclooctadiene [R, $R_1 = -(CH_2)_3-$], while it fails with the nonbridged one [R = R_1 = H or alkyl].[21]

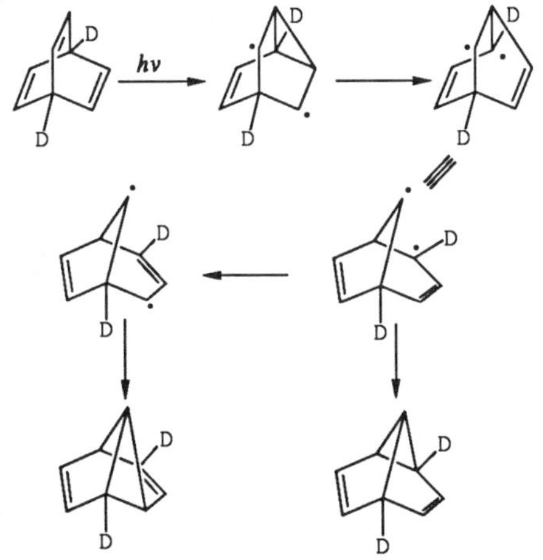

6.7. The exocyclic double bonds in 3,7-dimethylenebicyclo[3.3.1]nonane are so close that irradiation of its Cu(I) complex easily induces an intramolecular [2+2] cycloaddition.[22] The same is true for the C=O groups in bicyclo[3.3.1]nonane-3,7-dione (see 8.18).

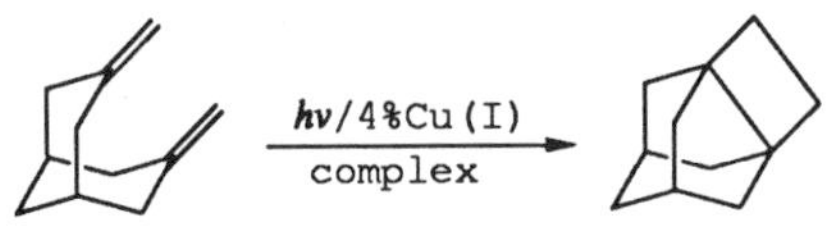

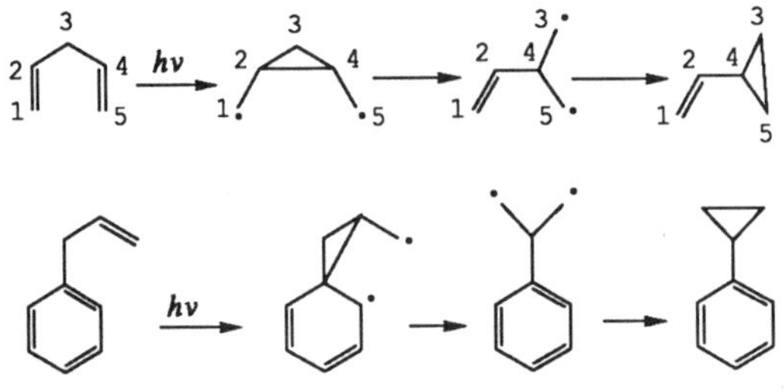

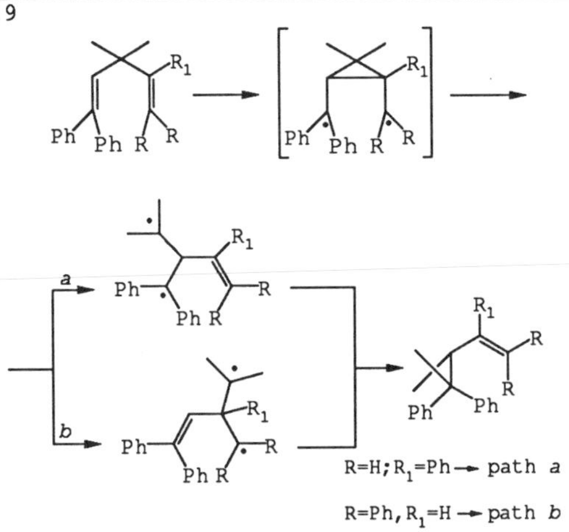

6.8. The most remarkable photoreaction of nonconjugated dienes is the *di-π-methane rearrangement (Zimmerman rearrangement)*[23] of dienes having π-systems separated by an sp^3 hybridized carbon atom (1,4-pentadienes) and also of 3-phenyl-alkenes in which one of the double bonds is replaced by a benzene ring. Irradiation induces rearrangement to a vinylcyclopropane or phenylcyclopropane derivative. The reaction formally consists of a [1,2] shift and closure to a three-membered ring. The rearrangement of monocyclic and acyclic systems occurs from the S_1 state, whereas the sensitized reaction leads instead to *cis-trans* isomerization or [2+2] cycloaddition.

6.9. This and the previous figure show 1,4- and 1,3-biradical intermediates only for ease of visualization and are not meant to imply their actual intermediacy. Nevertheless, the reaction path that would involve the more stable biradical intermediate is always followed. The rearrangement occurs by a concerted process, obeys the Woodward-Hoffmann rules, retains configuration on atoms 1 and 5 (a result of disrotatory ring closure between C-3 and C-5), and inverts it on atom 3 (*anti* disrotatory *mode*).[24,25]

6.10. Unlike linear and monocyclic 1,4-dienes, polycyclic dienes undergo the di-π-methane rearrangement from the T_1 state.[23,26,27] On the contrary, direct irradiation of barrelenes yields the corresponding cyclooctatetraenes.[28] These singlet vs. triplet differences are described by the effects of tight (singlet) or loose (triplet) geometries of the intermediate biradicals.[29,30]

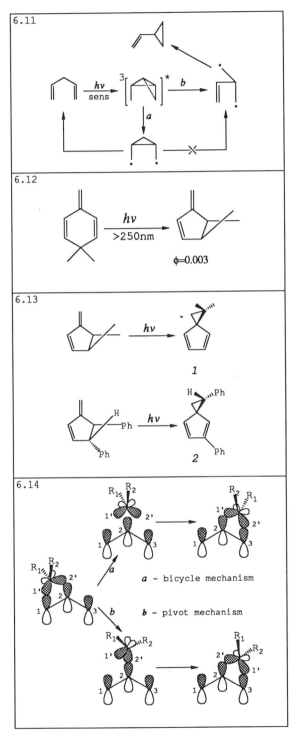

6.11. The cyclopropyldicarbinyl intermediate was originally assumed to play a role, at least in sensitized di-π-methane rearrangements. Recent experimental[31] and theoretical[32] studies have shown, however, that the triplet mechanism is efficient only if the cyclopropyldicarbinyl triplet intermediate is circumvented during the reaction so that the reverse reaction (*Grob fragmentation*) to the ground state of the reactant (a) is avoided. The singlet mechanism (b) is efficient if no barriers exist on the first singlet surface, e.g., with central dimethyl substitution. Then, cyclopropyldicarbinyl is not the intermediate.

6.12. Di-π-methane rearrangements also occur with disubstituted 1,4-cyclohexadienes and is particularly remarkable with 6,6-disubstituted 3-methylene-1,4-cyclohexadienes.[33] The products, 2-methylene-5,6-disubstituted bicyclo[3.1.0]hex-3-enes, are photoreactive and subsequently undergo a novel type of rearrangement: the *bicycle rearrangement*.

6.13. The bicycle rearrangement of 2-methylenebicyclo[3.1.0]hex-3-enes is a useful method for the synthesis of spiro[2.4]hepta-4,6-dienes. The reaction is stereospecific and occurs from the S_1 state. The *trans* isomer of the bicyclic compound 1 forms the *syn* isomer of the spirocyclic product 2. The quaternary carbon atom migrates along the π-system without inversion and without loss of its stereochemical integrity.[33-35]

6.14. This stereospecificity is well explained by a mechanism based on the motion of the C atom with two substituents and two hybrid AOs. The orbitals and the substituents represent the wheels and the handlebars of a bicycle traveling along the π-system (a). The fact that in some instances the reaction is not completely stereospecific led to the concept of a pivot mechanism (b).[36]

6.15

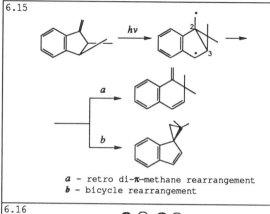

a - retro di-π-methane rearrangement
b - bicycle rearrangement

6.16

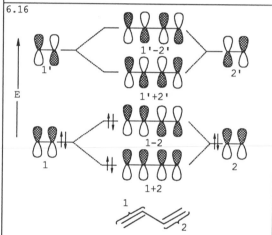

6.17   **trans-1,3-BUTADIENE LOW-LYING**

ELECTRONIC STATES, THEIR CONFIGURATION

AND CALCULATED[40] ENERGIES

STATE	ELECTRONIC CONFIGURATION	ENERGY kJ/mol (kcal/mol)
S_0 Ground (1^1A_g)	(π_1^2, π_2^2)	---
T_1 Lowest (^3B_u)	$^3(\pi_2^2 \to \pi_3^*)$	250 (60)
S_1 Doubly excited valence singlet (2^1A_g)	$\begin{cases} (\pi_2 \to \pi_3^*)^2 \\ (\pi_1 \to \pi_3^*) \\ (\pi_2 \to \pi_4^*) \end{cases}$	530 (127)
S_2 Singly excited valence singlet (^1B_u)	$^1(\pi_2 \to \pi_3^*)$	670 (160)

6.15. The bicycle process also competes with the retro di-π-methane rearrangement. If the cyclopropyldicarbinyl 1,4-biradical intermediate cannot be avoided, it is stabilized by homolytic splitting of 2,3 bond with the formation of a 1,4-pentadiene system (path a). If it is avoided, the bicycle process occurs (path b).[37]

6.16. The shape of MOs of dienes is visualized easily by means of a symmetric [1+2] or an antisymmetric [1−2] combination of the π orbitals of isolated double bonds Since the HOMO-LUMO gap decreases, the singlet state becomes lower in energy and absorption occurs at a longer wavelength. Because of delocalization, the central bond [C(2)—C(3)] has a small amount of double bond character in the S_0 state of butadiene ($\Psi_1^2\Psi_2^2\Psi_3^0\Psi_4^0$), which results in a low energy barrier (~21 kJ/mol, ~5 kcal/mol) for *s-cis→s-trans* isomer ization. According to the model shown, this barrier increases significantly in the excited state of electronic configuration $\Psi_1^2\Psi_2^1\Psi_3^1\Psi_4^0$ due to the increased double bond character of the central bond. This fact, together with the short lifetime of the singlet excited state, precludes the possibility of *cis-trans* isomerization in the excited state (see below).[38]

6.17. In addition, according to the model shown, the lowest π,π* state is formed by a practically pure HOMO-LUMO vertical transition. This was confirmed by *ab initio* calculations,[40] which show that there are four low-lying electronic states of *trans*-1,3-butadiene with planar geometry. These are the ground state (1^1A_g), π → π* singlet and triplet states ($1^{1,3}B_u$), and the (2^1A_g) singlet state formed by a combination of $\pi_1 \to \pi_3^*$, $\pi_2 \to \pi_4^*$, and the doubly excited ($\pi_2 \to \pi_3^*)^2$ transitions.[39] The latter state results from the coupling of two triplet states of each ethylene moiety.

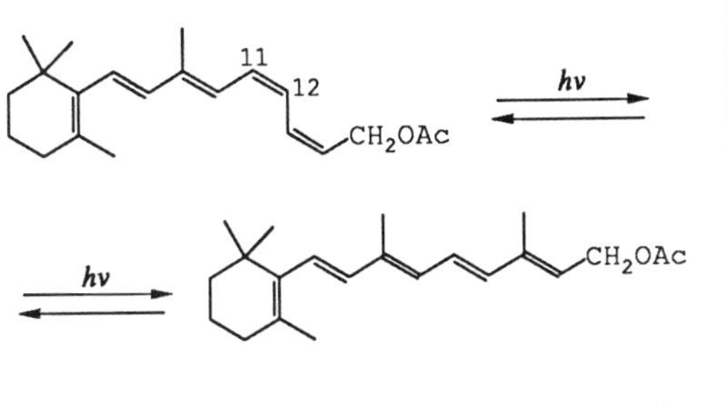

6.18. While the ground state of butadiene maintains planarity because of π electron delocalization, the equilibrium molecular structures in the excited states are not planar, reflecting differences in electronic character. Both triplet ($1^3A''$) and singlet ($1^1A'$) excited states arise from the B_u state of *trans*-butadiene and are described by a single electron excitation from π(HOMO) to π*(LUMO). While the triplet state is biradical, the corresponding singlet state is ionic. The lowest excited singlet state S_1 in the equilibrium is calculated to be the doubly excited state from the 2^1A_g state of *trans*-butadiene. This state is very flat.[40]

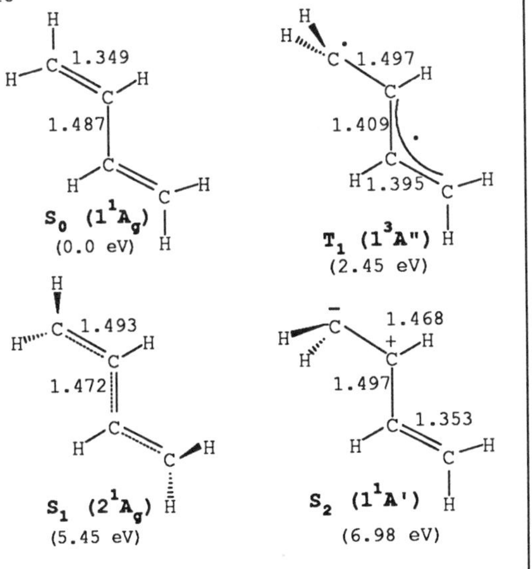

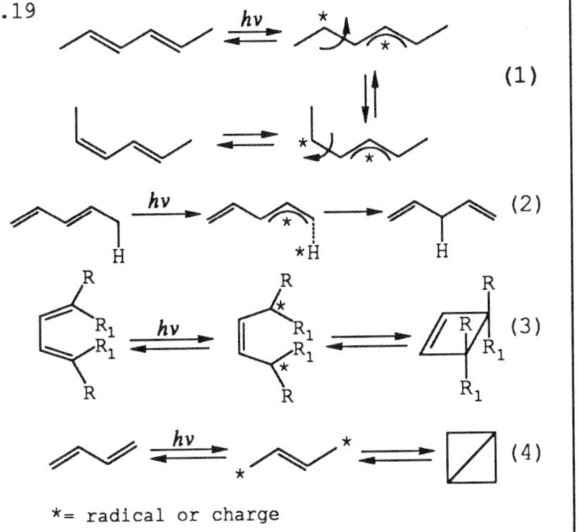

6.19. The most important concerted and nonconcerted phototransformations of dienes[41] described with biradical or zwitterionic intermediates are: (1) *cis-trans* isomerization; (2) sigmatropic shift; (3) disrotatory electrocyclic ring closure; (4) intramolecular $x[2+2]$ cycloaddition. *Cis-trans* isomerization of the central C—C bond is facile in both the doubly excited (S_1) state and the ground state.[40] Since the 1^1B_u state relaxes to a structure with a twisted geometry of the terminal C—C bond, the S_2 state seems to be unreactive toward ring closure. While the S_2—S_1 energy gap is small (see 6.17), internal conversion to S_1 is possible.

6.20. The mechanism of *cis-trans* isomerization of dienes and polyenes differs from that of simple alkenes (see 5.8).[42] While the S_1 state in simple alkenes slopes down to the twisted minimum, in dienes S_1 is flat and in polyenes there is a barrier in the reaction pathway. Since twisting about several double bonds is possible, that which has the lowest barrier leading to the funnel will occur preferentially (the deeper the funnel, the lower the barrier). The photoisomerization of 11-*cis*-retinyl acetate to the biologically active all-*trans* isomer, vitamin A acetate, is of industrial significance.[43]

6.21

SELECTION RULES FOR ELECTROCYCLIC REACTIONS OF POLYENES

NUMBER OF π ELECTRONS	ELECTROCYCLIC REACTION	
	THERMAL	PHOTOCHEMICAL
4q (q=1,2..)	*conr*[a]	*disr*[b]
4q+2	*disr*	*conr*

[a] *conr* = conrotatory mode

[b] *disr* = disrotatory mode

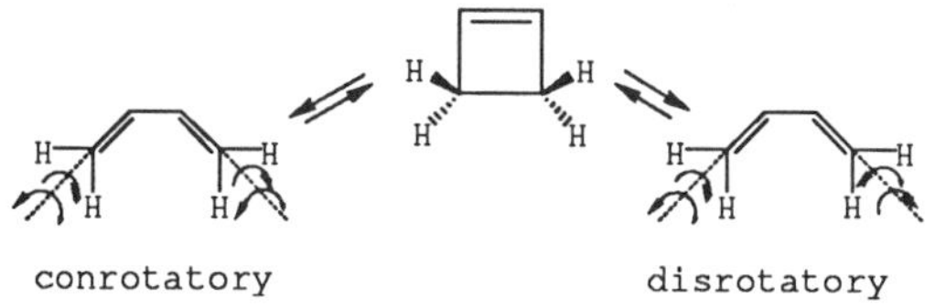

conrotatory   disrotatory

6.22

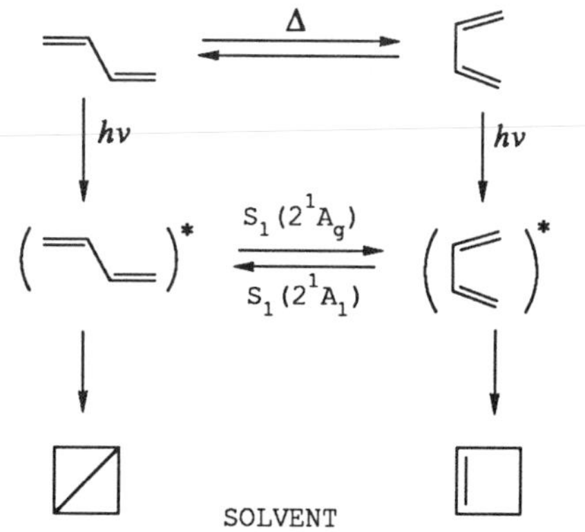

	SOLVENT	
1	Isooctane	16
1	Ether	7
1	Ether/CuCl$_2$	6

6.23

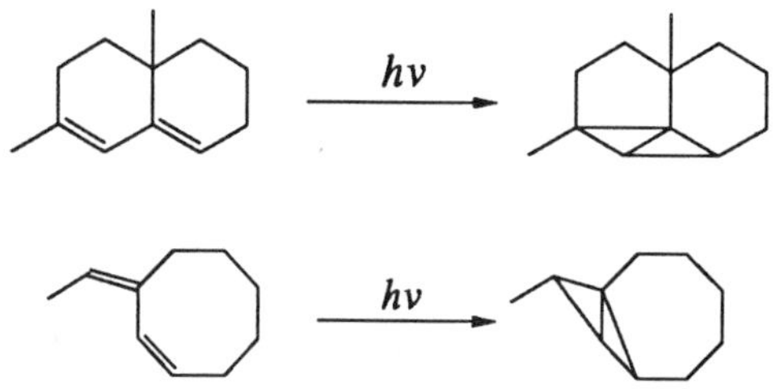

6.21. An example of a pericyclic reaction is the electrocyclic ring closure/opening reaction that obeys the Woodward–Hoffmann rules.[44] The stereochemistry of the product is directed by the mode of cyclization of ring opening, i.e., *conrotatory* or *disrotatory*, which in turn is determined by the electronic state in which the process occurs. For each ring closure there are two modes of twisting (rotation) about the π bond: *conrotatory mode*—clockwise-clockwise and counterclockwise-counterclockwise, and *disrotatory mode*—clockwise-counterclockwise and counterclockwise-clockwise. The conrotatory mode is represented by $[_\pi 2_s + _\pi 2_a]$ or $[_\pi 2_a + _\pi 2_s]$; the disrotatory mode by $[_\pi 2_s + _\pi 2_s]$ or $[_\pi 2_a + _\pi 2_a]$, where s = *supra* and a = *antara*. For a theoretical treatment of the butadiene $\leftrightarrow$ cyclobutene interconversion, see 4.9 and 4.12.

6.22. The photochemistry of butadiene provides a typical example of a photochemically initiated electrocyclic reaction. Irradiation of butadiene in cyclohexane produces cyclobutene from an electrocyclic reaction of the *s-cis* isomer, and bicyclo[1.1.0]butane, the product of an intramolecular $x[2+2]$ cycloaddition of the *s-trans* isomer, in a 6:1 ratio.[45,46] This ratio, which obviously does not reflect the relative population of the ground state conformations (~3% of the equilibrium mixture in the ground state at ambient temperature is the *cis* isomer), is explained by the difference in quantum yields, Φ_{cis} and Φ_{trans}.

6.23. In contrast to a concerted cyclobutene formation, bicyclobutane is probably the result of a nonconcerted process involving an intermediate biradical and is particularly important in systems that contain a rigid *s-trans*-diene gometry. However, it also occurs with *cis*-fused dienes, e.g., cyclopentadiene. Hydrocarbon diene systems with high E_T require the use of mercury as a sensitizer.[47]

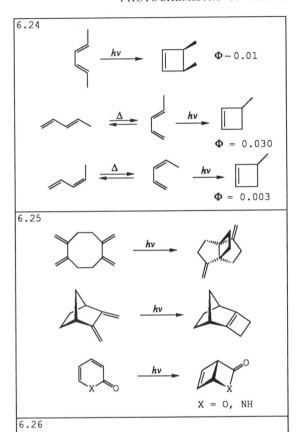

6.24. Although the electrocyclic ring closure of butadiene systems occurs only from the *s-cis* isomer, and *s-trans*-butadiene is thought to be the excited species (very low concentration of *s-cis* isomer in equilibrium), a mechanism for the *cis-trans* isomerization in the doubly excited S_1 state has been proposed.[40] This introduces other factors, such as a *steric* effect in addition to the orbital factors controlling the course of the disrotatory mode of the reactions. These additional factors have a significant effect on the efficiency of these reactions.[48]

6.25. The stereochemistry of butadiene (polyene) systems incorporated into acyclic, cyclic,[49] polycyclic,[50] and heterocyclic[51] molecules is controlled reliably by the Woodward–Hoffmann rules. Tetramethylenecyclooctane shown can be viewed as containing either two butadiene or two 1,6-heptadiene systems. According to the rule of five, a propellane derivative is formed, along with products of single- or double-electrocyclic reactions.

6.26. Despite the recently proposed energy level diagram for *E*-hexatriene,[52] there was some uncertainty in the assignment and ordering of its excited states,[53] particularly the A_g and B_u states (the former has been shown to be below the latter). Available experimental data for the E_{exc} of Z and E isomers of hexatriene shows no difference in the lower states of each multiplicity. It is known that the vertically allowed HOMO → LUMO transition in linear polyenes results in the formation of a polar (zwitterionic) excited state (^1B_u), the key intermediate in *cis-trans* isomerizations (see also 5.8),[54] which, after twisting, begins to mix with ^1A_g; at 90° of twist these states are thoroughly mixed and three low energy states are formed (two biradical and one zwitterionic species). No evidence exists for the doubly excited covalent state 2^1A_g of hexatriene, although it lies ~48 kJ/mol (~11.5 kcal/mol) below its zwitterionic ^1B_u state in octatetraene.[55]

Z AND E HEXATRIENE (HXT) VALENCE TRANSITION ENERGIES IN VAPOR PHASE [SOLUTION]

VALENCE STATE	Z - HXT E_{trans}	E - HXT E_{trans}
T_1	260[a] (62)	260 (62)
T_2	386 (92)	405 (97)
S_1	497[470] (119/112)	497[470] (119/112)
S_2	627[598] (150/143)	675 (161)

[a] In kJ/mol (kcal/mol) [ref. 53]

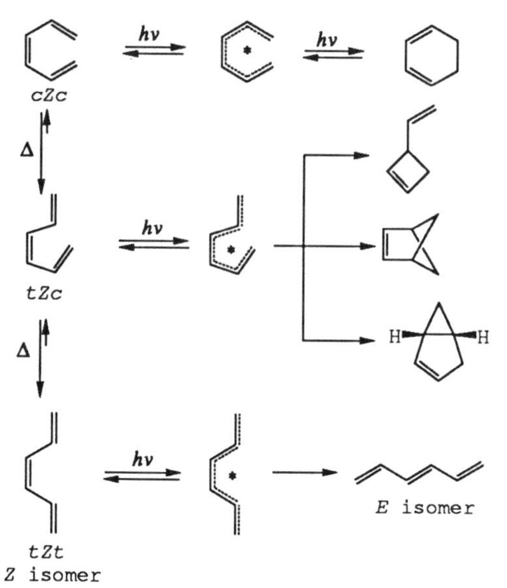

6.27. The photochemistry of hexatriene can also be explained by invoking biradical intermediates. It is assumed that because of the extremely short lifetimes of their excited states, the roughly planar conformers of hexatriene do not interconvert, and, consequently, each conformer yields its own photoproducts (NEER principle, see 15.8).[56] Therefore, the ground state conformations of hexatriene play an important role and, along with steric and orbital factors, determine the course of photoreactions.

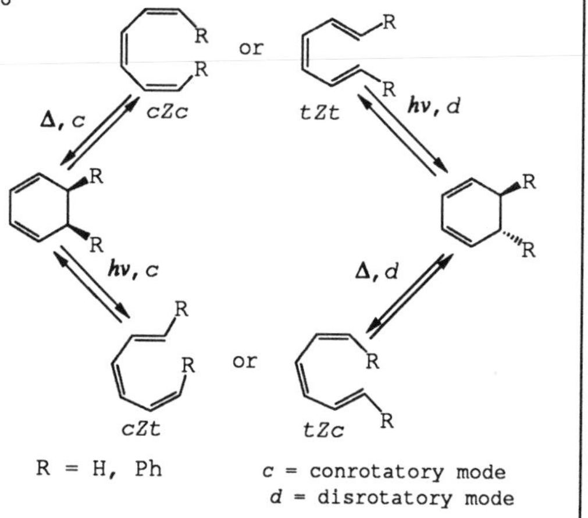

6.28. According to the Woodward–Hoffmann rules (see 6.21), the interconversion hexatriene $\rightleftharpoons$ 1,3-cyclohexadiene proceeds by a conrotatory mechanism. The ring opening is much more efficient than the ring closure due to the predominance of *tZt*-hexatriene in the conformational equilibrium. The Φ of vinylcyclobutene and the bicyclic products shown in Fig. 6.27 is very low compared to that of the *E–Z* photoisomerization.[57] Because of two possible conrotatory modes, the ring opening of 4,5-disubstituted cyclohexadiene generally leads to two different products. However, this mode is affected strongly by steric[58] (conformational[59]) effects so that even the less stable product is formed (R=Ph).[60]

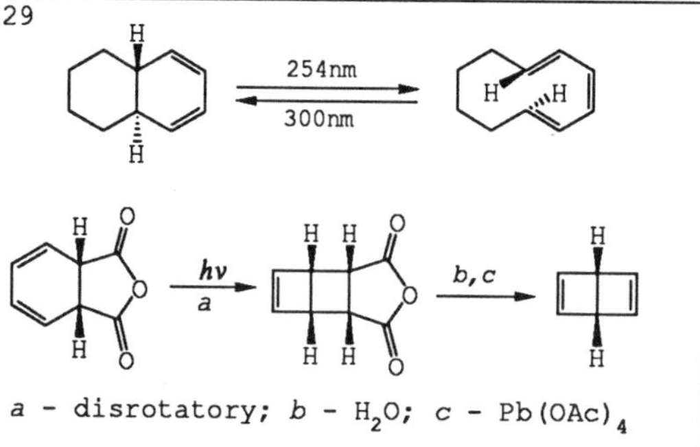

6.29. The conrotatory opening of a cyclohexadiene ring in a bicyclic system proceeds efficiently, especially if the resulting ring is sufficiently large to provide stability for the *E* double bond in the triene. Since the triene generally has a higher ϵ_{max} and broader absorption bands, longer wavelength irradiation is required for the reverse reaction. The ring opening/closure reactions are so exclusively conrotatory that the interconversion of optically active bicyclic diene and cyclic triene proceeds with chirality retention.[61] However, if a severely strained system results from ring opening, an energetically more favorable reaction path, e.g., diene → cyclobutene, is followed.[62]

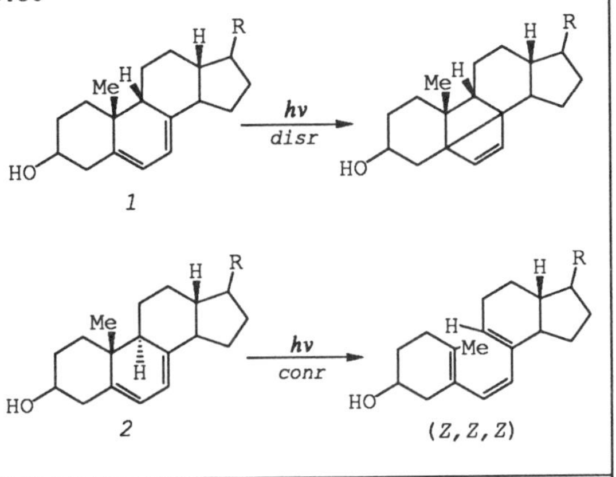

6.30. This is shown in the different photochemical behavior of ergosterol (2) and isopyrocalciferol (1).[63] 1 produces a system with considerable steric strain. The ring closure is more favorable than the ring opening since the latter would produce hexatriene with an E double bond in a cyclohexene ring. It is likely that the activation energy (E_a) for the opening of the cyclohexadiene ring is larger than that for the ring closure in the 4π electron system. In contrast, the conrotatory ring opening in 2 transforms the original *trans* configuration in the cyclohexadiene ring into c,Z,c or t,Z,t-hexatriene, which is strain free.

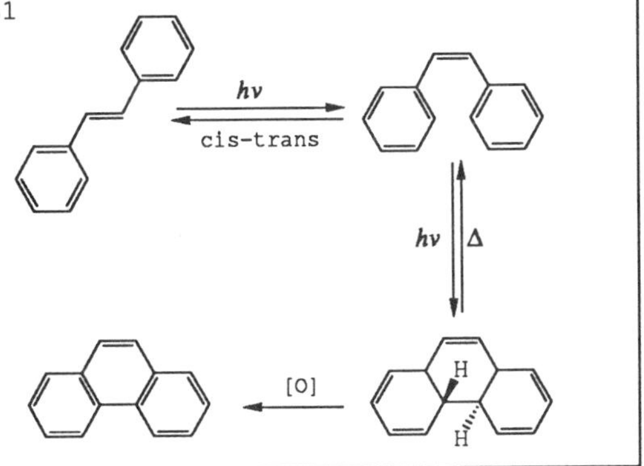

6.31. The *oxidative photocyclization* of *cis*-stilbene (formed by *trans* → *cis* isomerization) to phenanthrene proceeds in 50–90% yield and can be classified formally as an electrocyclic triene reaction. In accordance with theory, the dihydrophenanthrene formed has a *trans* configuration. If the irradiation is performed in the presence of an oxidizing agent (air, iodine, iron trichloride, etc.), the corresponding phenanthrene is isolated directly. The reaction occurs entirely from the S_1 (π, π^*) state and does not proceed with stilbenes that are substituted by electron acceptors or substituents that would introduce a low-lying n, π^* state.[64]

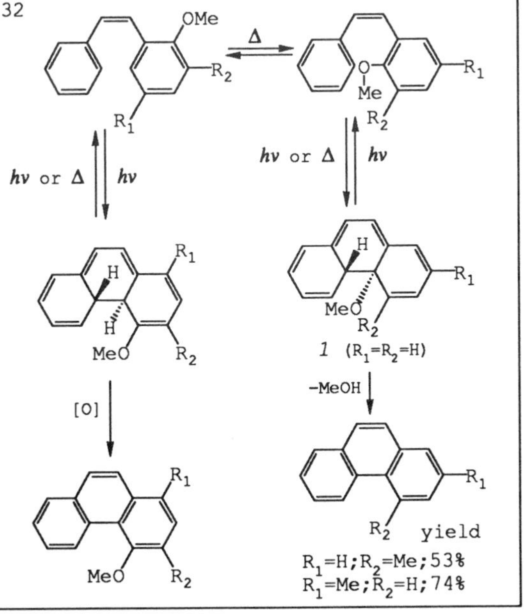

6.32. This reaction is a general method for the synthesis of 1-, 3-, and 9-substituted phenanthrenes in high yield from the corresponding o-, p-, or α-substituted stilbenes.[65] *m*-substituted stilbenes give a mixture of 2- and 4-substituted phenanthrenes, which are difficult to separate.[64c] An alternative method is the elimination of methanol from methoxydihydrophenanthrene (1), the product of cyclization of an o-methoxystilbene. The *eliminative photocyclization method* proceeds efficiently by trapping 1, using an acid catalyst and a solvent capable of proton transfer.[66]

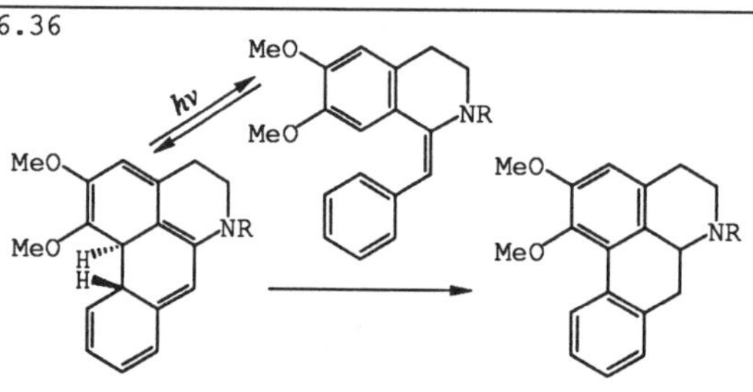

6.33. The utility of the original method is clearly demonstrated by the synthesis of helicenes. 1-methylhexahelicene was synthesized by a 20-step sequence with a total yield of 0.02%.[67] Using a double photocyclization-oxidation, the yield was 16%.[68] This method was reported for the synthesis of a series of helicenes, e.g., tridecahelicene,[69] benzohexahelicenes,[70] kekulene,[71] and double helicenes.[72]

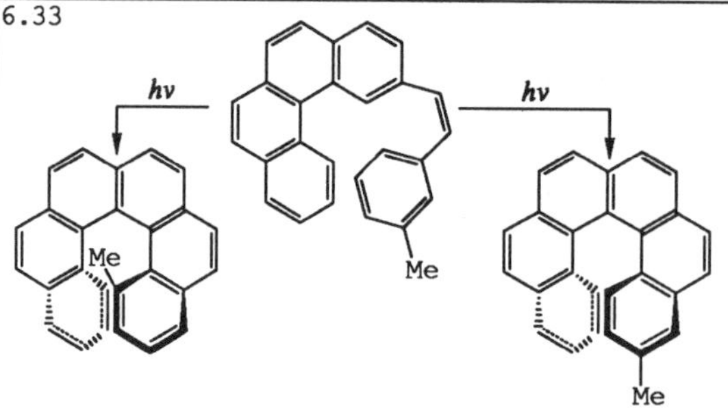

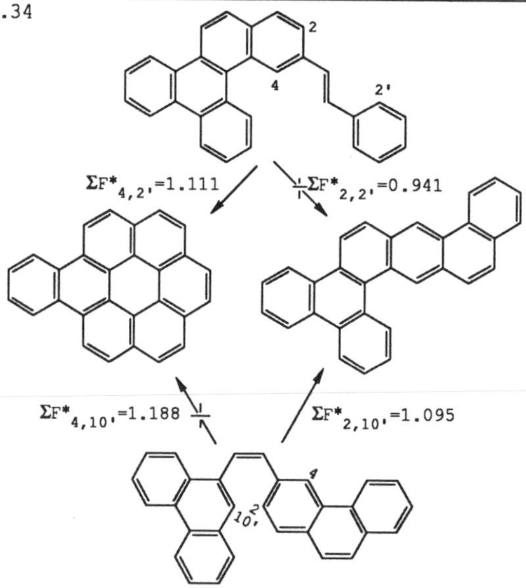

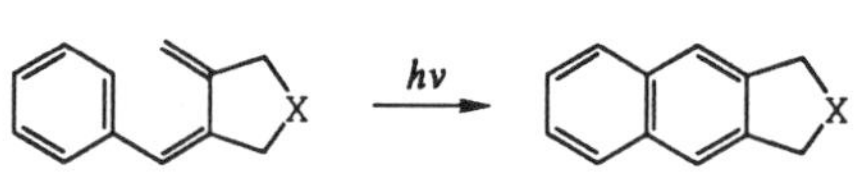

6.34. This reaction is a valuable synthetic method for many carbocyclic and heterocyclic systems, e.g., chrysene and benzo[c]phenanthrene from 1- and 2-styrylnaphthalene, respectively. It is surprisingly highly selective and usually gives a single product. The selectivity is controlled by three rules. (1) Photocyclization occurs only when the sum of the excited singlet state HMO free valence values in the two positions involved in the cyclization is 1 or higher ($\Sigma F_{ij}^* \geq 1$). (2) If several possibilities for ring closure exist, only one product is formed if the difference $\Delta F_{ij}^* > 0.1$; otherwise, several products result. (3) If both planar and nonplanar products are possible, the planar one always forms, unless prohibited by Rule (1).[70]

6.35. This reaction also occurs with various 1,2-diphenyl derivatives,[73] 1,4-diaryl-1,3-dienes,[74(a)] and 1,4-diaryl-1-buten-3-ynes.[75] Although the method fails with 1-phenyl-1,3-butadiene,[74(b)] it was successfully applied to its 2,3-alkylene derivatives where both double bonds are exocyclic.[74(c)] Azobenzene (benzalaniline) yields benzocinnoline[76] (phenanthridine,[77] see 11.8).

6.36. Oxidative photocyclization reactions are key in the synthesis of many alkaloids,[78] e.g., the aporphine alkaloid nucipherin.[79] An alternate synthesis is based on irradiation of the iodo or bromo precursor.[78(b),(c)] The reaction is feasible even if the central C=C bond of the triene is substituted by a $-CONH-$ group (see 10.27 through 10.30).

6.37

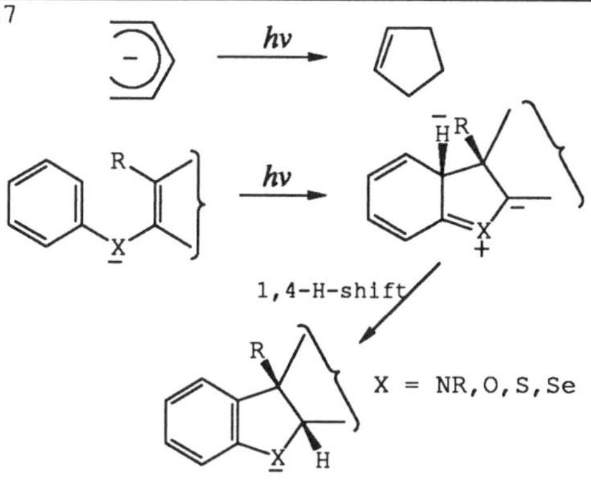

6.38

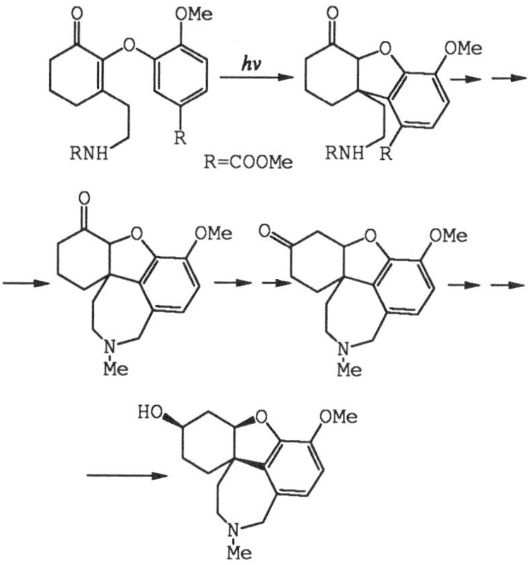

6.39

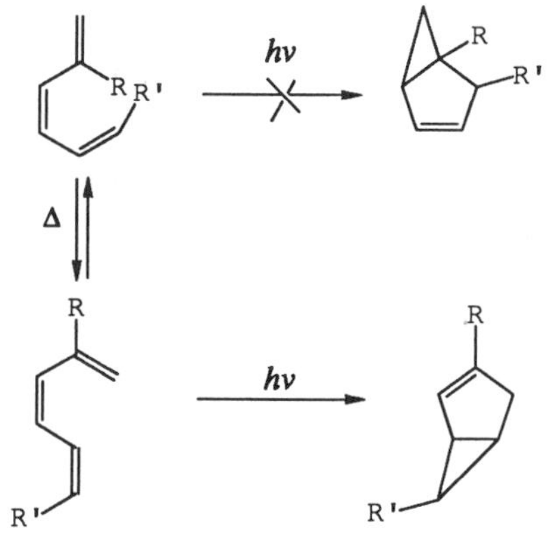

6.37. The nonoxidative photocyclization procedure is based on the fact that the pentadienyl anion is isoelectronic with hexatriene and undergoes 1,5-electrocyclic ring closure to yield the cyclopentenyl anion.[44] The heterosystems formed by substitution of the "carbanion" carbon atom with heteroatoms having an unshared electron pair (N, O, S, Se) are isoelectronic with the pentadienyl anion, are important in synthetic organic photochemistry, and undergo the six-electron heterocyclization reaction shown. The method is based on a conrotatory 1,5 electrocyclic reaction, yielding imminium-, carbonyl-, thiocarbonyl-, or selenocarbonyl- ylids. Subsequent 1,4-sigmatropic H-shifts or a sequence of two 1,2-shifts give the final products, i.e., derivatives of pyrrole, furan, thiophene, selenophene, or the corresponding benzo derivatives.[80]

6.38. This photochemical heterocyclization reaction is preparatively significant, especially for the synthesis of alkaloids.[81] The reaction is highly regioselective and although it is a triplet reaction, it is also highly stereoselective; chemical yields are 70–90%; quantum yields are of the order 0.25–0.5. Long wavelength irradiation in a Pyrex immersion reactor with a 450W discharge lamp of *oxygen-free* solutions (≥ 0.1 M) require about 24 hours.[82] The example illustrates the application of this reaction to the synthesis of *d,l*-lycoramine with a yield of about 10%, based on the starting 1,3-cyclohexadione.[83]

6.39. The isomerization of hexatrienes to bicyclo[3.1.0]hex-2-enes follows certain rules. (1) The triene must be in the cZt configuration. (2) The reaction is regiospecific and appears also to be stereospecific, so that it probably follows a concerted path. (3) The most substituted cyclopropane is usually formed.[84] This is a further example of the rule of five.

6.40

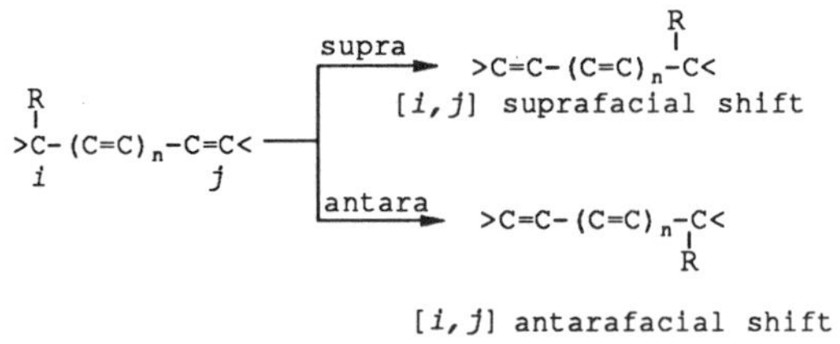

SELECTION RULES FOR [1,j] SIGMATROPIC SHIFTS

INDUCTION	$1+j = 4q$	$1+j = 4q + 2$
Thermal	supra+inversion[a]	supra+retention
	antara+retention	antara+inversion[a]
Photochemical	supra+retention	supra+inversion[a]
	antara+inversion[a]	antara+retention

[a]Inversion cannot occur with H-shifts

6.41

6.42

MA = maleic anhydride

6.40. An $[i, j]$ sigmatropic shift is a thermally or photochemically induced intramolecular pericyclic reaction in which a σ bond migrates along a π system from atom i to atom j ($[i, j]$-sigmatropic shift) or from atom i ($i \neq 1$) to atom j ($[i, j]$-sigmatropic shift) either suprafacially (at the same face of the π system) or antarafacially (at the opposite face of the π system). It proceeds in a concerted manner via a cyclic transition state and is controlled by orbital symmetry (Woodward–Hoffmann rules). The product(s) depend on the electronic state in which the shift occurs (S_0 or S_1) and the extent of the π system involved.[44] Sigmatropic shift systems generally yield different products by direct or sensitized irradiation. In organic photochemistry [1,3] hydrogen shifts in alkenes (see 5.13), and [1,5] and [1,7] sigmatropic shifts are the most important.

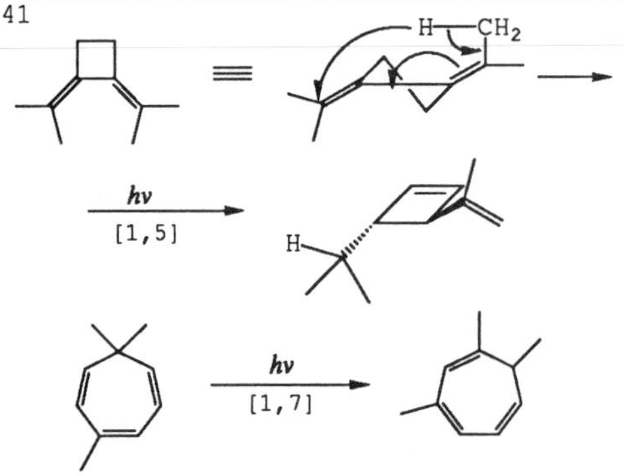

6.41. The migrating substituent is usually a hydrogen atom or an alkyl group. If the carbon atom of the alkyl group is chiral, it shifts with retention or inversion of its stereochemistry. The [1,5] hydrogen sigmatropic shift is the major characteristic of the photochemistry of 1,3-dienes and proceeds antarafacially.[85] The major feature of cycloheptatrienes is the [1,7] H shift, which occurs suprafacially.[86]

6.42. Photochemical [1,5] hydrogen sigmatropic shifts occur in 2-methyl-1-(alken-1-yl)cyclohexenes, which are isomerized to products with two exocyclic double bonds.[87] This reaction has been extended to o-methylstyrenes, which, upon irradiation, isomerize by a [1,5] H shift with the formation of a product with an o-xylylene structure. This very reactive 1,3-diene can be trapped by a dienophile, e.g., maleic anhydride.[88] The photoenolization of o-alkylphenones (see 8.18) and the deconjugation of enones (see 9.10) could be formally classified as [1,5] hydrogen sigmatropic shifts.

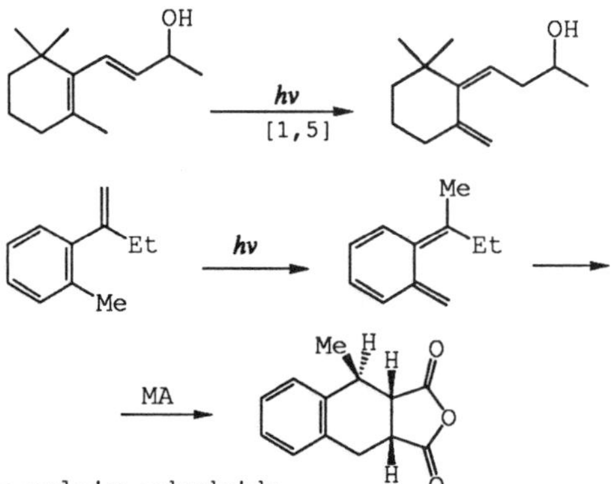

6.43

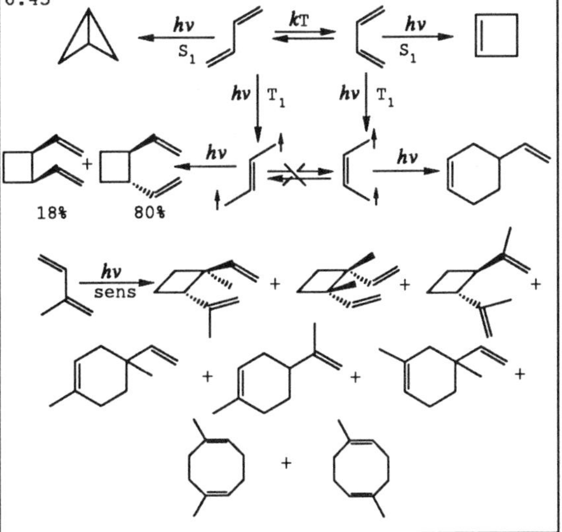

6.44

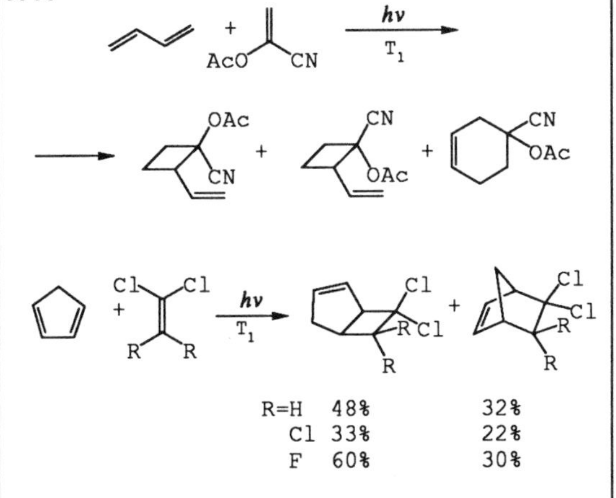

6.45

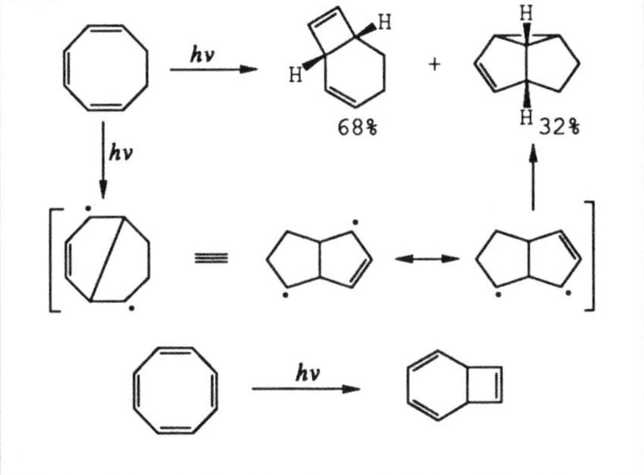

6.43. Direct irradation of 1,3-butadiene gives intramolecular cyclization reaction products (see 6.21),[89] while sensitization leads to both [2+2] and [4+2] dimerization products. The proportion of the products from a sensitized reaction depends on the amount of the individual conformers of butadiene in the irradiated mixture since each rotamer is a precursor for a different product.[90] The ratio of [2+2]:[4+2] photodimers depends on the triplet excitation energy of the sensitizer. With sensitizers of $E_T > 250$ kJ/mol (>60 kcal/mol), *s-trans*-butadiene, the major conformer in the equilibrium mixture, is excited to its T_1 state and [2+2] dimers are formed. With sensitizers of $E_T < 230$ kJ/mol (<55 kcal/mol) only *s-cis*-butadiene is the acceptor of triplet energy and, consequently, [4+2] dimers are formed. The sensitized irradiation of isoprene leads to [4+4] dimerization products, in addition to [2+2] and [4+2] dimers. The latter probably result from a thermal rearrangement of *cis*-divinylcyclobutane.[91] The ratio of [2+2]:[4+2] photodimers of cyclic dienes is independent of the E_T of the sensitizer because of the rigidity of the *cis*-diene system.

6.44. Similar to the dimerization of dienes, the cycloaddition of alkenes to dienes proceeds through the T_1 state of the diene, in which the more stable biradical is an intermediate. Both [2+2] and [4+2] cycloadducts are also formed.[92]

6.45. The disrotatory ring closure to cyclobutene is favored in an excited state reaction of a compound with a 1,3-diene system. Consequently, the principal products of the irradiation of cyclooctatriene[93] and cyclooctatetraene[94,95] are those of a [4n] electrocyclic reaction. However, 1,3,5-cyclooctatriene, which also contains a 1,5-hexadiene system, undergoes a cyclization process with preferential formation of a 1,5-bond as a minor reaction path, according to the rule of five (see 6.2).

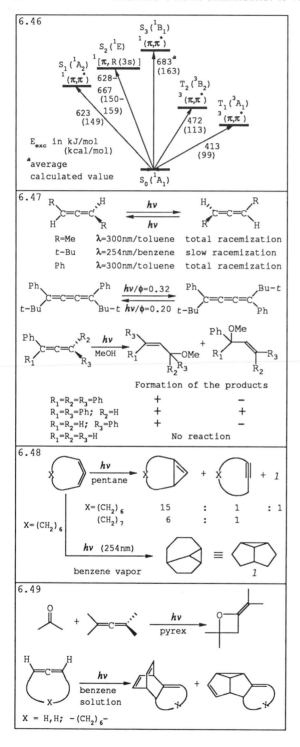

6.46. The UV spectrum of the simplest cumulene, allene, and its alkyl derivatives shows continuous weak absorption from 250–193 nm and numerous transitions at shorter wavelengths.[96,97] Thus, their excited states are complex and their spectral interpretations have generated controversy. However, theory and experiment are in agreement that the lowest singlet excited states of simple allenes are valence π,π^* states, rather than Rydberg $(\pi,3s)$ states. The figure shows the low-lying vertical excited states of allene.

6.47. The most important photoreactions of cumulenes[98] are: π bond rotation, protic solvent addition, rearrangement, and cycloaddition. With odd-carbon cumulenes π bond rotation results in interconversion of enantiomers, while even-carbon cumulenes undergo *cis-trans* isomerization. It is assumed that π bond rotation, a very efficient process, proceeds via the T_1 state with partial or total racemization, as shown.[99–101] Unlike odd-carbon cumulenes, *cis-trans* isomerization of butatrienes is of little interest. The addition of protic solvents to allenes appears to proceed via the S_1 state to form allyl ethers as major products.[102]

6.48. The ratio of rearranged products from irradiation of 1,2-cycloalkadienes is a function of ring size. The larger the ring, the higher the yield of alkyne.[103,104] When the starting material is an enantiomer, racemization occurs more rapidly than rearrangement. In the vapor phase a tricycloalkane is the only product.[105]

6.49. Photocycloaddition reactions of allenes[106] (see also 15.31) probably arise from the addition of allene in the ground state to another excited chromophore, e.g., ketone, enone, quinone, etc. A typical example of allene addition to a ketone is shown.[107] The photoaddition of allenes to aromatic compounds is different than that of alkenes [where the *meta*-cycloadduct predominates (see 7.29 through 7.31)]. Here *para*-cycloadducts are the major products.[108]

REFERENCES

(1)(a) Srinivasan, R. and Carlough, K. H. *J. Am. Chem. Soc.* **1967**, *89*, 4932. (b) For preparation of bicyclo[2.1.1]hexane, see Srinivasan, R. *Org. Photochem. Synth.* **1971**, *1*, 31. (c) For its 2-oxo derivative, see Bond, F. T., Jones, H. L., and Scerbo, L. *Org. Photochem. Synth.* **1971**, *1*, 33. (d) For its 2-methylene derivative, see Skattebol, L. and Peyman, J. *Org. Photochem. Synth.* **1971**, *1*, 77.

(2) Yamadada, C., Pakh, M. J., and Liu, R. S. H. *Chem. Commun.* **1970**, 882.

(3) Dauben, W. G. and Cargill, R. L. *J. Org. Chem.* **1962**, *27*, 1910.

(4)(a) Crowley, K. J. *Tetrahedron.* **1965**, *21*, 1001. (b) Crowley, K. J. *Proc. Chem. Soc.* **1962**, *245*, 334. (c) Saltiel, J. and Zafitious, O. C. *Mol. Photochem.* **1969**, *1*, 1.

(5)(a) Liu, R. S. H. and Hammond, G. S. *J. Am. Chem. Soc.* **1964**, *86*, 1892. (b) Liu, R. S. H. *Tetrahedron Lett.* **1966**, 2159. (c) For the preparation of 1, see Liu, R. S. H. *Org. Photochem. Synth.* **1971**, *1*, 48.

(6) Agosta, W. C. and Wolff, S. *J. Org. Chem.* **1980**, *49*, 3139.

(7) Lowry, T. H. and Richardson, K. S. *Mechanism and Theory in Organic Chemistry.* Harper & Row: New York, 1987, p. 805.

(8)(a) Gleiter, R. and Sander, W. *Angew. Chem., Int. Ed. Eng.* **1985**, *24*, 566. (b) Fischer, E. and Gleiter, R. *Angew Chem., Int. Ed. Engl.* **1989**, *28*, 925. (c) Gleiter, R. and Schafer, W. *Acc. Chem. Res.* **1990**, *23*, 374.

(9)(a) Srinivasan, R. and Hill, K. A. *J. Am. Chem. Soc.* **1965**, *87*, 4988. (b) For a preparative procedure, see Srinivasan, R. *Org. Photochem. Synth.* **1971**, *1*, 101.

(10)(a) Shani, A. *Tetrahedron Lett.* **1972**, 569. (b) For the 1-methyl derivative, see Heathcock, C. A., Badger, R. A., and Starkey R. A. *J. Org. Chem.* **1972**, *37*, 231.

(11) Gleiter, R. and Muller, G. *J. Org. Chem.* **1988**, *53*, 3912, and references therein.

(12) For a review, see (a) Dilling, W. L. *Chem. Rev.* **1966**, *66*, 373. (b) Jefford, C. W. *J. Chem. Educ.* **1976**, *53*, 477. (c) Prinzbach, H. *Pure Appl. Chem.* **1968**, *16*, 17.

(13)(a) Dauben, W. G. and Cargill, R. L. *Tetrahedron.* **1961**, *15*, 197. (b) Hammond, G. S., Turro, N. J., and Fischer, A. *J. Am. Chem. Soc.* **1961**, *83*, 4674. (c) Murrow, S. L. and Hammond, G. S. *J. Am. Chem. Soc.* **1968**, *90*, 2957.

(14)(a) For the preparative procedure, see Sonntag, F. I. and Srinivasan, R. *Org. Photochem. Synth.* **1971**, *1*, 97. (b) For acetoxy quadricyclane, see Gassman, P. G. and Patton, D. S. *Org. Photochem. Synth.* **1971**, *1*, 21. (c) For quadricyclanedicarboxylic acid, see Cristol, S. J. and Snell, R. L. *Org. Photochem. Synth.* **1971**, *1*, 94. (d) For the utilization of this reaction in the synthesis of 1H-azepines (and oxepines), see Ref. 13 and Kaupp, G. and Prinzbach, H. *Org. Photochem. Synth.* **1976**, *2*, 1, and references therein.

(15) Hammond, G. S., Wyatt, P., DeBoer, C. D., and Turro, N. J. *J. Am. Chem. Soc.* **1964**, *86*, 2532.

(16)(a) Eaton, P. E. *Acc. Chem. Res.* **1968**, *1*, 50. (b) Mehta, G., Srikrishna, A., Reddy, A. V., and Nair, M. S. *Tetrahedron.* **1981**, *37*, 4543. (c) For factors influencing the intramolecular [2+2] photochemical ring closure of a multiple-bridged diene, see Osawa, E., Rudzinski, J. M., and Xun, Y. M. *Struct. Chem.* **1990**, *1*, 333.

(17)(a) Schenck, G. O. and Steinmetz, R. *Chem. Ber.* **1963**, *96*, 520. (b) For the preparative procedure of anti-1,3-bis homocubanol, see Dilling, W. L. and Reineke, C. E. *Org. Photochem. Synth.* **1971**, *1*, 85.

(18)(a) Scharf, H.-D. *Tetrahedron.* **1967**, *23*, 3057. (b) Stedman, R. J. and Miller, L. S. *J. Org. Chem.* **1967**, *32*, 3544.

(19)(a) Mohler, D. L., Vollhardt, K. P. C., and Wolff, S. *Angew. Chem., Int. Ed. Engl.* **1990**, *29*, 1151. (b) For the preparation of pentacyclodecane-9,10-dicarboxylic acid anhydride, see Cuts, H. G., Cain, E. N., Westberg, H., and Masamune, S. *Org. Photochem. Synth.* **1971**, *1*, 83. (b) For a homocubane derivative, see Miller, R. D. and Dolce, D. L. *Org. Photochem. Synth.* **1976**, *2*, 17.

(20) Eaton, P. E. and Cole, T. W., Jr. *J. Am. Chem. Soc.* **1964**, *86*, 3157.

(21) Gleiter, R. and Karcher, M. *Angew. Chem., Int. Ed. Engl.* **1988**, *27*, 840.

(22)(a) Yurchenko, A. G., Voroshchenko, A. T., and Stepanov, F. N. *Zhurn. Org. Chim.* **1970**, *6*, 189; *Chem. Abstr.* **1970**, *72*, 89871[s]. (b) For another example, see Schriver, G. W. and Thomas, T. A. *J. Am. Chem. Soc.* **1987**, *109*, 4121. (c) For intramolecular [2+2] photocycloaddition of styrenes to cyclophanes, see Nishimura, J., Dol, H., Ueda, E., Ohbayashi, A., and Oku, A. *J. Am. Chem. Soc.* **1987**, *109*, 5293.

(23)(a) Dopp, D. and Zimmerman, H. E. In *Methoden der Organischen Chemie/(Houben-Weyl).* G. Thieme: Stuttgart, 1975, Vol. 4/5a: Photochemie, p. 413.

(b) Zimmerman, H. E. *Acc. Chem. Res.* **1982**, *15*, 312.   (c) Zimmerman, H. E. In *Rearrangements in Ground and Excited States*, De Mayo, P., ed. Academic Press: New York, 1980, Vol. 3, p. 131. (d) For references describing different features of di-π-methane rearrangements, see Zimmerman, H. E., Nuss, J. M., and Tantillo, A. W. *J. Org. Chem.* **1988**, *53*, 3793.

(24)(a) For R = Ph, R' = H, see Zimmerman, H. E. and Mariano, P. S. *J. Am. Chem. Soc.* **1969**, *91*, 1718. (b) For R = H, R' = Ph, see Zimmerman, H. E. and Baum, A. A. *J. Am. Chem. Soc.* **1971**, *93*, 3646.

(25)(a) For di-π-methane rearrangement of aza analogs, see Armesto, D., Horspool, W. M., and Langa, F. *J. Chem. Soc., Chem. Commun.* **1987**, 1874, and references therein.   (b) For the preparation of methoxyazabullvalene, see Paquette, L. A., Krow, G. R., and Barton, T. J. *Org. Photochem. Synth.* **1971**, *1*, 67.

(26)(a) Zimmerman, H. E. and Grunewald, G. L. *J. Am. Chem. Soc.* **1966**, *88*, 183.   (b) Zimmerman, H. E., Binkley, R. W., Givens, R. S., and Sherwin, M. A. *J. Am. Chem. Soc.* **1967**, *89*, 3939.   (c) Hixon, S. S., Mariano, P. S., and Zimmerman, H. E. *Chem. Rev.* **1973**, *73*, 531.   (d) Paquette, L. A. and Bay, E. J. *Am. Chem. Soc.* **1984**, *106*, 6693.   (e) For a chiral course of the di-π-methane photorearrangement, see Evans, S. V., Garcia-Garibay, M., Omkaram, N., Scheffer, J. R., Trotter, J., and Wireko, F. *J. Am. Chem. Soc.* **1986**, *108*, 5648.

(27)(a) Schroder, G., *Angew. Chem., Int. Ed. Engl.* **1963**, *2*, 481.   (b) For the dual-channeled triplet state di-π-methane rearrangement of benzonorbornadiene, see Paquette, L. A., Burke, L. D., Irie, T., and Tanida, H. *J. Org. Chem.* **1987**, *52*, 3246.

(28)(a) Pratapan, S., Ashok, K., Cyr, D. T., Das, P. K., and George, M. V. *J. Org. Chem.* **1987**, *52*, 5512. (b) For benzobarrelene, see Scheffer, J. R. and Yap, M. J. *J. Org. Chem.* **1989**, *54*, 2561, and references therein.

(29) Michl, J. *Mol. Photochem.* **1972**, *4*, 243, 257, 287.

(30) Zimmerman, H. E. and Cassel, J. M. *J. Org. Chem.* **1989**, *54*, 3800, and references therein.

(31)(a) Paquette, L. A., Varadarajan, A., and Burke, R. *J. Am. Chem. Soc.* **1986**, *108*, 8032.   (b) Adam, W., DeLucchi, O., and Dorr, M. *J. Am. Chem. Soc.* **1989**, *111*, 5209.   (c) Adam, W., Dorr, M., Dron, J., and Rosenthal, R. J. *J. Am. Chem. Soc.* **1987**, *109*, 7074.

(32) Jug, K. J., Iffert, R., and Muller-Remmers, P. L. *J. Am. Chem. Soc.* **1988**, *110*, 2045.

(33)(a) Tabata, T. and Hart, H. *Tetrahedron Lett.* **1969**, 4929.   (b) Zimmerman, H. E., Juers, D. F., McCall, J. M., and Schroder, B. *J. Am. Chem. Soc.* **1971**, *93*, 3662.

(34) For a review of bicycle rearrangement, see Zimmerman, H. E. *Chimia.* **1982**, *36*, 423.

(35) For a controversial mechanism, see Hamer, N. K. and Stubbs, M. *Chem. Commun.* **1970**, 1013.

(36) Figure adapted from Ref. 34.

(37) Zimmerman, H. E. and Factor, R. E. *J. Am. Chem. Soc.* **1980**, *102*, 3538.

(38) See Ref. 7, p. 79.

(39)(a) Koutecký, J. *J. Chem. Phys.* **1967**, *47*, 1501. (b) Dunning, T. H., Hosteny, R. P., and Shavitt, I. *J. Am. Chem. Soc.* **1973**, *95*, 5607, and references therein.   (c) Table adapted from Salem, L. *Electrons in Chemical Reactions, First Principles.* Wiley: New York, 1982, p. 132.

(40)(a) Schulten, K., Ohmine, I., and Karplus, M. *J. Chem. Phys.* **1976**, *64*, 4422.   (b) Aoyagi, M. and Osamura, Y. *J. Am. Chem. Soc.* **1989**, *111*, 470. (c) Aoyagi, M., Osamura, Y., and Iwata, S. *J. Chem. Phys.* **1985**, *83*, 1140.

(41) For reviews of conjugated diene photochemistry, see   (a) Courtot, P., Rumin, R., and Salaun, J.-Y. *Pure Appl. Chem.* **1977**, *49*, 317.   (b) Dilling, W. L. *Chem. Rev.* **1969**, *69*, 845.   (c) Saltiel, J., D'Agostino, J., Megarity, E. D., Metts, L., Neuberger, K. R., Wrighton, M., and Zarfiriou, O. C. *Org. Photochem.* **1973**, *3*, 1.   (d) Srinivasan, R. *Adv. Photochem.* **1966**, *4*, 113.   (e) Turro, N. J., Ramamurthy, V., Cherry, W., and Farneth, W. *Chem. Rev.* **1978**, *78*, 125.   (f) Warrener, R. N. and Bremner, J. B. *Pure Appl. Chem.* **1966**, *16*, 117.

(42) For the photoisomerization of *cis-trans*-1,3-cyclooctadiene, see (a) Liu, R. S. H. *J. Am. Chem. Soc.* **1967**, *89*, 112.   (b) Nebe, W. J. and Fonken, G. J. *J. Am. Chem. Soc.* **1969**, *91*, 1249.   (c) For preparative procedure, see Gassman, P. G. and Williams, E. A. *Org. Photochem. Synth.* **1971**, *1*, 44.

(43)(a) For this and other examples of industrial applications of *cis-trans* photoisomerization, see Fischer, M. *Angew Chem., Int. Ed. Engl.* **1978**, *17*, 16. (b) For pre-vitamin D_3, see Dauben, W. G. and Phillips, R. G. *J. Am. Chem. Soc.* **1982**, *104*, 5780. (c) For the preparation of *cis-β*-ionol from the *trans* isomer, see Ramamurthy, V. and Liu, R. S. H. *Org. Photochem. Synth.* **1976**, *2*, 70.

(44) Woodward, R. B. and Hoffman, R. *The Conservation of Orbital Symmetry.* Verlag Chemie International: Deerfield, Florida, 1970.

(45)(a) Srinivasan, R. *J. Am. Chem. Soc.* **1963,** *85,* 3048. (b) Srinivasan, R. and Sonntag, F. I. *J. Am. Chem. Soc.* **1965,** *87,* 3778.

(46) For a recent study of the thermal and photochemical cyclobutene/1,3-butadiene interconversion, see Clark, K. B. and Leigh, W. J. *J. Am. Chem. Soc.* **1987,** *109,* 6086.

(47)(a) Gassman, P. G. and Hymans, W. E. *Tetrahedron.* **1968,** *24,* 4437. (b) Dauben, W. G. and Poulter, C. D. *J. Am. Chem. Soc.* **1968,** *90,* 802.

(48)(a) Srinivasan, R. *J. Am. Chem. Soc.* **1968,** *90,* 4498. (b) Boue, S. and Srinivasan, R. *J. Am. Soc.* **1970,** *92,* 3226. (c) For a recent study of *s-cis* acyclic 1,3-dienes, see Squillacote, M. and Semple, T. C. *J. Am. Chem. Soc.* **1990,** *112,* 5546. (d) For the substituent and wavelength effects on the photochemical ring opening of monocyclic alkylcyclobutenes, see Leigh, W. J., Zheng, K., and Clark, K. B. *Can. J. Chem.* **1990,** *68,* 1988.

(49) Borden, W. T., Reich, I. L., Sharpe, L. A., and Reich, H. J. *J. Am. Chem. Soc.* **1970,** *92,* 3808.

(50) Ane, D. H. and Reynolds, R. N. *J. Am. Chem. Soc.* **1973,** *95,* 5027.

(51)(a) For X = NH, see Dilling, W. L. *Org. Photochem. Synth.* **1976,** *2,* 5. (b) For X = O, see Pirkle, W. H. and McKendry, L. H. *J. Am. Chem. Soc.* **1969,** *91,* 1179.

(52)(a) McDiarmid, R., Sabljic, A., and Doering, J. P. *J. Am. Chem. Soc.* **1985,** *107,* 826, and references therein. (b) See also Trulson, M. O., Dollinger, G. D., and Mathies, R. A. *J. Am. Chem. Soc.* **1987,** *109,* 586.

(53) For an excellent review of photophysics and photochemistry of simple trienes, see Jacobs, H. J. C. and Havinga, E. *Adv. Photochem.* **1979,** *11,* 305.

(54) Nebot-Gil, I. and Malrieu, J. P. *J. Am. Chem. Soc.* **1982,** *104,* 3320.

(55) Granville, M. F., Holton, G. R., and Kohler, B. E. *J. Chem. Phys.* **1980,** *72,* 4671, and references therein.

(56) See Ref. 53, p. 321.

(57) Minnaard, N. G. and Havinga, E. *Recl. Trav. Chim.* **1973,** *92,* 1179, 1315.

(58) Dauben, W. G., Rabinowitz, J., Vietmeyer, N. D., and Wendschuk, P. D. *J. Am. Chem. Soc.* **1972,** *94,* 4285.

(59) Vroegop, P. J., Lugtengurg, J., and Havinga, E. *Tetrahedron.* **1973,** *29,* 1393.

(60) Courot, P. and Rumin, R. *Tetrahedron. Lett.* **1970,** 1849.

(61) Matuszewski, B., Burgstahler, A. W., and Givens, R. S. *J. Am. Chem. Soc.* **1982,** *104,* 6874.

(62)(a) This reaction path has been employed in the synthesis of Dewar benzene, see van Tamelen, E. E., Pappas, S. P., and Kirk, K. *J. Am. Chem. Soc.* **1971,** *93,* 6092. For other examples of this path, see (b) Bremmer, J. B. and Warrener, R. N. *Chem. Commun.* **1967,** 926. (c) Holovka, J. M. and Gardner, P. D. *J. Am. Chem. Soc.* **1967,** *89,* 6390. For the change in reactivity with substitution, see (d) Jefford, C. W. and Delay, F. *J. Am. Chem. Soc.* **1975,** *97,* 2272. (e) Bellus, D., Helferich, G., and Weis, C. D. *Helv. Chim. Acta.* **1971,** *54,* 463. (f) Goldschmidt, Z. and Genizi, E. *Tetrahedron Lett.* **1987,** *28,* 4867.

(63)(a) Meier, H. In *Methoden der Organischen Chemie/(Houben-Weyl).* G. Thieme: Stuttgart, 1975, Vol. 4/5a: Photochemie, p. 262, and references therein. (b) Havinga, E. *Experientia.* **1973,** *29,* 1181, and references therein.

(64) For authoritative reviews of photocyclization-oxidation reactions, see (a) Laarhoven, W. H. *Recl. J. R. Neth. Chem. Soc.* **1983,** *102,* 185, 241. (b) Laarhoven, W. H. *Pure Appl. Chem.* **1984,** *56,* 1225. (c) Mallory, F. B. and Mallory, C. W. *Org. React.* **1984,** *30,* 1. (d) See Ref. 53, p. 511. (e) For an extensive study of valence isomerization of *cis*-stilbene to 4a,4b-dihydrophenanthrene, see Muszkat, K. A. and Fischer, E. *J. Chem. Soc. B.* **1967,** 662. (f) For a recent study of this reaction, see Prinsen, W. J. C. and Laarhoven, W. H. *J. Org. Chem.* **1989,** *54,* 3689. (g) For bimolecular reactions of *trans*-stilbene, see Lewis, F. D. *Acc. Chem. Res.* **1979,** *12,* 152, and references therein.

(65) For the preparation of 3-fluorophenanthrene, see Mallory, C. W. and Mallory, F. B. *Org. Photochem. Synth.* **1971,** *1,* 55. (b) For the preparation of an azaphenanthrene, see Dybas, R. A. *Org. Photochem. Synth.* **1971,** *1,* 25.

(66) Mallory, F. B., Rudolph, M. J., and Oh, S. M. *J. Org. Chem.* **1989,** *54,* 4619, and references therein.

(67) Knauer, B. *Diss. Abstr. Int. B.* **1969,** *30,* 1042.

(68) Laarhoven, W. H. and Veldhuis, R. G. M. *Tetrahedron.* **1972,** *28,* 1811.

(69) Martin, R. H., Morren, G., and Schurter, J. J. *Tetrahedron. Lett.* **1969**, 3683.

(70) Laarhoven, W. H., Cuppen, T. J. H. M., and Nivard, R. F. J. *Tetrahedron.* **1970**, *26*, 4865.

(71) Diedrich, F. and Staab, H. A. *Angew. Chem., Int. Ed. Engl.* **1978**, *17*, 372.

(72)(a) Laarhoven, W. H. and Veldhuis, R. G. M. *Tetrahedron.* **1972**, *28*, 1823. (b) Laarhoven, W. H., Cuppen, T. J. H. M., and Nivard, R. F. J. *Tetrahedron.* **1974**, *30*, 3343. (c) Martin, R. H., Eyndels, C., and Defay, N. *Tetrahedron.* **1974**, *30*, 3339.

(73) Stahlke, K. R., Heine, H. G., and Hartmann, W. *Liebigs Ann. Chem.* **1972**, *764*, 116.

(74)(a) Leznoff, C. C. and Hayward, R. J. *Can. J. Chem.* **1970**, *48*, 1842. (b) Baldry, P. J. *J. Chem. Soc., Perkin Trans. 2.* **1980**, 805. (c) Olsen, R. J., Minniear, J. C., Overton, W. M., and Sherrick, J. M. *J. Org. Chem.* **1991**, *56*, 989. (d) For a recent review of photocyclizations and intramolecular photocycloadditions of conjugated arylolefins and related compounds, see Laarhoven, W. H. *Org. Photochem.* **1990**, *10*, 163.

(75)(a) Tinnemans, A. H. A. and Laarhoven, W. H. *J. Am. Chem. Soc.* **1974**, *96*, 4617. (b) For the preparation, see Tinnemans, A. H. A. and Laarhoven, W. H. *Org. Photochem. Synth.* **1976**, *2*, 93.

(76)(a) Lewis, G. E. *J. Org. Chem.* **1960**, *25*, 2193. (b) Badger, G. M., Drewer, R. J., and Lewis, G. E. *Austr. J. Chem.* **1963**, *16*, 1042; **1964**, *17*, 1036.

(77)(a) El'Tsov, A. V., Studzinskii, O. P., Ogol'tsova, N. V. *Zhurn. Org. Zhim.* **1970**, *6*, 405 (*Chem. Abstr.* **1970**, *72*, 111266x). (b) For the first oxidative photocyclization of pyrroles, see Perrine, D. M., Kagan, J., Huang, D. B., Zeng, K., and Teo, B. K. *J. Org. Chem.* **1987**, *52*, 2213.

(78)(a) For a review of applications of this cyclization to alkaloid synthesis, see Kametani, T. and Fukumoto, K. *Acc. Chem. Res.* **1972**, *5*, 512. For the recent application of the intramolecular coupling photodehydrohalogenation, see (b) Bringmann, G. and Jansen, J. R. *Tetrahedron Lett.* **1988**, *29*, 2987. (c) Hoshino, O., Ogasawara, H., Takahashi, A., and Umezawa, B. *Heterocycles.* **1987**, *25*, 155.

(79) Cava, M. P., Mitchell, M. J., Havlicek, S. C., Lindert, A., and Spangler, R. J. *J. Org. Chem.* **1970**, *35*, 175.

(80) For authoritative reviews of this cyclization, see (a) Schultz, A. G. *Acc. Chem. Res.* **1983**, *16*, 210. (b) Schultz, A. G. and Motyka, L. *Org. Photochem.* **1983**, *6*, 1.

(81) For a recent paper, see Gramain, J. C., Husson, H. Q., and Troin, Y. *J. Org. Chem.* **1985**, *50*, 2323.

(82) For preparative examples of this reaction, see Schultz, A. G. and DeTar, M. B. *Org. Photochem. Synth.* **1976**, *2*, 47, 101.

(83)(a) Schultz, A. G., Yee, Y. K., and Berger, M. H. *J. Am. Chem. Soc.* **1977**, *99*, 8065. (b) For a photochemical heterocyclization of functionalized dienamines, see Gelas-Mialhe, Y., Mabiala, G., and Verriere, R. *J. Org. Chem.* **1987**, *52*, 5395.

(84) Courtot, P. and Rumin, R. *Bull. Soc. Chem. Fr.* **1972**, 4238.

(85) Kiefer, E. G. and Fukunaga, J. Y. *Tetrahedron Lett.* **1969**, 933.

(86) Jones, L. B. and Jones V. K. *J. Am. Chem. Soc.* **1968**, *90*, 1540.

(87) Roof, A. A. M., van Wageningen, A., Kruk, C., and Cerfontain, H. *Tetrahedron. Lett.* **1972**, 367.

(88) Hornback, J. M. and Barrow, R. D. *J. Org. Chem.* **1982**, *47*, 4285.

(89) However, irradiation of concentrated solutions of butadiene in isooctane with 254 nm produces a mixture of dimers. See Srinivasan, R. and Sonntag, F. I. *J. Am. Chem. Soc.* **1965**, *87*, 3778.

(90)(a) Hammond, G. S., Turro, N. J., and Liu, R. S. H. *J. Org. Chem.* **1963**, *28*, 3297. (b) Liu, R. S. H. and Gale, C. M. *J. Am. Chem. Soc.* **1968**, *90*, 1897. (c) DeBoer, C. D., Turro, N. J., and Hammond, G. S. *Org. Synth.* **1967**, *47*, 64.

(91) Liu, R. S. H., Turro, N. J., and Hammond, G. S. *J. Am. Chem. Soc.* **1965**, *87*, 3406, and references therein.

(92)(a) Dilling, W. L. *J. Am. Chem. Soc.* **1967**, *89*, 2742. (b) Turro, N. J. and Bartlett, P. D. *J. Org. Chem.* **1965**, *30*, 1849, 4396.

(93) Roth, W. R. and Peltzer, B. *Angew. Chem., Inst. Ed. Engl.* **1964**, *76*, 378.

(94) Yamazaki, H. and Shida, S. *J. Chem. Phys.* **1956**, *24*, 1278.

(95) For the photochemistry of polycyclic polyenes see (a) Dressel, J., Pansegrau, P. D., and Paquette, L. A. *J. Org. Chem.* **1988**, *53*, 3996. (b) Anthony, I. J., Byrne, L. T., McCulloch, R. K., and Wege, D. *J. Org. Chem.* **1988**, *53*, 4123.

(96) Runge, W. In *The Chemistry of the Allenes*, Landor, S. R., ed. Academic Press: 1982, Vol. 3, p. 777.

(97) Rabalais, J. W., McDonald, J. M., Scherr, V., and McGlynn, S. P. *Chem. Rev.* **1971,** *71*, 73.

(98) Johnson, R. P. *Org. Photochem.* **1985,** *7*, 75.

(99) See Ref. 98, p. 89.

(100) Rodriguez, O. and Morrison, H. *Chem. Commun.*, **1971,** 679.

(101) Hornback, J. M. *17th Annual Report on Research Under Sponsorship of the Petroleum Research Fund.* American Chemical Society, 1972, p. 121.

(102) Kuhn, R. and Schulz, B. *Chem. Ber.* **1965,** *98*, 3218.

(103) Stierman, T. J. and Johnson, R. P. *J. Am. Chem. Soc.* **1985,** *107*, 3971.

(104) Price, J. D. and Johnson, R. P. *J. Am. Chem. Soc.* **1985,** *107*, 2187.

(105) For a review, see Landor, S. R., ed., *The Chemistry of the Allenes.* Academic Press: New York, 1982, Vols. 1–3.

(106) Hopf, H. In Ref. 105, Vol. 2, Ch. 5, p. 525.

(107) Gotthardt, H., Steinmetz, R., and Hammond, G. S. *J. Org. Chem.* **1968,** *33*, 2774.

(108) Berridge, J. C., Forrester, J., Foulger, B. E., and Gilbert, A. *J. Chem. Soc., Perkin Trans. 1*, **1980,** 2425.

Chapter 7
PHOTOCHEMISTRY OF AROMATIC COMPOUNDS

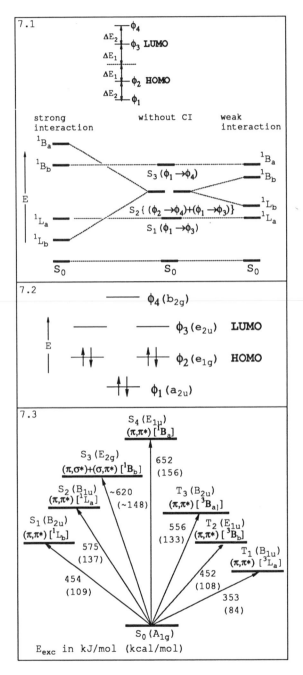

7.1 strong interaction · without CI · weak interaction

E_{exc} in kJ/mol (kcal/mol)

7.1. Since benzenoid hydrocarbons are *alternant*[1] with symmetrical arrangements of bonding and antibonding π orbitals, the S_2 state is degenerate at the SCF level.[2] This degeneracy is removed by the introduction of configuration interaction (CI). If the interaction is weak, the formation of the lowest singlet state can be described as the transition of an electron from the HOMO to the LUMO (an ^1L_a type transition, according to Platt). This type of interaction is typical of polyacenes beginning with anthracene, and the transition from the ground state to the excited state is allowed.[3] If the interaction is stronger, the states cross and the formation of the lowest singlet state can be described as a mixture of $(\phi_2 \rightarrow \phi_4)$ + $(\phi_1 \rightarrow \phi_3)$ electron transitions [an ^1L_b type transition (Platt)], which is forbidden.

7.2. Singly excited configurations of benzene are produced by transition of an electron from one of the two degenerate e_{1g} HOMOs to one of the two degenerate e_{2u} LUMOs.[4] Therefore, at the *Hückel* level there are four singlet and four triplet states (all degenerate) with the electron configuration $(a_{2u})^2$ $(e_{1g})^3(e_{2u})$.[1] The degeneracy is partially removed by CI, which produces three excited singlet states with energy $^1B_{2u}$ < $^1B_{1u}$ < $^1E_{1u}$[4] and three excited triplet states with energy $^3B_{1u}$ < $^3E_{1u}$ < $^3B_{1u}$.[5]

7.3. Although considerable spectroscopic[6] and theoretical[7] data for benzene is available for the prediction of its photochemical behavior, some uncertainty still exists regarding the assignment of an $^1A_{1g} \rightarrow {}^1E_{2g}$ transition to the 205 or 190 nm band and the location of the $^1E_{2g}$ state (above[8] or below[9] $^1B_{1u}$). This results from the sensitivity of the $^1E_{2g}$ state to electron interaction (CI). The electronic spectrum of benzene is shown in 2.1.

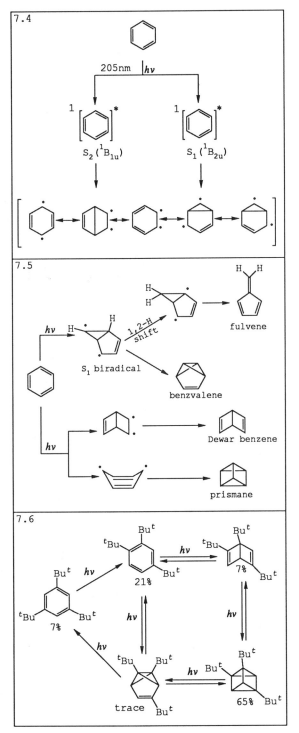

7.4. The following photoreactions of aromatic compounds will be discussed: valence isomerization; *ortho* (1,2), *meta* (1,3), and *para* (1,4) cycloaddition reactions of alkenes and alkynes to arenes, and aromatic photosubstitution (S Ar*) of arenes, especially nucleophilic substitution reactions by S_N2 Ar* or $S_{RN}1$ mechanisms. While selective excitation into S_1 gives preferentially *meta* (and *ortho*) products,[10] excitation into S_2 gives *para* bonded products by valence isomerization.[11] These processes are described by biradical intermediates for ease of visualization.

7.5. Benzene has long been considered photochemically unreactive because its valence isomers are formed with very low quantum yields. Irradiation of liquid benzene under nitrogen at 254 nm causes excitation to the S_1 state and the products, benzvalene and fulvene, are formed via a 1,3-biradical.[10, 13] The formation of *Dewar benzene* occurs in the S_2 state[11, 14] upon short wavelength irradiation (205 nm). It has been confirmed that Dewar benzene is converted to prismane.[15] The formation of benzvalene can be viewed formally as an $x[2+2]$ cycloaddition reaction, whereas Dewar benzene results from disrotatory ring closure of a "butadiene" subsystem.

7.6. The individual valence isomers of benzene are photoreactive at 254 nm and isomerize to benzene and fulvene. Photolysis of benzvalene and Dewar benzene in the gas phase (10 Pa) yields benzene with $\Phi = 0.7$ and 0.9, respectively. As a result, the irradiation of benzene, and especially its more photoreactive alkyl derivatives, leads to a photostationary mixture (see 5.9 and 5.10). Because of steric strain, benzene derivatives with bulky substituents, e.g., 1,3,5-tri-*t*-butyl benzene, are particularly prone to valence photoisomerizations.[16, 17]

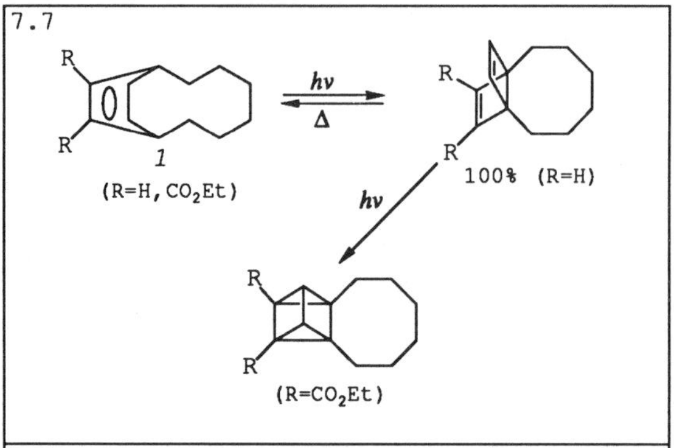

7.7. As shown in 7.6, valence isomerization is particularly facile and efficient when the planarity of the aromatic system is perturbed by strain. The product is thermally reisomerized to the starting cyclophane (R=H).[18] This thermal reversion has been used as a key step in the formation of [4] and [5]paracyclophanes.[19] Some derivatives of *1* (R=COOEt) give the corresponding prismane as the only product.[20]

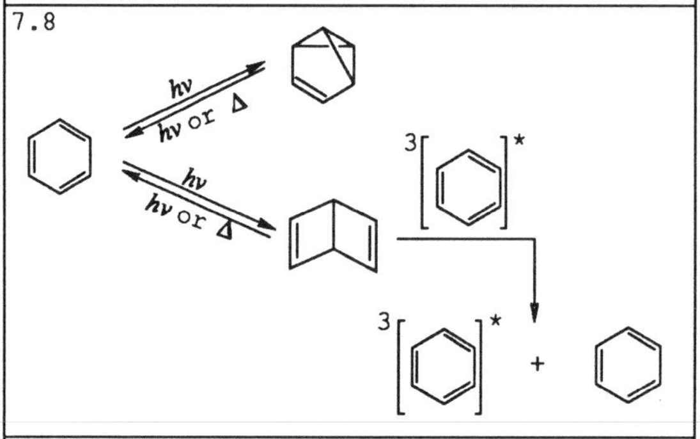

7.8. Both benzvalene and Dewar benzene reisomerize thermally or photochemically to benzene. For some fraction of molecules the thermal reversion of Dewar benzene proceeds via triplet benzene[21] and produces chemiluminescence in very low yield. No chemiluminescence was observed from benzvalene. The triplet-sensitized (E_T > 280 kJ/mol, 67 kcal/mol) reversion of Dewar benzene leads to the formation of triplet benzene (E_T > 357 kJ/mol, 85 kcal/mol), which then sensitizes the reversion of another molecule of Dewar benzene (chain reaction).[22]

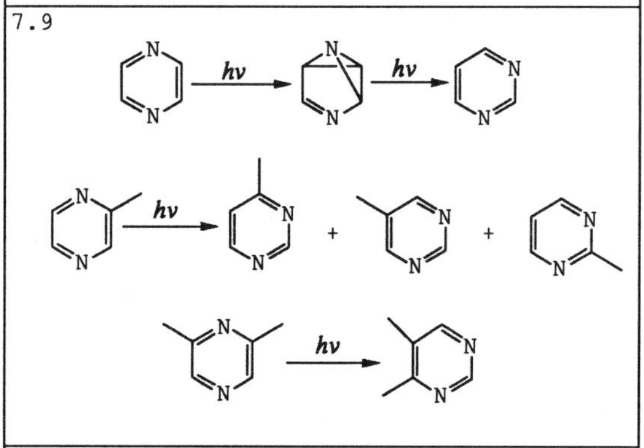

7.9. Azabenzenes (pyridine, pyrazine, etc.) exhibit electronic spectra similar to benzene. However, their S_1 state is frequently n,π^*, while the π,π^* state is lowest for T_1. Their photochemical valence isomerization occurs by a mechanism similar to that of benzene $\rightarrow$ benzvalene. The initially formed diazabenzvalenes from pyrazine and its methyl derivatives are unstable and isomerize immediately to give the corresponding pyrimidines.[23] The analogous Dewar pyridines and/or azaprismanes[24] are formed from pyridine[25] and its perfluoroalkyl derivatives.[26]

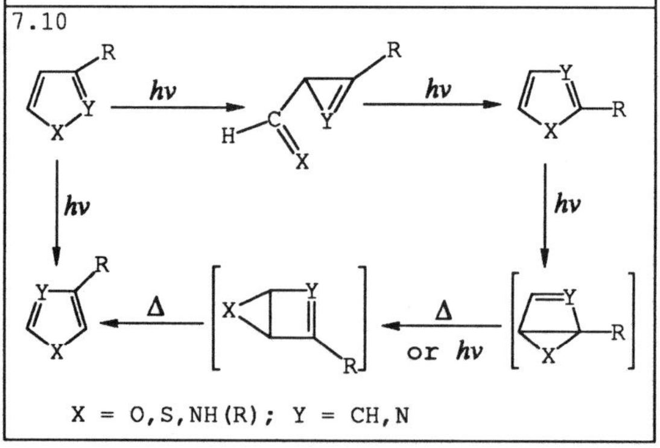

7.10. A shift of substituents occurs by the irradiation of monohetero compounds, such as furans,[27] thiophenes,[28] and pyrroles.[29] The interconversion of 1,2 into 1,3 isomers in dihetero compounds, i.e., oxazole to isoxazole,[30] thiazole to isothiazole,[31] pyrazole to imidazole,[32] and their benzoderivatives[33] was observed.

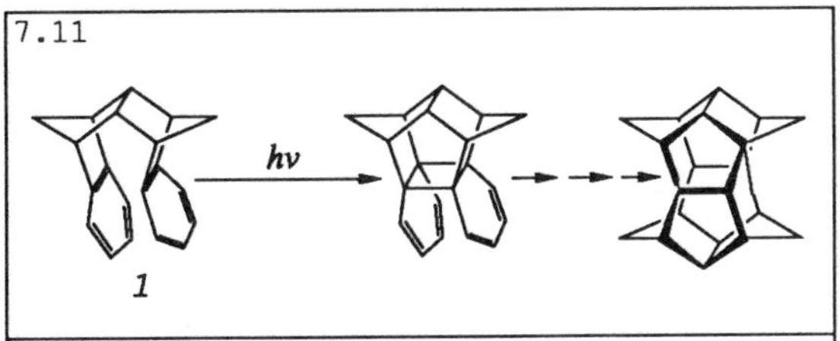

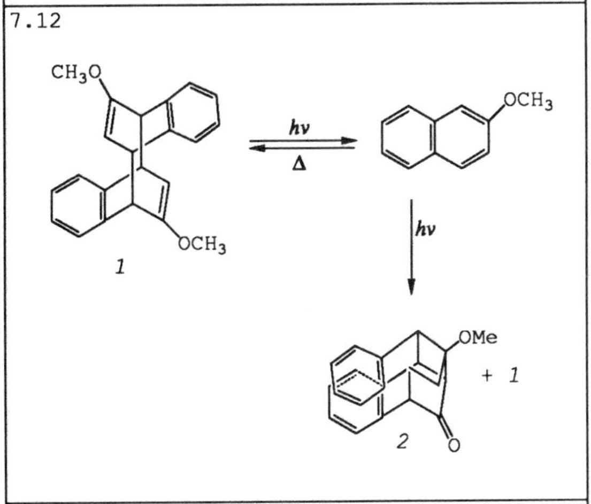

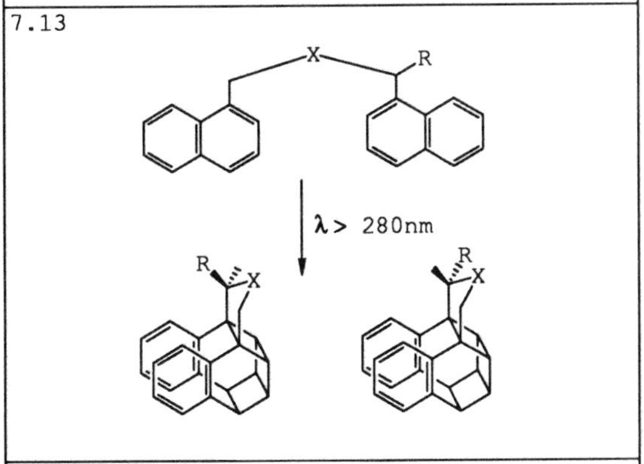

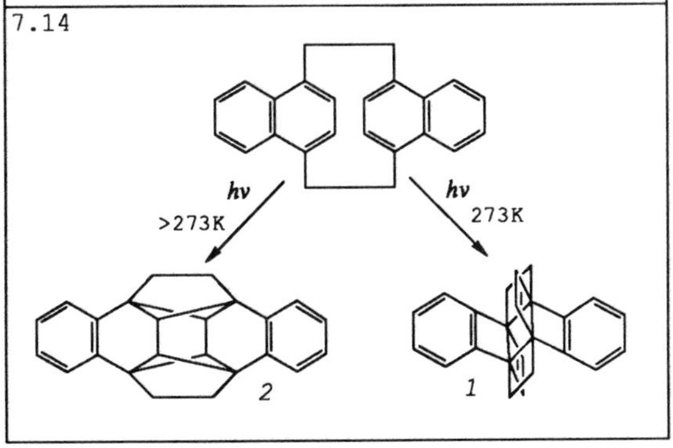

7.11. Photodimerization of the parent benzene is not known. Its [2+2] dimer has been synthesized,[34] but it rapidly dissociates to benzene. Intramolecular [2+2] photodimerization of the benzene rings in the bichromophoric compound 1, a key step in the synthesis of pagodane,[35] has been successfully achieved. Also, a double [2+2] photodimerization of the benzene rings in a layered, multiply linked dithiacyclophane was observed.[36]

7.12. Some derivatives of naphthalene,[37] e.g., 2-alkoxynaphthalenes,[37(c)] (but not 1-alkoxynaphthalenes) undergo photodimerization. Depending on the solvent, either the *exo* [4+4] dimer *1*, or both *1* and the cage dimer *2* (resulting from the *endo* [4+4] dimer) are formed. The yield of dimer was 45% after 14 days of irradiation; after two months it was 90%.

7.13. Although 1-methylnaphthalene, as well as amino-, bromo-, and hydroxynaphthalenes, do not photodimerize in a bimolecular process, bichromophoric naphthalenes made by linking two naphthalene molecules by a three-atom chain facilitates the intramolecular [4+4] *endo*-photodimerization reaction (see also 7.16). After a thermal [3.3] sigmatropic shift (e.g., X=CH$_2$, R=H, OH)[39(a),(b)] and, in some cases, X=CH$_2$, R=OH),[39(b)] the entire process is terminated by a [2+2] cycloaddition reaction. Both *endo* and *exo* [4+4] photodimers are formed from the corresponding oxa analog (X=O, R=H).[40] The intramolecular cycloaddition reaction also occurs if two different chromophores, e.g., naphthalene and anthracene, are linked by a three-atom chain. The cycloaddition can also occur intermolecularly.[41]

7.14. Irradiation of another bichromophoric compound, e.g., the *exo* isomer of 1,1′,4,4′-[2.2]paracyclonaphthane at 273K, leads to the thermally unstable product *1* by a [4+4] intramolecular dimerization. Dibenzoequinene (*2*), the cage product of a double [4+2] cycloaddition, is produced at higher temperatures.[42]

7.15

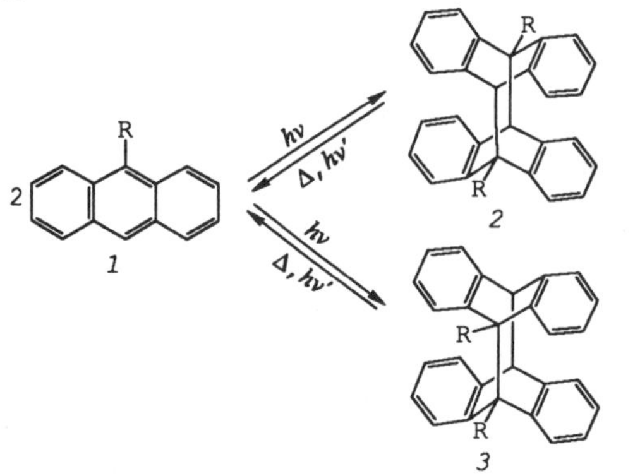

7.16

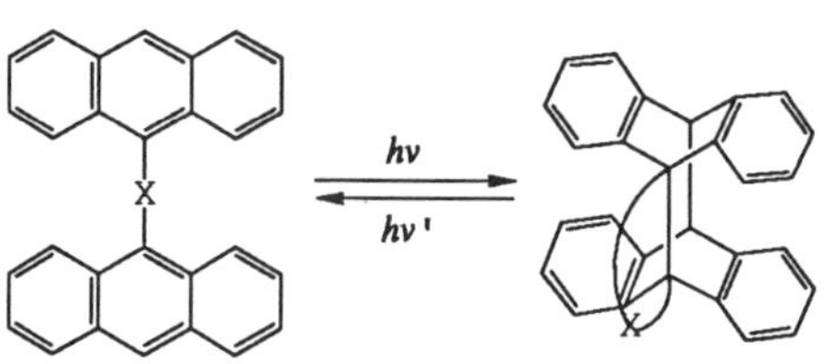

X= —(CH$_2$)$_{10}$ Φ=0.0004   X—CH$_2$CH$_2$CH—   Φ=0.14
                                          |
                                          OH

X=—(CH$_2$)$_2$·   Φ=0.24   X—CH$_2$CHCH$_2$-   Φ=0.046
                                    |
                                    OH

X—CH$_2$CH—   Φ=0.35   X—CH$_2$CH$_2$C—   Φ=0.65
       | ||
       OH   O

X=—(CH$_2$)$_3$·   Φ=0.14

7.17

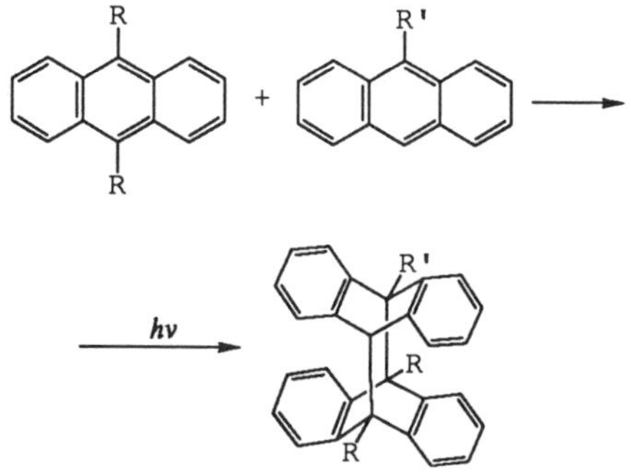

e.g.,   R=CH$_3$, OCH$_3$; R'=CN

7.15. The [4+4] photodimerization of anthracene occurs via a singlet excimer at positions 9 and 10.[43] If the excimer does not collapse to the photodimer, competing excimer fluorescence occurs. The ratio of these two processes is a function of substitution in positions 9 and 10, which decreases the efficiency of dimerization and increases excimer fluorescence, probably because of steric strain. Although 9-substituted anthracenes *1* in polar solvents yield primarily the expected HT-dimers (head-to-tail) *2*,[44] irradiation of *1* in nonpolar solvents and in micellar solution dramatically increases the proportion of the head-to-head (HH) dimer *3*.[45] Substitution in both positions 9 and 10 can completely suppress photodimerization.

7.16. Bichromophoric systems having a linking chain of atoms often leads to photodimerization or cycloaddition reactions that do not occur by bimolecular processes or can enhance the yields of some that have been observed. The length of the linking chain affects not only the excimer (exciplex) intermediate formation, but also its collapse to the product. The quantum yields for the [4+4] cycloaddition of the 1, ω-di-9-anthryl-alkane series vary with the length of the linkage. The highest quantum yield (0.65) observed is a result of the longer lifetime of the triplet state through which the reaction proceeds. It is apparent that excimer fluorescence competes with photodimerization.[46]

7.17. Cross-dimerization reactions are particularly efficient if one of the components is an anthracene derivative incapable of dimerization, e.g., 9,10-dimethylanthracene, or if one of the components contains an electron donor and the other, an acceptor (CT interaction).[47] An increase in the polarity of the medium enhances the yield of cross dimerization between 9,10-dimethylanthracene and 9-anthracenocarbonitrile.[48]

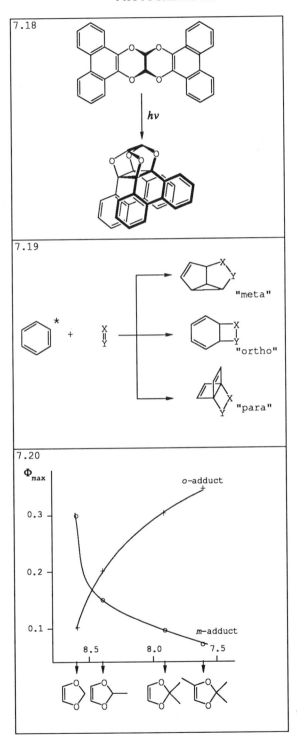

7.18. Although the photodimer of phenanthrene has been synthesized by an intramolecular [2+2] photocycloaddition from the irradiation of a macrocyclic diene, phenanthrene, like benzene and naphthalene, does not undergo photodimerization. However, some of its derivatives, e.g., 9-phenanthrenocarbonitrile,[49] 4H-cyclopenta-[d,e,f]phenanthracene,[50] and the bichromophoric glyoxal derivative of 9,10-dihydroxyphenanthrene[51] (shown in the figure) do undergo [2+2] photodimerization at positions 9 and 10.

7.19. Benzene and its derivatives in the S_1 state (the T_1 state of aromatic compounds is usually quenched by the transfer of excitation energy to the alkene) undergo cycloaddition reactions with various π systems, especially alkenes, but also azocompounds,[52] alkynes, and dienes. Under certain structural and electronic circumstances a reverse model, i.e., S_1 alkene + S_0 arene, prevails. Three different modes of cycloaddition reactions have been observed: 1,2 or *ortho*-, 1,3 or *meta*-, and 1,4 or *para*-, of which the *para* is the least efficient. These reactions occur either bimolecularly or intramolecularly.[53,54]

7.20. Alkenes with ionization potentials (IP) close to that of benzene (9.24 eV) preferentially undergo *meta*-cycloaddition (see 7.29). Electron-poor alkenes, such as acrylonitrile, maleic anhydride, and those containing electron donors, e.g., vinyl ethers, preferentially form *ortho*-, in addition to some *para*-products (these may be secondary products due to the photolability of *ortho*-products). The formation of the *ortho*-cycloadduct is generally favored by donor-acceptor interaction. The more polar the interactions (charge transfer) in the exciplex, the larger the proportion of *ortho*- and *para*-cycloadducts.[53,55] Selection rules have been proposed for the prediction of the regio- and stereoselectivity of these reactions.[53,56]

7.21

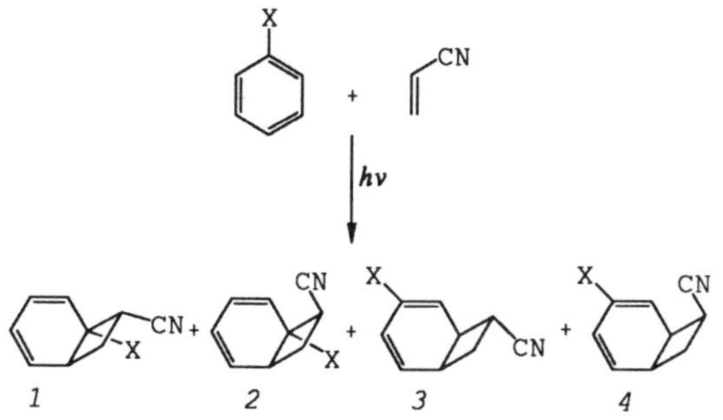

X	1	2	3+4
H	73%	17%	
CH$_3$	57%	14%	29%
OCH$_3$		100%	

7.22

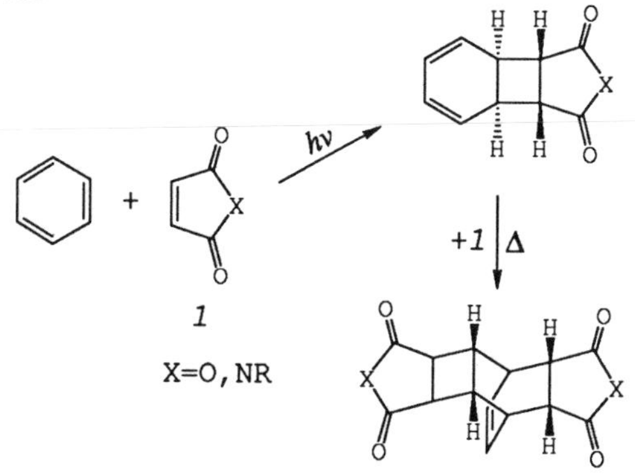

1

X=O, NR

7.23

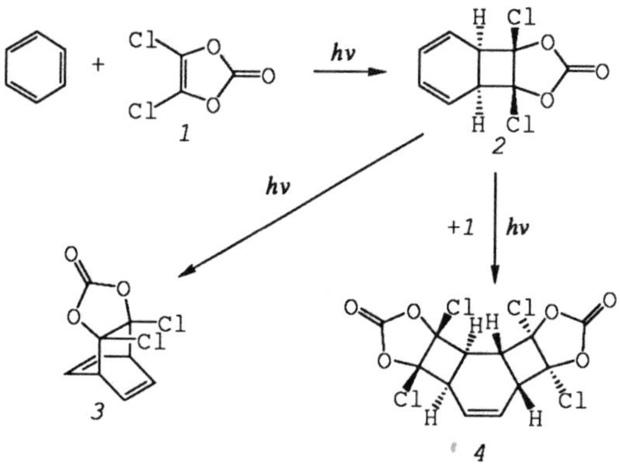

7.21. Benzene and its derivatives generally form CT complexes with dienophiles and, upon irradiation, give *ortho*-cycloadducts, usually with *exo*-orientation.[57] The CT complexes absorb at longer wavelengths than the individual components, and the cycloaddition reaction occurs even with radiation corresponding in energy to the CT absorption band, which is lower in energy than the E_{exc} of the benzene derivative. The substituents control the regioselectivity of the *ortho*-cycloaddition reaction (the stronger the donor, the greater the control).

7.22. Irradiation of benzene in the presence of maleic anhydride yields, via the S_1 state of a CT complex, an *exo* 1:2 cycloadduct, as a result of a photochemical [2+2] *ortho*-cycloaddition reaction followed by a thermal [4+2] cycloaddition.[58,59] The reaction also proceeds with sensitizers ($E_T \geq$ 285 kJ/mol, 68 kcal/mol) with benzene and substituted benzenes (toluene, xylenes, *p*-t-butylbenzene, chlorobenzene, etc.).[60] The cycloaddition of maleimides by direct or sensitized irradiation in strongly dependent on the substitutent on the nitrogen atom. Whereas H and alkyl groups yield the desired 2:1 photoproduct, N-aryl derivatives yield the corresponding *anti* dimer in sensitized reactions.[61]

7.23. In contradiction to the selection rules (see 7.20), 1,3-dioxol-2-one yields *meta*-products[62] upon irradiation with benzene because of the photolability of the initially formed *ortho*-cycloadduct and the reaction conditions (T_1 process). This view is supported by the sensitized (E_T sens $\geq$ 170 kJ/mol, 40 kcal/mol) reaction of the dichloro-1,3-dioxol-2-one (*1*) with benzene, in which the *ortho*-cycloadduct 2 rearranges to the *para*-cycloadduct 3 or undergoes a further [2+2] cycloaddition reaction with the formation of the 2:1 adduct *4*.[63] With naphthalene (N), the *ortho*-product is formed by an N(S_1) + *1*(S_0) interaction, while the *para*-product results from an N(S_0) + *1*(S_1) mechanism.[63(c)]

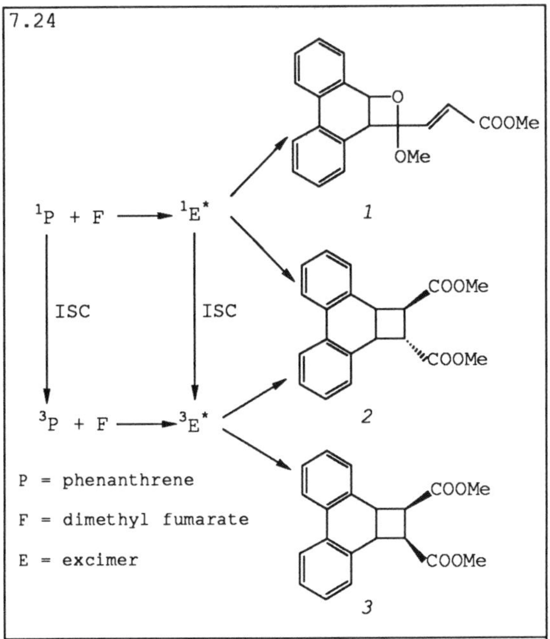

7.24

P = phenanthrene

F = dimethyl fumarate

E = excimer

7.24. Electron-poor alkenes, e.g., acrylonitrile,[64] methyl cinnamate, maleate, and fumarate, undergo cycloaddition to phenanthrene (P), exclusively at the 9,10 double bond. The formation of *endo*-cycloadducts with methyl cinnamate provides evidence of the existence of an exciplex (E*) intermediate (strong π orbital overlap between the aromatic rings of the two components).[65] In the more complex cycloaddition of dimethyl fumarate (F) to phenanthrene, oxetane *1* and, partially, product *2*, are formed via a singlet exciplex.[66] However, the nonstereospecific formation of *3* and the major part of *2* are the result of a triplet reaction in which the triplet is formed by ISC of either ^{1}E* or ^{1}P, prior to exciplex formation, as shown.[67]

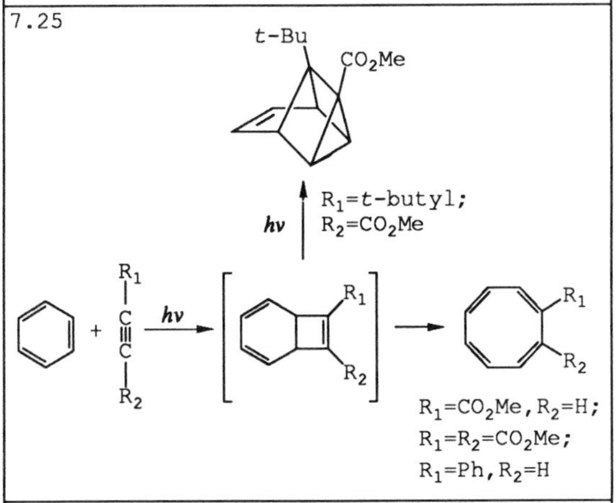

7.25. The primary products from irradiation of alkynes, especially those with electron-withdrawing groups (dienophilic alkynes), with benzene are the corresponding *ortho*-cycloadducts. These are often unstable at ambient temperature and by ring opening provide substituted cyclooctatetraenes[68,69] or a cage structure by a subsequent photoreaction.[70] Either singlet[71] or triplet[72] states may be involved in which the alkyne is usually the excited species.[68(d)] The cycloaddition is more efficient by an intramolecular process, in which the alkyne moiety is linked to the arene by at least a two-atom chain and the cycloaddition proceeds via the arene singlet state.[73]

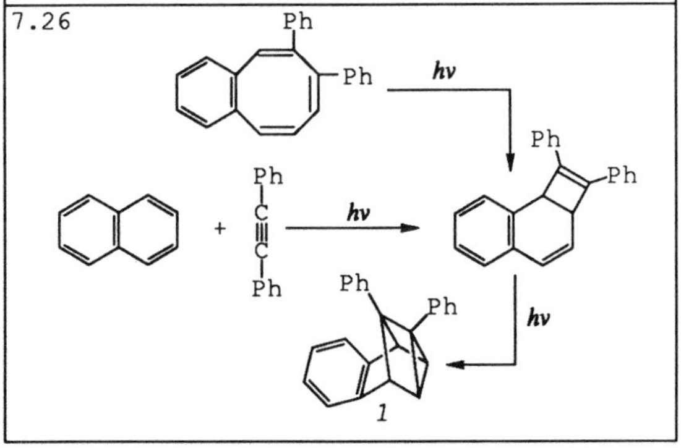

7.25

$R_1 = t$-butyl; $R_2 = CO_2Me$

$R_1 = CO_2Me$, $R_2 = H$;
$R_1 = R_2 = CO_2Me$;
$R_1 = Ph$, $R_2 = H$

7.26. The final product of the irradiation of tolane with naphthalene is the cage compound *1*.[74] The exciplex formed by naphthalene in its S_1 state and tolane in its S_0 state yields an *ortho*-product which, upon further irradiation, gives the cage compound by an intramolecular [2+2] cycloaddition. The reaction takes place with a wide range of substituted naphthalenes[75] and diarylacetylenes.[76] Cage compound *1* is also obtained by the irradiation of 6,7-diphenylbenzocyclooctatetraene.[74]

7.26

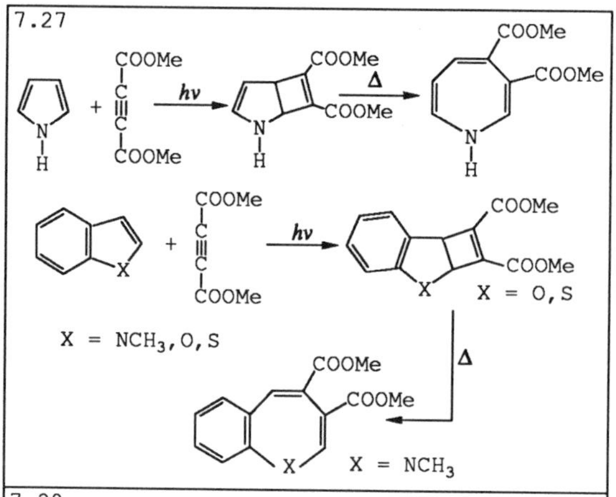

7.27. The sensitized irradiation of dimethyl acetylenedicarboxylate with pyrrole[77] or N-methylindole affords azepine[77] or benzazepine[78] derivatives (as one of the products). These are thought to arise from the allowed photochemical disrotatory ring opening of the primary product, the *ortho*-cycloadduct. With benzofuran or benzothiophene, the primary product can be isolated.[79]

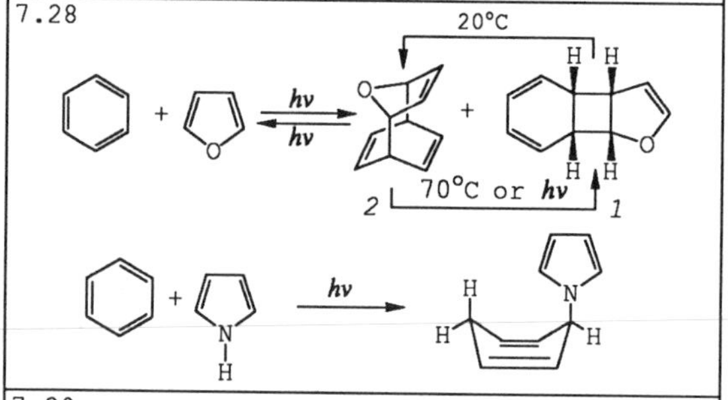

7.28. Irradiation of a furan-benzene mixture at 254 nm yields two 1:1 cycloadducts: *ortho*-cycloadduct *1* and *para*-cycloadduct *2*, which rearranges thermally or photochemically to *1*. Sensitized irradiation of either cycloadduct causes cycloreversion.[80] Unlike furan, pyrrole adds to excited benzene to form a 1,4-cyclohexadiene derivative ($\Phi = 0.2$).[81] Pyrrole and indole react similarly[82] with naphthalene, apparently via an electron transfer mechanism involving an S_1 exciplex.[83]

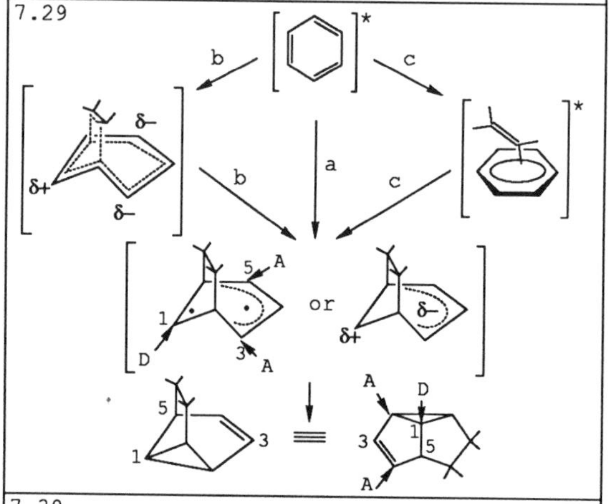

7.29. *Meta*-cycloaddition occurs by the interaction of an alkene in the S_0 state with an arene in the S_1 state, only if the IP of the alkene closely matches that of the arene, i.e., < 0.4 eV difference. Otherwise, *ortho*- and/or *para*-cycloaddition occurs. Of the mechanisms shown: (a) is concerted, with the formation of a biradical or zwitterionic intermediate. These can be formed via either a charge-transfer (b) or an exciplex (c). Recent studies substantiate paths (b) and (c).[84]

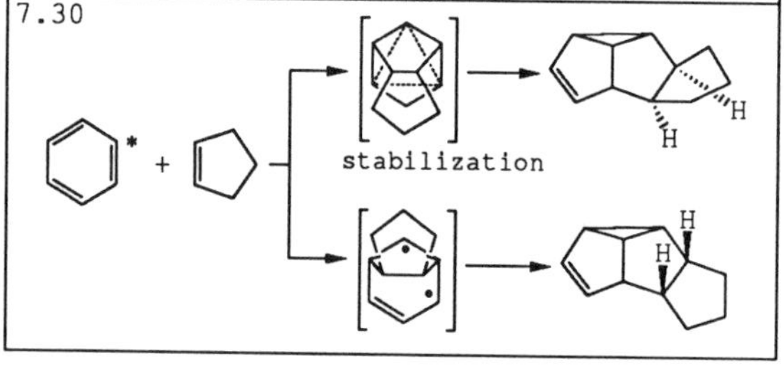

7.30. Irradiation (254 nm) of simple aliphatic or cyclic alkenes with benzene yields *meta*-cycloadducts, preferentially with an *endo*-configuration, as major products.[85] For example, cyclopentene and benzene form an *endo*-*meta*-cycloadduct ($\Phi = 0.17$), while the *exo*-cycloadduct is formed with a quantum yield of $\Phi = 0.02$.[86] The reaction course via the stabilized exciplex favors *endo*-stereochemistry, while the most stable biradical intermediate would favor the formation of the *exo*-isomer.

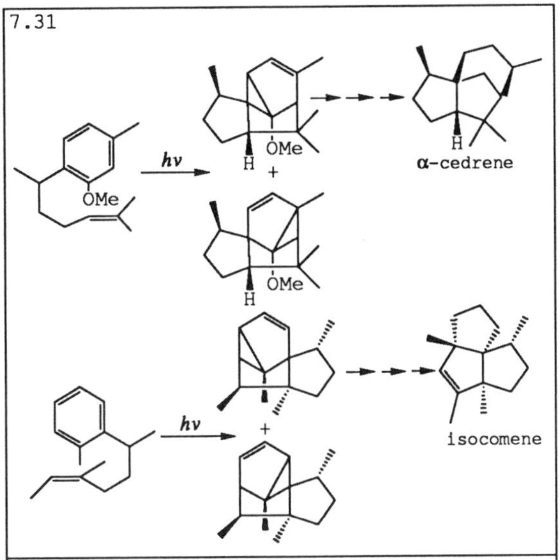

7.31. *meta*-Cycloadditions of substituted benzenes with alkenes are highly regioselective in which the position of the substituent in the *meta*-cycloadduct depends on whether the substituent is an electron donor or acceptor, e.g., the alkene bonds to the benzene ring in positions *ortho* to an electron donor substituent. The intramolecular (bichromophoric) version of the reaction is also particularly efficient, however, only when both chromophores are separated by three connecting atoms.[87] Regioselectivity is so pronounced that the intramolecular version of the photo-*meta*-cycloaddition was used as the key step in the synthesis of a number of natural products, e.g., the sesquiterpene hydrocarbons, α-cedrene,[88] and isocomene.[89,90]

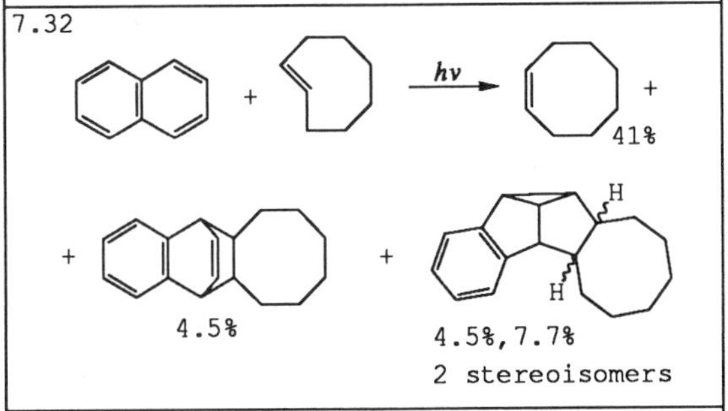

7.32. The products of the irradiation (300 nm) of naphthalene with *trans*-cyclooctene are *cis*-cyclooctene (41%), two *meta*-cycloadducts (12.2%) of the cis isomer, whose configurations have not yet been resolved, and a *para*-cycloadduct (4.5%).[91] Under these conditions *trans*-cyclooctene does not form any cycloadducts, although in the presence of benzene it does give an *endo-meta*-cycloadduct with $\Phi = 0.31$ and an *exo-meta*-cycloadduct with $\Phi = 0.02$, as well as an *ortho*-cycloadduct with $\Phi = 0.09$.

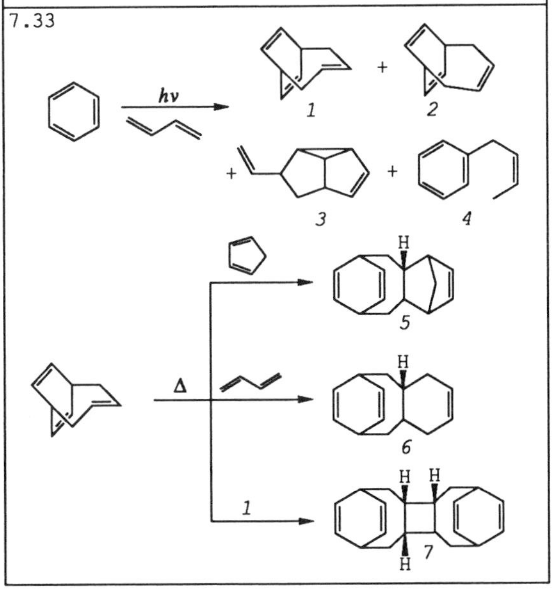

7.33. Irradiation of benzene with 1,3-butadiene gives a mixture in which *para*-cycloadducts prevail. The major product (67% of the reaction mixture) is the tricyclic triene *1* with one double bond in a *trans*-configuration. This very reactive alkene easily undergoes dimerization or forms [2+2] cycloadducts. The configuration of dimer *7* is in agreement with orbital conservation rules (see 5.16), according to which the configuration of the product of a thermal dimerization reaction should be the result of a concerted $[_\pi 2_s + _\pi 2_a]$ process.[92]

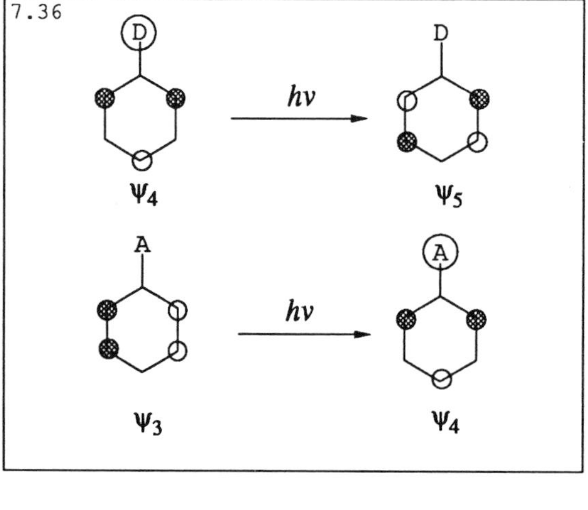

7.34. Photocycloaddition of 1,3-cyclohexadiene to naphthalene yields two photochemically allowed $[_\pi 4_s + _\pi 4_s]$ products (56% *syn* and 5.5% *anti* isomer) and, in addition, two products of a forbidden $[_\pi 4_s + _\pi 2_s]$ cycloaddition reaction. The latter apparently resulted from the decomposition of the exciplex into a vibrationally excited hot ground state, followed by a thermally allowed cycloaddition.[93,94] The *syn* isomer can undergo a subsequent photosensitized [2+2] cycloaddition to yield a cage compound. The [4+4] photocycloaddition reaction of 5,6-disubstituted cyclohexadienes has been utilized in the synthesis of the mixed [4+4] adducts (heterodimers) of benzene and naphthalene[95] or anthracene.[96]

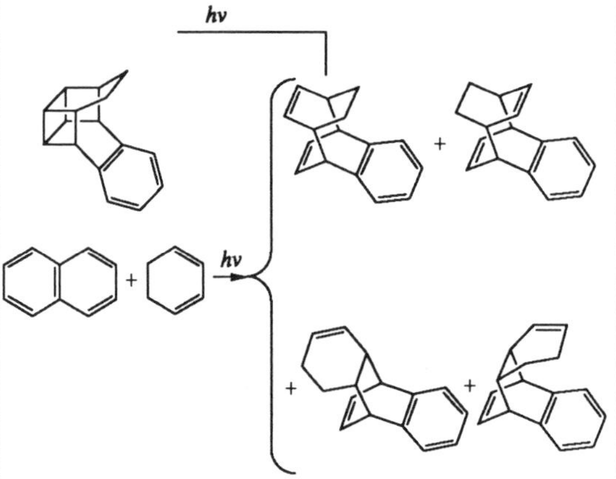

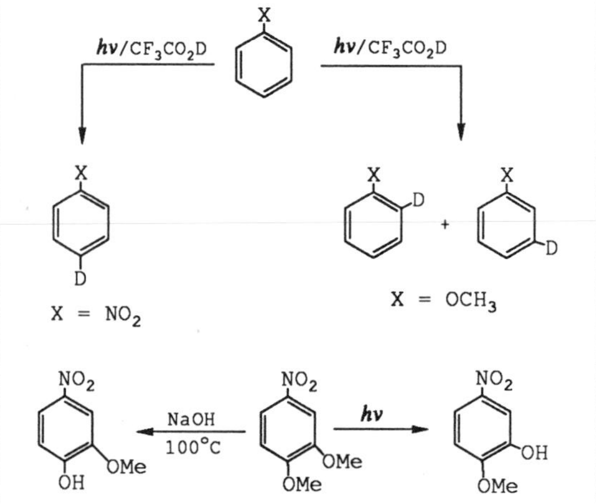

7.35. In aromatic photosubstitution reactions (S Ar*), electrophilic (S_E Ar*), or nucleophilic (S_N Ar*), the directing effects of the substituents are exactly opposite to those in the thermal aromatic S_E or S_N reactions. Although only a few examples of S_E Ar* reactions are known (perhaps an implication of the electron transfer from the π,π^* excited state of the substrate to the electrophile),[97] numerous examples of S_N Ar* reactions have been reported,[98] many of which proceed with a high Φ.

7.36. The different reactivity can be explained in terms of changes in the electron population of the ground and excited states, particularly changes in the electron density at the sites of nucleophilic substitution, which result from the frontier orbitals of benzene derivatives. Benzene derivatives can be divided into two groups: those substituted by an electron donor D (HOMO = Ψ_4 and LUMO = Ψ_5) and those substituted with an electron acceptor A (HOMO = Ψ_3 and LUMO = Ψ_4). For the S_N Ar* reaction, the sites are those which have a maximum population of electrons in HOMO (indicated by circles) and a minimum in LUMO. These are positions 4 in D and 3 and 5 in A. For the S_E Ar* reaction, the positions are 2 and 3 in D and positions 2 and 4 in A.

7.37

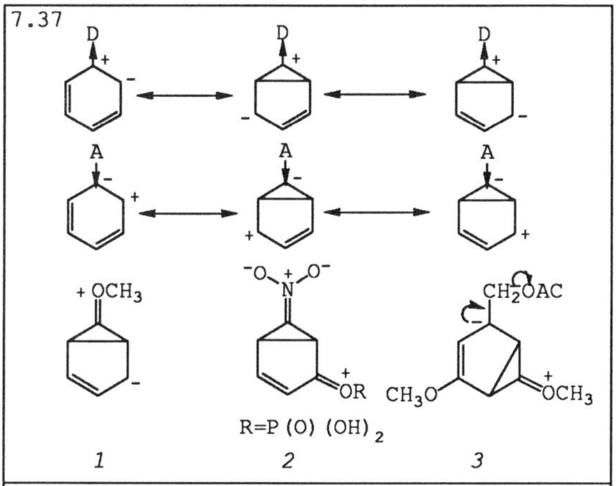

7.38

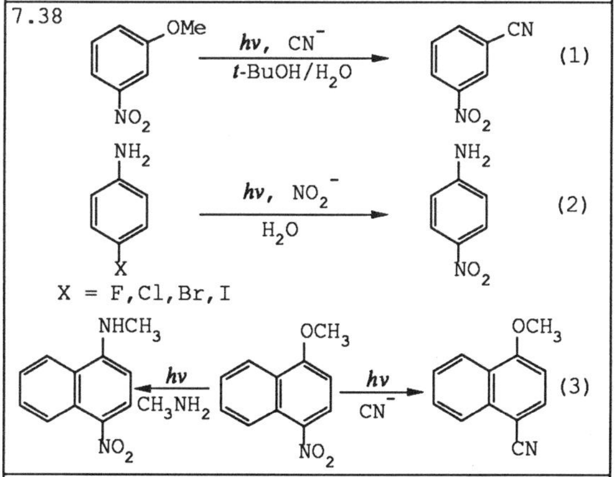

7.39

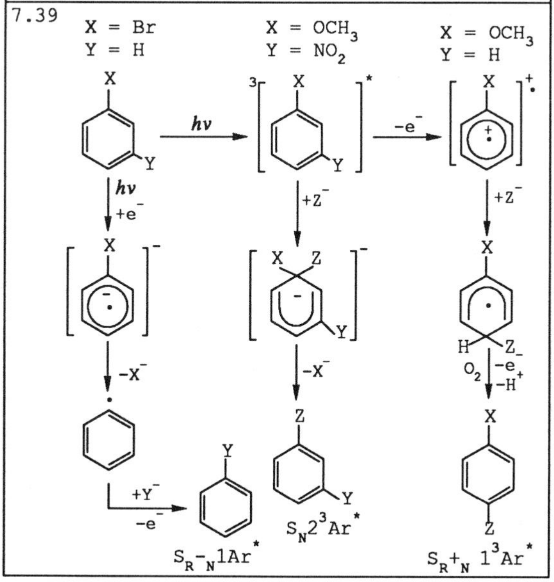

7.37. The above results can also be expressed in terms of the valence bond (VB) theory using VB structures for the π, π^* states of benzene derivatives in which, in the presence of a donor, the sites of highest π electron density are the *ortho*- and *meta*-positions. At these positions the S_E Ar* reaction occurs. With an acceptor, the *meta*-position is the center of lowest π electron density where the S_N Ar* reaction proceeds. With these structures it is possible to explain the electrophilic photodeuterium exchange (*1*) in anisoles,[99] as well as the facile photosolvolysis of *m*-nitrophenylphosphate (*2*)[100] and 3,5-dimethoxybenzylacetate (*3*).[101]

7.38. Both procedures shown, together with the empirical rules given below, rationalize the preference of a given reaction path, at least qualitatively. The site of an S_N Ar* reaction is: (1) the position *meta* to the electron-attracting substituent (NO_2, CN, $COCH_3$); or (2) the position *ortho* or *para* to the electron-donating group (OH, OCH_3, NR_2); or (3) the α position of the *cata*-condensed polynuclear arenes; or (4) the substituent of the same electronic properties as the substitute (nucleophile), i.e., *the merging resonance stabilization rule.*[102]

7.39. The mechanism of the S_N Ar* reaction, however, is complex, and three different pathways have been proposed:[102(a)] (1) direct displacement at the $^3(\pi, \pi^*)$ state of the arene (S_N2 ^{3}Ar*); (2) electron transfer from the $^3(\pi, \pi^*)$ state of the arene to an acceptor, followed by the attack of a nucleophile on the aromatic radical-cation with the formation of a σ complex ($S_{R+N}1$ ^{3}Ar*); (3) electron transfer from the nucleophile to the (π, π^*) state of the aromatic substrate to form a radical-anion ($S_{R-N}1$ Ar*). The S_N2 ^{3}Ar* reaction competes with the S_{R-N} Ar* reaction, particularly with amine nucleophiles of lower ionization potential (secondary amines).[102(b)]

$S_{R+N}1$

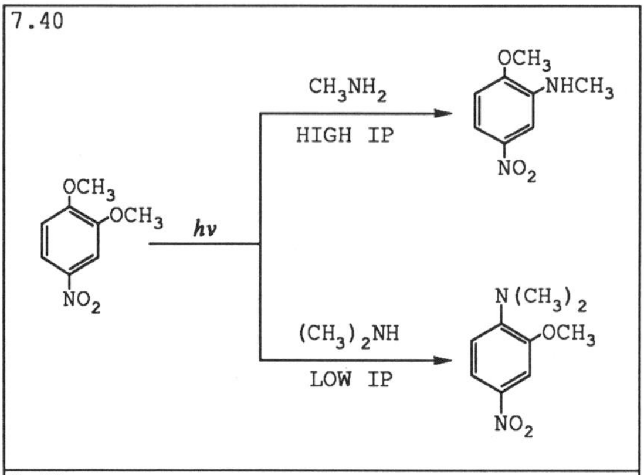

7.40

7.41

σ-complex   70%

7.42

X = (CH$_2$)$_n$

X=(CH$_2$)$_n$, yield 21-33%

7.40. With primary amines (higher ionization potential) no electron transfer from the nucleophile to the $^1(\pi,\pi^*)$ or the $^3(\pi,\pi^*)$ state of the aromatic substrate takes place. Thus, S$_N$2 Ar* chemistry (*meta* regioselectivity) is observed, e.g., with 4-nitroveratrole. With secondary amines (lower IP), *para* regioselectivity is obtained, arising from a radical-ion pair via electron transfer from the amine to the $^3(\pi,\pi^*)$ state of the aromatic substrate.[103]

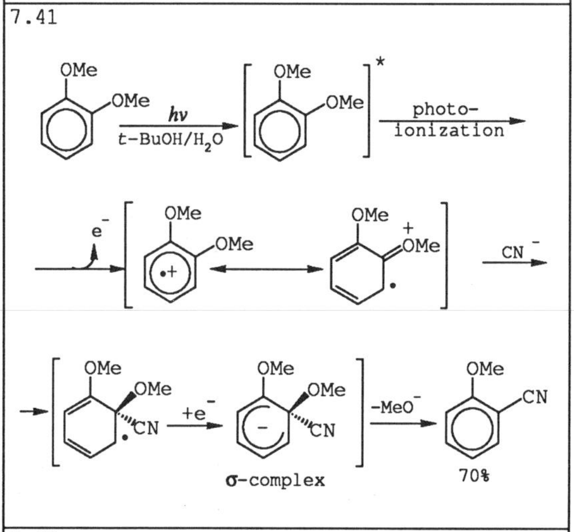

7.41. In S$_{R^+N}$ ^{3}Ar* reactions, an electron can be expelled from the π,π^* triplet state of the aromatic substrate and transferred, e.g., to a solvent molecule. The addition of a nucleophile to the ionized molecule (radical-cation) then takes place. The photosubstitution of the methoxy group by a nitrile group in veratrole illustrates this reaction.[104]

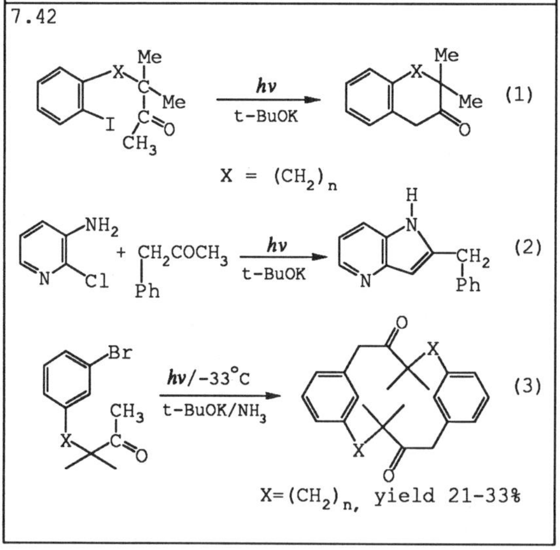

7.42. The S$_{R-N}$1 Ar* mechanism is a well-known process, referred to as a radical-chain mechanism of nucleophilic substitution, and involves radical-anions and radicals as intermediates.[105] Many reactions of, e.g., nonactivated aryl halides with various nucleophiles, proceed smoothly under photostimulation[106] and have recently been a subject of attention for their synthetic value. This reaction with an enolate as the nucleophile is particularly attractive, both in inter- and intramolecular arrangements. The reaction proceeds in good yield and offers a convenient method for the synthesis of benzocycloalkanones (*1*),[107] indole derivatives (*2*),[108] benzofurans,[109] etc. This method has recently been used for a facile synthesis of [m.m]metacyclophanediones (*3*) by irradiation of o-(*m*-bromophenyl)-3,3-dimethylalkan-2-ones in the presence of *t*-BuOK in liquid ammonia.[110] These cyclization reactions are successful only when the internal H atom transfer from position 8 to the benzene ring is hindered, e.g., by substitution.

REFERENCES

(1) Greenwood, H. H. *Computing Methods in Quantum Organic Chemistry.* Wiley-Interscience: London, 1972, p. 44.

(2) Koutecký, J. *J. Chem. Phys.* **1966**, *44*, 3702.

(3) See Ref. (1), pp. 84, 174.

(4) Parr, R. C., Craig, D. P., and Ross, I. C. *J. Chem. Phys.* **1950**, *18*, 1561.

(5) Kearns, D. *J. Chem. Phys.* **1962**, *36*, 1608.

(6) For an authoritative review of excitation and deexcitation of benzene, see Cundall, R. B., Robinson, D. A., and Pereira, L. C. *Adv. Photochem.* **1977**, *10*, 147.

(7) Buma, W. J., van der Waals, J. H., and van Hemert, M. C. *J. Am. Chem. Soc.* **1989**, 111, 86.

(8) Peyerinhoff, S. D. and Buenker, R. J. *Theor. Chim. Acta.* **1970**, *19*, 1.

(9) Bloor, J. E., Lee, J., and Gartside, S. *Proc. Chem. Soc.* **1960**, 413.

(10) Kaplan, L., Rausch, D. J., and Wilzbach, K. E. *J. Am. Chem. Soc.* **1972**, *94*, 8638.

(11) Bryce-Smith, D., Gilbert, A., and Robinson, D. A. *Angew. Chem., Int. Ed. Engl.* **1971**, *10*, 745.

(12) For a review of benzene valence photoisomerization, see (a) Bryce-Smith, D. and Gilbert, A. In *Rearrangements in Ground and Excited States*, De Mayo, P., ed. Academic Press: New York, 1980, Vol. 3, p. 349. (b) Fischer, M. In *Methoden der organischen Chemie/(Houben-Weyl).* G. Thieme: Stuttgart, 1975, Vol. 4/5a: Photochemie, p. 472. (c) Noyes, W. A. and Al-Ani, K. E. *Chem. Rev.* **1974**, *74*, 29. (d) Van Tamelen, E. E. *Acc. Chem. Res.* **1972**, *5*, 186.

(13)(a) Kaplan, L., Ritscher, J. S., and Wilzbach, K. E. *J. Am. Chem. Soc.* **1966**, *88*, 2881. (b) Farenhorst, E. and Bickel, A. F. *Tetrahedron Lett.* **1966**, 5911. (c) For the matrix-controlled photochemistry of benzene and pyridine, see Johnstone, D. E. and Sodeau, J. R. *J. Phys. Chem.* **1991**, *95*, 165.

(14) For different Dewar arenes, see (a) Ward, H. R. and Wishnok, J. S. *J. Am. Chem. Soc.* **1968**, *90*, 5353. (b) hexafluoro: Barlow, M. G., Haszeldine, R. N., and Dingwall, J. G. *J. Chem. Soc., Perkin Trans. 1.* **1973**, 1542. (c) perfluoroalkyl: Barlow, M. G., Haszeldine, R. N., and Kershaw, M. J. *J. Chem. Soc., Perkin Trans. 1.* **1975**, 2005. (d) Dewar naphthalene and anthracene: Meador, M. A. and Hart, H. *J. Org. Chem.* **1989**, *54*, 2336, and references therein.

(15)(a) Haller, I. *J. Chem. Phys.* **1967**, *47*, 1117. (b) Halevi, E. A. *Nouveau J. Chim.* **1977**, *1*, 229.

(16)(a) van Tamelen, E. E., Pappas, S., and Kirk, K. L. *J. Am. Chem. Soc.* **1971**, *93*, 6092. (b) For the detection of the benzvalene intermediate, see Den Besten, I. E., Kaplan, L., and Wilzbach, K. E. *J. Am. Chem. Soc.* **1968**, *90*, 5868.

(17) Wilzbach, K. E. and Kaplan, L. *J. Am. Chem. Soc.* **1965**, *87*, 4004.

(18)(a) Kamula, S. L., Iroff, L. D., Jones, M., Jr. van Straten, J. W., deWolf, W. H., and Bickelhaupt, F. *J. Am. Chem. Soc.* **1977**, *99*, 5815. (b) For the photoreactivity of [5]metacyclophanes, see Jenneskens, L. W., de Boer, H. J. R., de Wolf, W. H., and Bickelhaupt, F. *J. Am. Chem. Soc.* **1990**, *112*, 8941.

(19)(a) Kostermans, G. B. M., Bobelkijk, M., deWolf, W. H., and Bickelhaupt, F. *J. Am. Chem. Soc.* **1987**, *109*, 2471. (b) Tsuji, T. and Nishida, S. *J. Am. Chem. Soc.* **1988**, *110*, 2157, and references therein.

(20) Liebe, J., Wolff, C., and Tochtermann, W. *Tetrahedron Lett.* **1982**, 2439.

(21) Turro, N. J., Schuster, G., Pouliquen, J. Pettit, R., and Mauldin, C. *J. Am. Chem. Soc.* **1974**, *96*, 6797.

(22) Turro, N. J., Ramamurthy, V., and Katz, T. J. *Nouveau J. Chim.* **1977**, *1*, 363.

(23)(a) Lahmani, F., Ivanoff, N., and Magat, M. *Compt. Rend.* **1966**, *263C*, 1005. (b) Lahmani, F. and Ivanoff, N. *Tetrahedron Lett.* **1967**, 3913.

(24) For a review of photoisomerizations of heterocyclic compounds, see Zanker, V. In *Methoden der organischen Chemie/(Houben-Weyl).* G. Thieme: Stuttgart, 1975, Vol. 4/5a: Photochemie, p. 546. (b) For a review of photoisomerizations of five-membered heterocyclic compounds, see Padwa, A. In *Rearrangements in Ground and Excited States.* De Mayo, P., ed. Academic Press: New York, 1980, Vol. 3, p. 501.

(25) Wilzbach, K. E. and Rausch, D. J. *J. Am. Chem. Soc.* **1970**, *92*, 2178.

(26) Barlow, M. G., Haszeldine, R. N., and Dingwall, J. G. *J. Chem. Soc., Perkin Trans. I.* **1973**, 1542.

(27) van Tamelen, E. E. and Whitesides, T. H. *J. Am. Chem. Soc.* **1971**, *93*, 6129, and references therein.

(28)(a) Wynberg, H. and van Driel, H. *J. Am. Chem. Soc.* **1965**, *87*, 3998. (b) Wynberg, H., Kellogg, R. M., van Driel, H., and Beekhuis, G. E. *J. Am. Chem. Soc.* **1966**, *88*, 5047. (c) For the SINDO1 study of the photoisomerization mechanism of thiophenes, see Jug, K. and Schluff, H. P. *J. Org. Chem.* **1991**, *56*, 129.

(29)(a) Hiralka, H. *J. Chem. Soc., Chem. Commun.* **1971,** 1610. (b) Barltrop, J., Day, A. C., Moxon, P. D., and Ward, R. W. *J. Chem. Soc., Chem. Commun.* **1975,** 786.

(30)(a) Ullman, E. F. and Singh, B. *J. Am. Chem. Soc.* **1967,** *89,* 6911. (b) Albanesi, S., Marchesini, A., and Gioia, B. *Tetrahedron Lett.* **1979,** 1875.

(31) LaBlanche-Combier, A. and Pollet, A. *Tetrahedron.* **1972,** *28,* 3141.

(32) Beak, P. and Messer, W. *Tetrahedron.* **1969,** *25,* 3287, and references therein.

(33)(a) Tiefenthaler, H., Dorschelen, W., Goth, H., and Schmid, H. *Helv. Chim. Acta.* **1967,** *50,* 2244. (b) Goth, H. and Schmid, H. *Chimia.* **1966,** *20,* 148.

(34)(a) Oth, J. F. M., Rottele, H., and Schroder, G. *Tetrahedron Lett.* **1970,** 61. (b) Berson, J. A. and Davis, R. F. *J. Am. Chem. Soc.* **1972,** *94,* 3658.

(35)(a) Fessner, W.-D., Sedelmeier, G., Spurr, P. R., Rihs, G., and Prinzbach, H. *J. Am. Chem. Soc.* **1987,** *109,* 4627. (b) For a theoretical study of the [2+2] photodimerization of benzene, see Engelke, R., Hay, P. J., Kleier, D. A., and Wadt, W. R. *J. Am. Chem. Soc.* **1984,** *106,* 5439.

(36) Higuchi, H., Kobayashi, E., Sakata, Y., and Misumi, S. *Tetrahedron.* **1986,** *42,* 1731.

(37) For the photodimerization of　(a) 2-cyano-naphthalene, see Mattingly, T. W., Jr., Lancaster, J. E., and Zweig, A. *Chem. Commun.* **1971,** 595. (b) Methyl 2-naphthoate and dimethyl 1,8-naphthalenedicarboxylate: Collin, P. J., Roberts, D. B., Sugowdz, G., Wells, D., and Sasse, W. H. F. *Tetrahedron Lett.* **1972,** 321. (c) 2-methoxynaphthalene: Bradshaw, J. S. and Hammond, G. S. *J. Am. Chem. Soc.* **1963,** *89,* 3953; Teitei, T., Wells, D., and Sasse, W. H. F. *Austral. J. Chem.* **1976,** *29,* 1783. (d) Naphthalene end-labeled poly(ethylene glycol) oligomers: Tung, C.-H. and Wang, Y.-M. *J. Am. Chem. Soc.* **1990,** *112,* 6322.

(38)(a) Bradshaw, J. S., Nielson, N. B., and Rees, D. P. *J. Org. Chem.* **1968,** *33,* 259. (b) Collin, P. J. *Austral. J. Chem.* **1974,** *27,* 227.

(39)(a) Chandross, E. A. and Dempster, C. J. *J. Am. Chem. Soc.* **1970,** *92,* 704. (b) Castellan, A., Desvergne, J.-P., and Bouas-Laurent, H. *Chem. Phys. Lett.* **1980,** *76,* 390.

(40) Todesco, R., Gelan, J., Martens, H., and Put, J. *Tetrahedron.* **1983,** *39,* 1407.

(41)(a) For intramolecular cycloaddition, see Desvergne, J. P., Bitit, N., Castellan, A., and Bouas-Laurent, H. *J. Chem. Soc., Perkin Trans. 2.* **1983,** *2,* 109.

(b) For intermolecular cycloaddition, see Albini, A. and Fasani, E. *J. Am. Chem. Soc.* **1988,** *110,* 7760.

(42)(a) Kaupp, G. and Zimmerman, I. *Angew. Chem., Int. Ed. Engl.* **1976,** *15,* 441. (b) Wasserman, H. H. and Keehn, P. M. *J. Am. Chem. Soc.* **1969,** *91,* 2374.

(43) For a review of photodimerization of anthracene and related compounds, see　(a) Stevens, B. *Adv. Photochem.* **1971,** *8,* 161. (b) Bouas-Laurent, H., Castellan, A., and Desvergne, J.-P. *Pure Appl. Chem.* **1980,** *52,* 2633. (c) Cowan, D. O. and Drisko, R. L. In *Elements of Organic Photochemistry.* Plenum: New York, 1976, Ch. 2.

(44) Kaupp, G. and Teuffel, E. *Chem. Ber.* **1980,** *113,* 3669, and references therein.

(45)(a) Wolff, T., Muller, N., and von Bunau, G. *J. Photochem.* **1983,** *22,* 61. (b) For a discussion of the role of a pericyclic intermediate in HT-dimer formation and the excimer intermediate in the HH-dimer formation, see Bonačić-Koutecký, V., Koutecký, J., and Michl, J. *Angew. Chem., Int. Ed. Engl.* **1987,** *26,* 170. (c) For the role of triplet-triplet annihilation in the dimerization of anthracene, see Nickel, B. and Roden, G. *Chem. Phys.* **1982,** *66,* 365.

(46)(a) Becker, H.-D. *Pure Appl. Chem.* **1982,** *54,* 1589. (b) Becker, H.-D and Amin, K. A. *J. Org. Chem.* **1989,** *54,* 3182, and references therein. (c) For the involvement of an ion-pair character intermediate in the photodimerization of linked anthracenes, see Manring, L. E., Peters, K. S., Jones, G., II and Bergmark, W. R. *J. Am. Chem. Soc.* **1985,** *107,* 1485.

(47) Castellan, A., Lapouyade, R., Bouas-Laurent, H., and Lallemand, J. Y. *Tetrahedron Lett.* **1975,** 2467.

(48) Bouas-Laurent, H. and Lapouyade, R. *J. Chem. Soc., Chem. Commun.* **1969,** 817.

(49)(a) Sargent, M. V. and Timmons, C. J. *J. Chem. Soc.* **1964,** 5544. (b) Caldwell, R. A., Ghali, N. I., Chien, C. K., DeMarco, D., and Smith, L. *J. Am. Chem. Soc.* **1978,** 2857.

(50) Sugowdz, G., Collin, P. J., and Sasse, W. H. F. *Tetrahedron Lett.* **1969,** 3843.

(52) For one of the rare cases where *para*-cycloaddition occurs, see Hamrock, S. J. and Sheridan, R. S. *J. Am. Chem. Soc.* **1989,** *111,* 9247.

(53)(a) Koch, H., Runsink, J. and Scharf, H.-D. *Tetrahedron Lett.* **1983,** 3217. (b) For an authoritative review of benzene photocycloaddition reactions, see Gilbert, A. *Tetrahedron.* **1976,** *32,* 1309; **1977,** *33,* 2459. (c) For a review of thermal and photochemical additions of dienophiles to (hetero)arenes, see Wagner-Jauregg, T. *Synthesis.* **1980,** 165. (d) For the substituent effects on the photocycloaddition of phenol to benzonitrile with the formation of substituted 1,2-dihydroazocin-2-ones, see Al-Jalal, N. A. *J. Heterocycl. Chem.* **1990,** *27,* 1323.

(54) For recent reviews, see (a) Wender, P. A., Siggel, L., and Nuss, J. M. *Org. Photochem.* **1989,** *10,* 357. (b) Wender, P. A. and von Geldern, T. W. In *Photochemistry in Organic Synthesis.* Coyle, J. D., ed. The Royal Society of Chemistry: Burlington House, London, 1986, p. 226. (c) Wender, P. A. In *Selectivity, A Goal for Organic Chemistry.* Trost, B. M., ed. Verlag Chemie: Weinheim, 1984. (d) McCullough, J. J. *Chem. Rev.* **1987,** *87,* 811.

(55) Mattay, J. *Tetrahedron.* **1985,** *41,* 2392, 2405.

(56)(a) Houk, K. N. *Pure Appl. Chem.* **1982,** *54,* 1633. (b) Gilbert, A. and Yianni, P. *Tetrahedron.* **1981,** *37,* 3275. (c) Gilbert, A. *Pure Appl. Chem.* **1980,** *52,* 2669. (d) Bryce-Smith, D., Foulger, B., Forrester, J., Gilbert, A., Orger, B. H., and Tyrrell, H. M. *J. Chem. Soc., Perkin Trans. 1.* **1980,** 55, and references therein.

(57) Photocycloaddition of acrylonitrile (AN) to: (a) Benzene; Job, B. E. and Littlehailes, J. D. *J. Chem. Soc. [C].* **1968,** 886. For increase in yield using ZnCl₂-AN complexes, see Ohashi, M., Yoshino, A., Yamasaki, K., and Yonezawa, T. *Tetrahedron Lett.* **1973,** 3395. (b) Toluene: see Ref. 56(b), (c). (c) Anisole: Ohashi, M., Tanaka, Y., and Yamada, S. *Tetrahedron Lett.* **1977,** 3629.

(58)(a) Bryce-Smith, D. *Pure App. Chem.* **1968,** *16,* 47. (b) Angus, H. J. F. and Bryce-Smith, D. *J. Chem. Soc.* **1960,** 4791. (c) Bryce-Smith, D. and Hems, M. A. *Tetrahedron Lett.* **1966,** 1895.

(59)(a) For the preparative procedure, see Grovenstein, E., Jr., Rao, D. V., and Taylor, J. W. *Org. Photochem. Synth.* **1976,** *2,* 97. For the photocycloaddition of maleic anhydride to (b) phenanthrene: Bryce-Smith, D. and Vickery, B. *Chem. Ind.* **1961,** 429. (c) Acenaphthylene: see 15.15.

(60) Schenck, G. O. and Steinmetz, R. *Tetrahedron Lett.* **1960,** 1.

(61) Bryce-Smith, D. and Gilbert, A. *J. Chem. Soc.* **1965,** 918.

(62) Heine, H.-G and Hartmann, W. *Angew. Chem., Int. Ed. Engl.* **1975,** *14,* 698.

(63)(a) Scharf, H.-D. and Klar, R. *Chem. Ber.* **1972,** *105,* 575. (b) Lechtken, P. and Hesse, G. *Liebigs Ann. Chem.* **1971,** *754,* 1. (c) Scharf, H.-D., Leisman, H., Erb, W., Gaidetzka, H. W., and Aretz, J.

Pure Appl. Chem. **1975,** *41,* 581, and references therein.

(64) Miyamoto, T., Mori, T., and Odaira, Y. *J. Chem. Soc., Chem. Commun.* **1970,** 1598.

(65) Williams, J. L. R., Farid, S. Y., Doty, J. C., Specht, D. P., Searle, R., Borden, D. G., and Chang, H. J. *Pure Appl. Chem.* **1977,** *49,* 523.

(66)(a) Farid, S., Hartman, S. E., Doty, J. C., and Williams, J. L. R. *J. Am. Chem. Soc.* **1975,** *97,* 3697. (b) Farid, S., Dory, J. C., and Williams, J. L. R. *J. Chem. Soc., Chem. Commun.* **1972,** 711.

(67)(a) Creed, D. and Caldwell, R. A. *J. Am. Chem. Soc.* **1974,** *96,* 7369. (b) Creed, D., Caldwell, R. A., and Ulrich, M. M. *J. Am. Chem. Soc.* **1978,** *100,* 5831.

(68)(a) Bryce-Smith, D. and Lodge, J. E. *J. Chem. Soc.* **1963,** 695. (b) Grovenstein, E., Jr. and Rao, D. V. *Tetrahedron Lett.* **1961,** 148. (c) Grovenstein, E., Jr., Campbell, T. C., and Shibata, T. *J. Org. Chem.* **1969,** *34,* 2418. (d) Bryce-Smith, D., Gilbert, A., and Grzonka, J. *J. Chem. Soc., Chem. Commun.* **1970,** 498. (e) Paquette, L. A., Oku, M., Heyo, W. E., and Meisinger, R. H. *J. Am. Chem. Soc.* **1974,** *96,* 5818. (f) For the isolation of the cycloadduct of hexafluorobenzene, see Sket, B. and Zupan, M. *J. Am. Chem. Soc.* **1977,** *99,* 3504.

(69) For preparative procedures: (a) R₁=H, R₂=COOMe: Grunewald, G. L. and Grindel, J. M. *Org. Photochem. Synth.* **1976,** *2,* 20. (b) R₁=R₂=COOMe: Paquette, L. A. and Beckley, R. S. *Org. Photochem. Synth.* **1976,** *2,* 45.

(70)(a) Bryce-Smith, D. and Lodge, J. E. *Proc. Chem. Soc.* **1963,** 695. (b) Hanzawa, Y. and Paquette, L. A. *Synthesis.* **1982,** 661. For other acetylenes, see (c) Bryce-Smith, D., Gilbert, A., and Grzonka, J. *Chem. Commun.* **1970,** 498. (d) Paquette, L. A. *Tetrahedron.* **1975,** *31,* 2855 (review).

(71) Morrison, H., Nylund, T., and Palensky, F. *J. Chem. Soc., Chem. Commun.* **1976,** 4.

(72) Miller, R. D. and Abraityl, V. Y. *Tetrahedron Lett.* **1971,** 891.

(73) Lippke, W., Ferree, W. I., Jr., and Morrison, H. *J. Am. Chem. Soc.* **1974,** *96,* 2134.

(74)(a) Sasse, W. H. F., Collins, P. J., Roberts, D. B., and Sugowdz, G. *Aust. J. Chem.* **1971,** *24,* 2339. (b) Collins, P. J. and Sasse, W. H. F. *Aust. J. Chem.* **1971,** *24,* 2325. (c) For a preparative procedure, see Sasse, W. H. F. *Org. Photochem. Synth.* **1976,** *2,* 74.

(75) Sasse, W. H. F., Collins, P. J., Roberts, D. B., and Sugowdz, G. *Aust. J. Chem.* **1971,** *24,* 2151.

(76)(a) Teitei, T., Collins, P. J., and Sasse, W. H. F. *Aust. J. Chem.* **1972**, *25*, 171. (b) Teitei, T. and Sasse, W. H. F. *Aust. J. Chem.* **1973**, *26*, 2129.

(77) Gandhi, R. P. and Chadha, V. K. *Ind. J. Chem.* **1971**, *4*, 305.

(78)(a) Davis, P. D. and Neckers, D. C. *J. Org. Chem.* **1980**, *45*, 456. (b) Davis, P. D., Neckers, D. C., and Blount, J. R. *J. Org. Chem.* **1980**, *45*, 462.

(79)(a) For benzofuran, see Tinnemans, A. H. A. and Neckers, D. C. *J. Org. Chem.* **1978**, *43*, 2493. (b) Sindler-Kylyk, M., Neckers, D. C., and Blount, J. R. *Tetrahedron.* **1981**, *37*, 3377.

(80) Berridge, J. C., Bryce-Smith, D., Gilbert, A., and Cantrell, T. S. *J. Chem. Soc., Chem. Commun.* **1975**, 611.

(81) Bellas, M., Bryce-Smith, D., and Gilbert, A. *Chem. Commun.* **1967**, 263.

(82) McCullough, J. J., Wu, W. S., and Huang, C. W. *J. Chem. Soc., Perkin Trans. 2* **1972**, 370.

(83) Meng, J.-B., Ito, Y., and Matsuura, T. *Tetrahedron Lett.* **1987**, *28*, 6665.

(84)(a) Mattay, J., Rumbach, T., and Runsink, J. *J. Org. Chem.* **1990**, *55*, 5691, and references therein. (b) Bryce-Smith, D., Gilbert, A., Orger, B. H., and Tyrell, H. M. *J. Chem. Soc., Chem. Commun.* **1974**, 334.

(85) Wilzbach, K. E. and Kaplan, L. *J. Am. Chem. Soc.* **1971**, *93*, 2073, and references therein.

(86)(a) Wilzbach, K. E. and Kaplan, L. *J. Am. Chem. Soc.* **1966**, *88*, 2066. (b) Cornelisse, J. and Srinivasan, R. *Chem. Phys. Letters.* **1973**, *20*, 278. Merritt, V. Y., Cornelisse, J., and Srinivasan, R. *J. Am. Chem. Soc.* **1973**, *95*, 8250. (c) For the *meta*-cycloaddition of cyclopentene to anisole, see Srinivasan, R., Merritt, V. Y., and Subrahmanyam, G. *Tetrahedron Lett.* **1974**, 2715. (d) For the preparative procedure of (c), see Subrahmanyam, G. *Org. Photochem. Synth.* **1976**, *2*, 99.

(87) Ferree, W. I., Jr., Grutzner, J. B., and Morrison, H. *J. Am. Chem. Soc.* **1971**, *93*, 5502, and references therein.

(88) Wender, P. A. and Howbert, J. J. *J. Am. Chem. Soc.* **1981**, *103*, 688.

(89) Wender, P. A. and Dreyer, G. B. *Tetrahedron.* **1981**, *37*, 4445.

(90) For the utilization of this reaction in natural product syntheses, see Wender, P. A. and von Geldern, T. W. In *Photochemistry in Organic Synthesis.* Coyle, J.

D., ed. The Royal Society of Chemistry: Burlington House, London, 1986, p. 243. (b) For the use of this reaction in the synthesis of (±)-laurenene, see Wender, P. A., von Geldern, T. W., and Levine, B. H. *J. Am. Chem. Soc.* **1988**, *110*, 4858.

(91) Yoshihisa, I., Nishida, K., Ishibe, K., Hakushi, T., and Turro, N. J. *Chem. Lett.* **1982**, 471.

(92) Kraft, K. and Koltzenburg, G. *Tetrahedron Lett.* **1967**, 4357.

(93) Kimura, M., Sagara, S., and Morosawa, S. *J. Org. Chem.* **1982**, *47*, 4344, and references therein.

(94) Smothers, W. K., Meyer, M. C., and Saltiel, J. *J. Am. Chem. Soc.* **1983**, *105*, 545 and references therein.

(95) Mak, K. T., Srinivasachar, K., and Yang, N. C. *J. Chem. Soc., Chem. Commun.* **1979**, 1038.

(96)(a) Yang, N. C., Chem, M.-J., Chen, P., and Mak, K. T. *J. Am. Chem. Soc.* **1982**, *104*, 853. (b) For a recent paper in this field, see Kimura, M., Okamoto, H., Kura, H., Okazaki, A., Nagayasu, E., Fukazawa, M., Abdel-Halim, H., and Cowan, D. O. *J. Org. Chem.* **1988**, *53*, 3908, and references therein. (c) For the pericyclic chemiluminescence of the [4+4] heterodimers of benzene and substituted anthracenes, see Yang, N. C. and Yang, X. *J. Am. Chem. Soc.* **1987**, *109*, 3804. (d) For the cycloaddition of alkenes to phenanthrene, see Caldwell, R. A., Mizuno, K., Hansen, P. E., Vo, L. P., Frentrup, M., and Ho, C. D. *J. Am. Chem. Soc.* **1981**, *103*, 7263, and references therein.

(97)(a) Lodder, G. and Havinga, E. *Tetrahedron.* **1972**, *28*, 5583. (b) Spillane, W. J. *Tetrahedron.* **1975**, *31*, 495. (c) Fischer, F. C. and Havinga, E. *Rec. Trav. Chim.* **1974**, *93*, 21. (d) Shizuka, H. and Tobita, S. *J. Am. Chem. Soc.* **1982**, *104*, 6919.

(98) For reviews, see (a) Havinga, E. and Cornelisse, J. *Pure Appl. Chem.* **1976**, *47*, 1. (b) Cornelisse, J. and Havinga, E. *Chem. Rev.* **1975**, *75*, 353. (c) Döpp, D. O. *Topics Curr. Chem.* **1975**, *55*, 49. (d) Cornelisse, J., de Grinst, G. P., and Havinga, E. *Adv. Phys. Org.* **1976**, *11*, 225. (e) For the dynamic behavior of photoexcited solutions of 4-nitroveratrole containing OH⁻ or amines, see Van Eijk, A. M. J., Huizer, A. H., Varma, C. A. G. O., and Marquet, J. *J. Am. Chem. Soc.* **1989**, *111*, 88, and references therein.

(99) Shizuka, H. and Tobita, S. *J. Am. Chem. Soc.* **1982**, *104*, 6919, and refrences therein.

(100) Havinga, E., deJongh, R. O., and Dorst, W. *Rec. Trav. Chim.* **1956**, *75*, 378.

(102) For recent reports on this subject, see (a) van Riel, H. C. H., Lodder, G., and Havinga, E. *J. Am. Chem. Soc.* **1981,** *103*, 7257. (b) Bunce, N. J., Cater, S. R., Scaiano, J. C., and Johnston, L. J. *J. Org. Chem.* **1987,** *52*, 4214. (c) Wubbels, G. G., Susens, D. P., and Coughlin, E. B. *J. Am. Chem. Soc.* **1988,** *110*, 2538, and references threin.

(103)(a) Cervello, J., Figueredo, M., Marquet, J., Moreno-Manas, M., Bertran, J., and Lluch, J. M. *Tetrahedron Lett.* **1984,** *25*, 4147. (b) Cantos, A., Marquet, J., and Moreno-Manas, M. *Tetrahedron Lett.* **1987,** *28*, 4191.

(104) See Ref. 48(a), p. 8.

(105) Bunnett, J. F. *Acc. Chem. Res.* **1978,** *11*, 413.

(106) Rossi, R. A. and Bunnett, J. F. *J. Org. Chem.* **1973,** *38*, 1407.

(107)(a) Semmelhack, M. F. and Bargar, T. M. *J. Org. Chem.* **1977,** *42*, 1481. (b) Semmelhack, M. F. and Bargar, T. M. *J. Am. Chem. Soc.* **1980,** *102*, 7762.

(108)(a) Wolfe, J. F., Sleevi, M. C., and Goehring, R. R. *J. Am. Chem. Soc.* **1980,** *102*, 3646, (b) Beugelmans, R., Boudet, B., and Quintero, L. *Tetrahedron Lett.* **1980,** *21*, 1943.

(109) Beugelmans, R. and Ginsburg, H. *J. Chem. Soc., Chem. Commun.* **1980,** 508.

(110) Usui, S. and Fukazawa, Y. *Tetrahedron Lett.* **1987,** *28*, 91.

Chapter 8
PHOTOCHEMISTRY OF CARBONYL COMPOUNDS

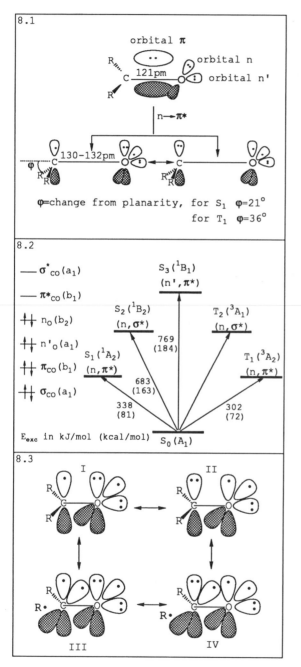

8.1 orbital π, orbital n, orbital n'

$n \rightarrow \pi^*$

121pm

130-132pm

φ=change from planarity, for S_1 φ=21°
for T_1 φ=36°

8.2

$\sigma^*_{CO}(a_1)$

$\pi^*_{CO}(b_1)$

$n_O(b_2)$

$n'_O(a_1)$

$\pi_{CO}(b_1)$

$\sigma_{CO}(a_1)$

$S_3(^1B_1)$
(n',π^*)

$S_2(^1B_2)$
(n,σ^*)

$T_2(^3A_1)$
(n,σ^*)

769
(184)

$S_1(^1A_2)$
(n,π^*)

683
(163)

$T_1(^3A_2)$
(n,π^*)

338
(81)

302
(72)

E_{exc} in kJ/mol (kcal/mol) $S_0(A_1)$

8.3

I   II

III   IV

8.1. Carbonyl compounds have two nonbonding orbitals located essentially on the O atom and one π orbital extending over the C and O atoms. A nonbonding p orbital (n) is higher in energy than an sp orbital (n') since the latter has a large s character contribution. The lowest S → S and S → T transitions are $n \rightarrow \pi^*$, which form either the $^1(n,\pi^*)$ or $^3(n,\pi^*)$ state. The main features of the n,π^* states are that the carbonyl group is not as strongly polarized as in the ground state and that the O atom, radical like in character, bears an odd electron.

8.2. Carbonyl compounds are characterized by low-lying S_1 and T_1 states whose population from the ground state ($n \rightarrow \pi^*$ transition, see 2.46) is forbidden by local symmetry (low-absorption intensity, $\epsilon \sim 10$–30 in the 280–320 nm region).[1] These compounds have a small singlet-triplet splitting, and a high probability of intersystem crossing between excited singlet and triplet states exists. As a result, carbonyl compounds are good triplet sensitizers.[2] The major photoreactions of carbonyl compounds involve radical species rather than the electron-paired charged species associated with thermal reactions.

8.3. The S_1 or T_1 state can be expressed by four valence structures. Therefore, the most important reactions are those in which the excited carbonyl compound completes the nonbonding electron pair on O, either by path 1—the addition of a suitable radical—or path 2—transfer of an electron from the C—R σ bond, which is hyperconjugated with the n orbital. Path 1 leads to various intra- or intermolecular *hydrogen abstractions* or to 1,4-biradical intermediates in *oxetane* formation. Path 2 involves expulsion of radical R·, which undergoes further reactions, such as *α-cleavage*.

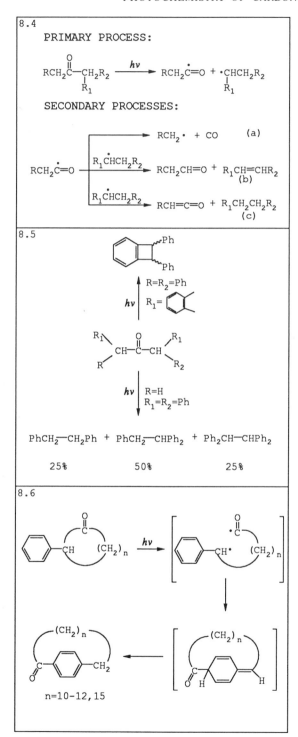

8.4. The primary photoreaction of a saturated carbonyl compound, particularly in the gas phase, is its fragmentation into an acyl and an alkyl radical by cleavage of the α bond (α-*cleavage* or *Norrish Type I reaction*).[3,4] The α bond that produces the more stable radical cleaves preferentially in unsymmetrically substituted ketones.[5] The initially formed acyl radical is stabilized by one of the secondary processes shown [(a)–(c)], whereas the alkyl radical can be stabilized by recombination or disproportionation. α-cleavage occurs from both $^1(n,\pi^*)$ and $^3(n,\pi^*)$ states. The latter is more efficient because of its lower activation barrier and endothermicity. However, α-cleavage in aryl alkyl ketones is less efficient in compounds whose (n,π^*) state is not the lowest excited state. In this case, there is a larger barrier on the reaction coordinate.

8.5. α-cleavage is efficient in the liquid phase only if a stable radical is formed (allylic, benzylic, or acyl radical by resonance stabilization, or *t*-butyl by steric stabilization).[6] Confirmation of the radical nature of the cleavage is derived from an analysis of the proportions of the products from unsymmetrical ketones. These products are the result of a statistical combination of radicals.[7–9] However, when the photolysis is performed in a micellar solution,[9] essentially a single product is formed. Molecules of the ketone and the photolytically produced radical pairs are enclosed in the micelle cavity, out of which they diffuse at a much lower rate than that of recombination (see also 15.33).

8.6. α-cleavage has been utilized in the synthesis of strained and energy-rich compounds (see 1.11[10(a)]) and natural products. An example is the synthesis of paracyclophanes by irradiation of macrocyclic phenylcycloalkanones (n = 10–12, 15) via an intramolecular acyl-alkyl biradical, stabilized by recombination.[10(b)]

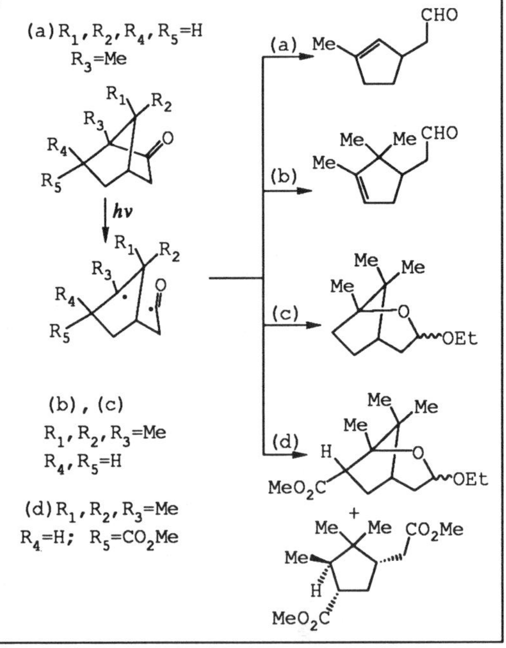

8.7. The efficiency of α-cleavage in cyclic ketones increases with steric strain, e.g., by a factor of 10 for cyclobutanone over cyclopentanone in the $^1(n,\pi^*)$ state.[11] Although the quantum yields of decarbonylation are low in strained systems, such as $(-)$-thujone, the chemical yield is quantitative.[12] Photodecarbonylation provides an efficient route to *anti*-Bredt compounds with large internal strain, e.g., the 9,9'-dehydrodimer of anthracene.[13]

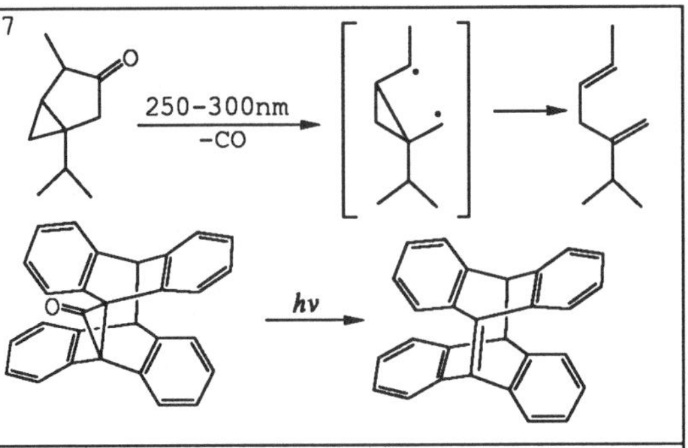

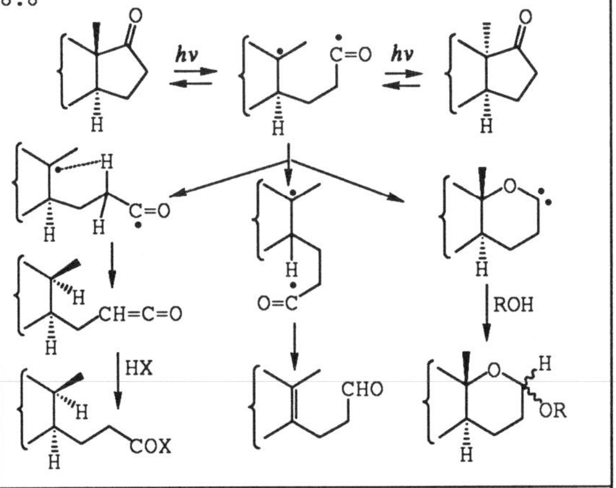

8.8. Although α-cleavage is a primary process in the photolysis of cyclic ketones in the liquid phase, the alkylacyl biradical pair is usually not stabilized by decarbonylation, but rather by an intramolecular H atom transfer. This occurs either by: (1) a shift of an αH atom to the alkyl radical with the formation of ketene, which then yields a carboxylic acid or its derivative in the presence of a protic nucleophile (*photochemical hydrolysis* of cyclic ketones), or (2) a shift of the γH atom to the acyl radical to produce an unsaturated aldehyde.[14] However, when the cyclopentanone ring is part of a strained polycyclic system or where structural features prohibit aldehyde and ketene formation, ring expansion occurs.[15] This ring strain factor is especially pronounced in cyclobutanones (see 8.12).

8.9. A necessary but not sufficient condition for the formation of a ketene is the presence of an αH atom in the cyclic ketone. The photolysis is affected by other factors, such as strain in the six-membered cyclic transition state. A general rule can be formulated: in the photolysis of bicyclo-[2.2.1]heptane-2-ones in solution, a δH atom shift occurs and an unsaturated aldehyde is formed; with bicyclo-[3.2.1]octane-3-ones, an αH atom shift takes place and the product is a ketene. However, as shown in paths (a) and (b), there must be either *syn*-C(7) or *endo*-C(6) H atoms available. Otherwise, path (c) and/or (d) will occur.[16]

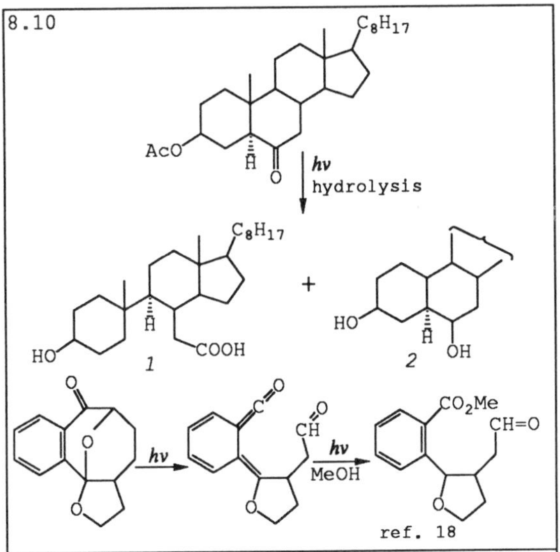

8.10

+

1

2

ref. 18

8.11

slow

$\dot{C}=O$

cis

fast

$\dot{C}=O$

trans

8.12

$S_1(n, \pi^*)$

(1) −CO

(2)

(3)

ROH

8.10. The photohydrolysis of cyclic ketones[17] is a mild degradation method used in steroid and terpene chemistry. The yields vary from 30–70% and are affected by side products from reactions of the carbonyl group. However, this can be suppressed by the use of suitable solvents. The photohydrolysis of 3β-acetoxy-6-oxo-5α-cholestane in dilute acetic acid gives 3β-hydroxy-5,6-secocholestane-6-olic acid (*1*) in 54% yield. In dioxane-water, the yield is reduced to 30% and the reduction product *2* is formed.[18]

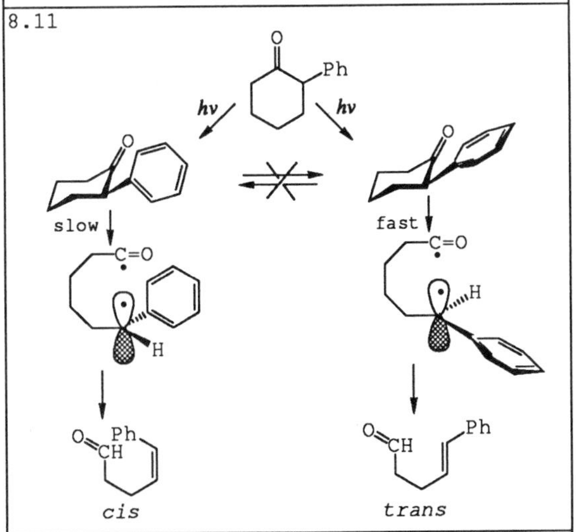

8.11. A study of the photolysis of 2-phenylcyclohexanone indicates that the products, *trans*-6-phenyl-5-hexanal (the major product), and the *cis* isomer are formed from triplet states T_{cis} and T_{trans}, which have different lifetimes. This occurs when the initially formed 1,6-biradicals have short lifetimes and are not in conformational equilibrium, i.e., the rate at which they convert to products is higher than their interconversion.[19]

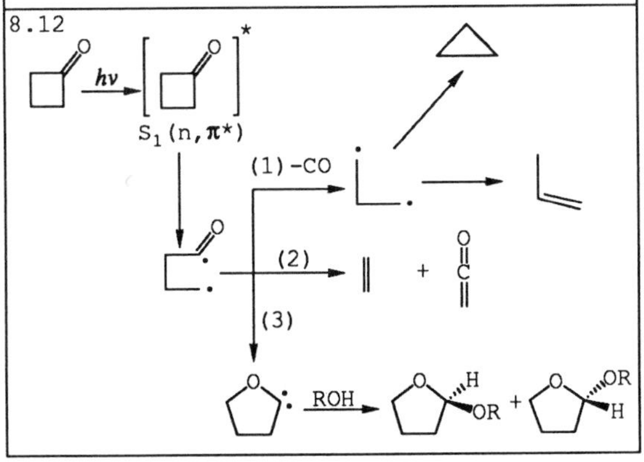

8.12. The photochemistry of cyclobutanones, although it begins with α-cleavage, differs significantly from the photochemistry of other cyclic ketones both in mechanism and in the reaction pathways following α-cleavage.[16,20] Unlike other ketones, cleavage occurs from $S_1(n, \pi^*)$ and leads to the formation of a 1,4-acyl-alkyl biradical. There are three different pathways for stabilization of the biradical: (1) loss of CO and the formation of a 1,3-biradical that undergoes either recombination to cyclopropane or an H shift to form propene; (2) by a subsequent β-cleavage (see 8.19) ethylene and ketene are formed (the latter can be trapped by a protic nucleophile); (3) by rebonding to O, an *oxacarbene* is formed (ring expansion). The oxacarbene can be trapped by an alcohol to give a pair of acetals (2-alkoxy-tetrahydrofurans). Low temperature suppresses the decarbonylation and β-cleavage processes.

8.13

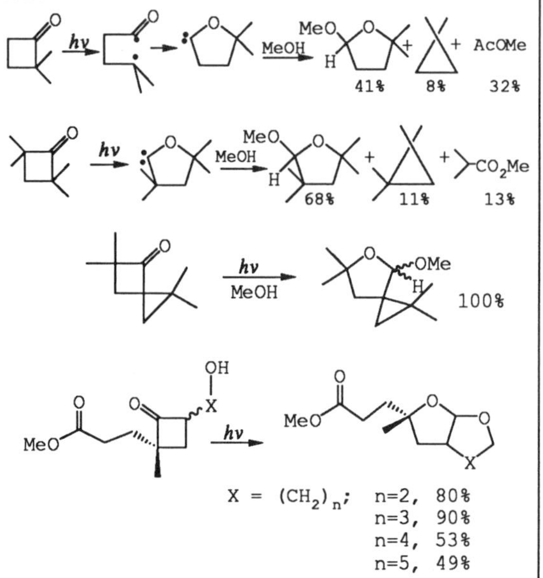

41%   8%   32% + AcOMe

68%   11%   13% + CO$_2$Me

100%

X = (CH$_2$)$_n$;   n=2, 80%
                    n=3, 90%
                    n=4, 53%
                    n=5, 49%

8.14

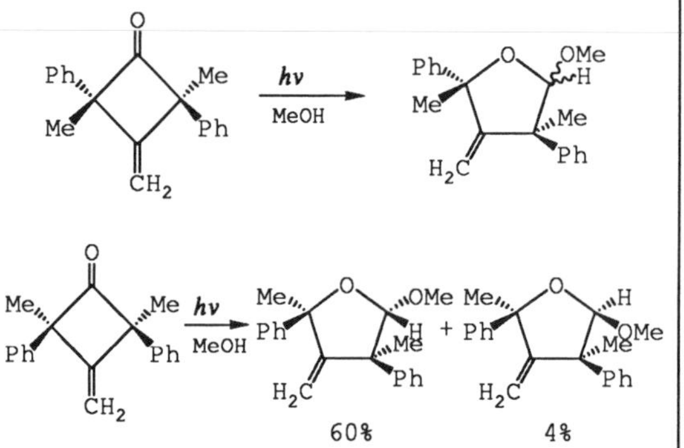

60%   4%

8.15

hv, ROH

1   2   3

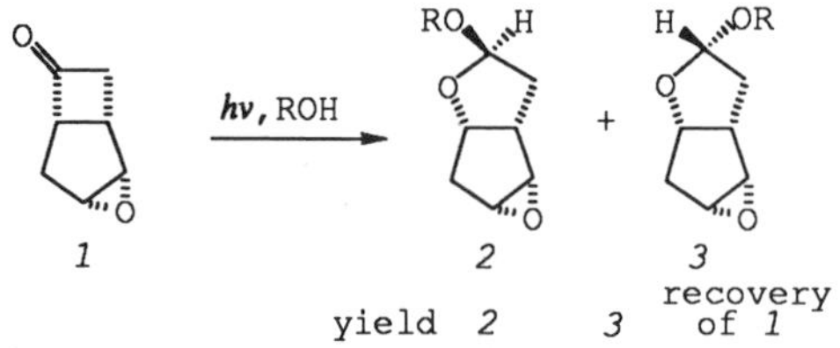

	yield 2	3	recovery of 1
(1) Q, MeOH	20%	3%	10%
(2) Q, MeOH, NaHCO$_3$	33%	10%	15%
(3) Q, MeOH, NaHCO$_3$, DHD	37%	10%	21%
(4) Q, EtOH, NaHCO$_3$, DHD	21%	0	20%
(5) Q, t-BuOH, NaHCO$_3$, DHD	11%	0	22%

8.13. Experimental results support the mechanism shown. Ring expansion of unsymmetrical cyclobutanones is highly regioselective, which indicates the need for a nucleophilic alkyl radical to attack the O atom of the acyl radical. Indeed, alkyl substitution does increase the yield of ring-expanded products.[15,21] The ring contraction product from the irradiation of cyclobutanone accounts for only 8% of the yield. Both 2,2-dimethyl- and 2,2,4,4-tetra-methylcyclobutanone form ring-expansion products predominantly. Bicyclic ring systems with five- to eight-membered rings containing O, N, or S have been prepared by an intramolecular process from heteroalkycyclobutanones.[22]

8.14. The stereospecific course (retention of configuration on the C atom that migrates to the carbonyl O atom) seems to provide evidence against the participation of the 1,4-acyl-alkyl biradical intermediate.[23] However, this apparent conflict was resolved by an experiment in which the biradicals were independently prepared from the corresponding diazene. When the biradicals are formed in MeOH, the resulting tetrahydrofuryl ethers exhibit retention of configuration.[24]

8.15. The cyclobutanone-oxacarbene ring-expansion reaction has been utilized in the synthesis of natural products, e.g., prostaglandins.[25] The best yields are obtained: (1) in a photoreactor with a quartz (Q) immersion well (Example 1) (see 14.29); (2) by adding NaHCO$_3$ to neutralize acidic byproducts (Example 2); (3) in the presence of 2,5-dimethyl-2,4-hexadiene (DHD), a triplet quencher (Example 3); and (4) by using deoxygenated solutions to avoid butyrolactone formation. Acetal yields decrease with increasing nucleophilicity of the alcohol (Examples 3–5). This method is the key step in the total synthesis of muscarines (alkaloids of the red fly agaric mushroom, *Amanita muscaria*).[26]

8.16

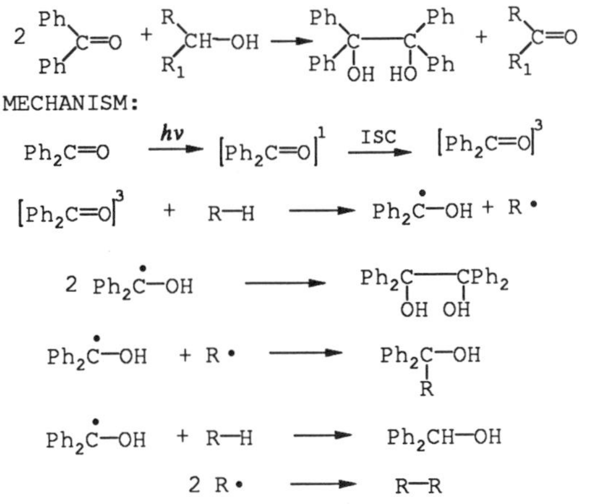

MECHANISM:

8.17

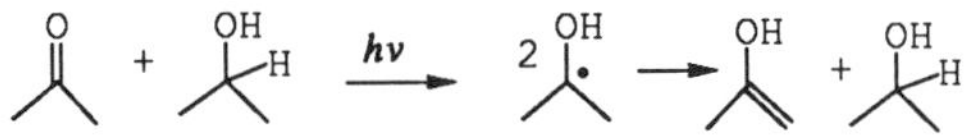

system		pK$_{ENOL}$
$CH_3-CH=O$ ⇌ $CH_2=CH-OH$		6.23
$(CH_3)_2CH-CH=O$ ⇌ $(CH_3)_2C=CH-OH$		3.84
$(CH_3)_2C=O$ ⇌ $CH_3-\underset{OH}{\overset{\vert}{C}}=CH_2$		8.33
$(CH_3)_2\underset{\overset{\vert}{CH_3}}{CH}C=O$ ⇌ $(CH_3)_2C=\underset{OH}{\overset{\vert}{C}}-CH_3$		7.33
		6.38

8.18

100%

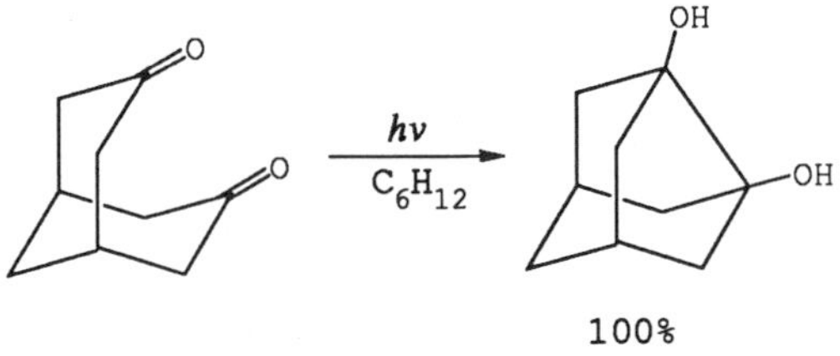

8.16 Abstraction of a hydrogen atom from a suitable donor by the electrophilic n orbital of an n,π^* excited carbonyl group is one of the best-known photoreactions. It proceeds either *inter*molecularly or *intra*molecularly (Norrish Type II reaction, see 8.19)[27] A correlation diagram analysis favors coplanar H abstraction in the $^{1,3}(n,\pi^*)$ states and excludes the involvement of $^{1,3}(\pi,\pi^*)$ states (see 4.20). H abstraction by benzophenone is a typical and well-studied example of this photoreaction. Among the useful H atom donor solvents for photopinacolization reactions are secondary alcohols (especially 2-propanol), alkylbenzenes (toluene, cumene), and tributyl hydride. The initially formed ketyl radicals combine to form a pinacol. Photopinacolization is a general reaction that occurs with all carbonyl compounds whose lowest excited states are n,π^* triplet or singlet.[28-30]

8.17. The hydrogen abstraction reaction can be viewed as either a photoreduction of ketones or a photooxidation of secondary alcohols. Using an appropriate ketone/secondary alcohol pair, identical ketyl radicals can be produced. Chemically induced dynamic polarization (CIDNP), a spectroscopic method for studying radical pairs, indicates that these ketyl radicals can undergo disproportionation to form enols.[31] Thus, the reaction can be used to determine keto-enol equilibrium constants (pK$_{enol}$). The results indicate that alkyl substitution on the carbonyl group decreases the amount of enol (aldehyde vs. ketone) while alkyl substitution on the β-carbon atom enhances it (isobutyraldehyde, cyclic ketones).[32,33]

8.18. Intramolecular photopinacolization reactions occur readily. In bicyclo[3.3.1]nonane-3,7-dione, in which both cyclohexane rings are in the chair conformations, the ketyl radicals are in such close proximity that the pinacol is formed quantitatively.[34]

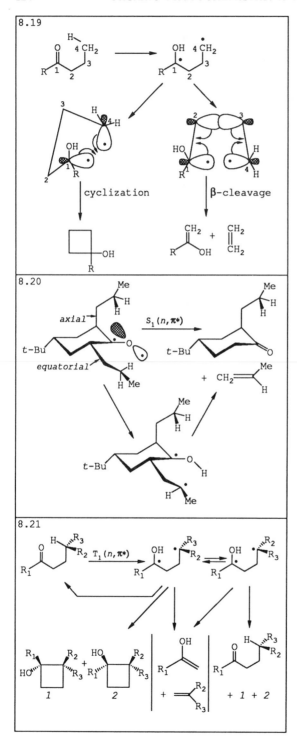

8.19. $^{1,3}(n,\pi^*)$ excited carbonyl compounds having a γ hydrogen atom undergo a characteristic 1,5-hydrogen atom transfer by an intramolecular cyclic process with the formation of a ketyl-like 1,4-biradical. Depending on the conformation of the initially formed 1,4-biradical, two different pathways to stabilization are possible.[35] (1) If only the *sp* orbitals of the radical centers can overlap, a cyclobutanol is the product. (2) If the *sp* orbitals of the radical centers are parallel to the β bond, they participate in the formation of two double bonds (one is the enol double bond), a result of the cleavage of the β bonds (Norrish Type II reaction).[36,37]

8.20. Although the reaction occurs from both the singlet and triplet n,π^* states, the quantum yields from the singlet state are generally lower than from the triplet state.[38] Because aromatic ketones can undergo rapid intersystem crossing, the reaction has not been observed from the singlet state with arylalkyl ketones. The requirement that the *p* orbitals are parallel to the β bond is equivalent to the need for a planar transition state (in agreement with the correlation diagram analysis). In 4-*t*-butyl-2,6-di-*n*-propylcyclohexanone, this requirement is fulfilled only by the equatorial propyl group, which is, in fact, cleaved in the reaction.[39(a)] However, there are many exceptions.[39(b),(c)]

8.21. Participation of a 1,4-biradical intermediate in the Norrish Type II reaction has been confirmed by trapping experiments and spectroscopic techniques.[40] That β-cleavage, as well as the hydrogen atom abstraction-cyclization reaction (often termed the *Type II photocyclization*), proceeds through a biradical intermediate has also been proven chemically. Photoracemization of a ketone with a chiral δ-carbon atom and loss of the chirality in the product was observed.[41]

8.22

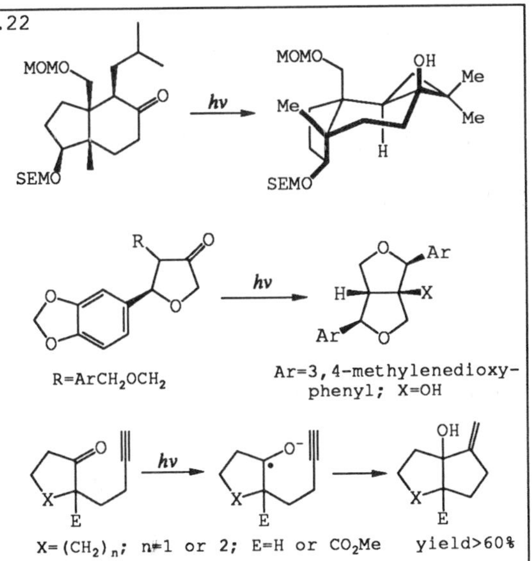
MOMO
SEMO
hν
Me
OH
H
R
O
Ar
X
R=ArCH$_2$OCH$_2$
Ar=3,4-methylenedioxy-phenyl; X=OH
E
X=(CH$_2$)$_n$; n=1 or 2; E=H or CO$_2$Me
yield>60%

8.22. Type II photocyclization has not been widely used in the synthesis of natural products. One exception is its use to create a cyclobutane ring in *punctatin*.[42] This reaction proceeds via a 1,5 hydrogen transfer involving a six-member cyclic transition state. A 1,6 H atom transfer is less probable and occurs only if the 1,5 H transfer pathway is inhibited,[43] e.g., in the key step in the total synthesis of the lignan *paulownin*.[44] Photocyclization to a five-membered ring can also be achieved by the photoreduction of δ,ϵ-unsaturated ketones via a photochemically induced electron transfer from triethylamine to the unsaturated ketone.[45]

8.23

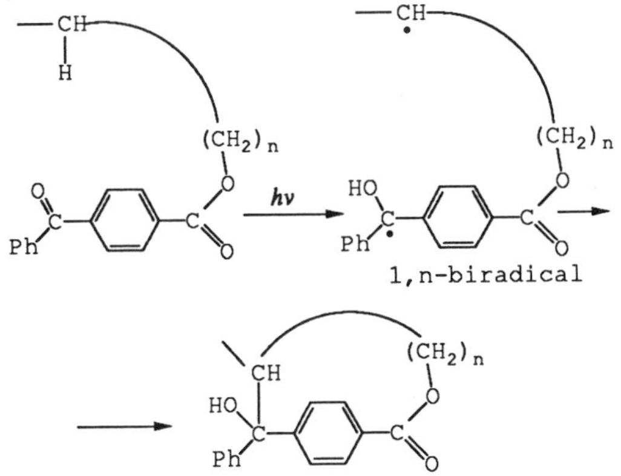
CH
H
(CH$_2$)$_n$
hν
HO
Ph
1,n-biradical

8.23. Intramolecular γ-hydrogen abstraction by an excited carbonyl group is approximately 20 times faster than δ-hydrogen abstraction. Nevertheless, the tendency of the half-occupied n orbital of the carbonyl oxygen to abstract a hydrogen atom from other positions is considerable. A necessary condition is the planarity of the transition state. In the photolysis of alkyl esters of benzophenone-4-carboxylic acid, the planar transition state is possible only for n $\geq$ 10, which leads to the formation of a $\geq$ 1,18-biradical whose recombination produces a paracyclophane.[46] Intramolecular abstraction of distant hydrogen atoms has been used in biomimetic photoreactions (see 15.42).

8.24

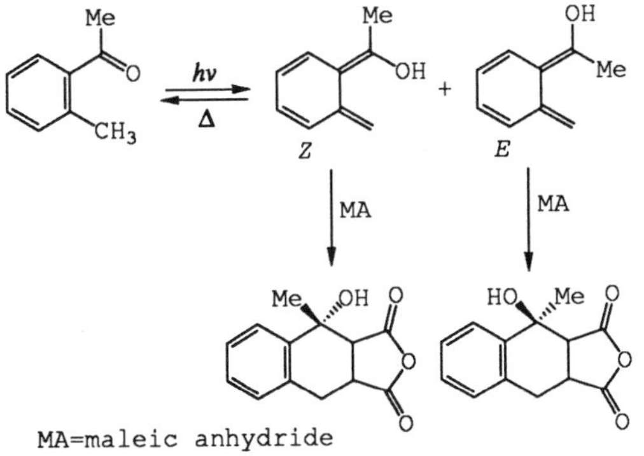
Me
CH$_3$
hν
Δ
OH
Z
E
MA
MA=maleic anhydride

8.24. If the carbonyl group and the carbon atom bearing the γ hydrogen are ortho to each other in the aralkyl ketone, e.g., 2-methylacetophenone, irradiation gives a mixture of Z- and E-photoenols with an *o*-xylylene (*o*-quinonedimethane) structure. The enol form of the starting ketone undergoes thermal reisomerization or, in the presence of a dienophile, a Diels–Alder reaction.[47–49] This photoenolization of a phenone is a key step in the synthesis of estrone (see 15.55).

8.25

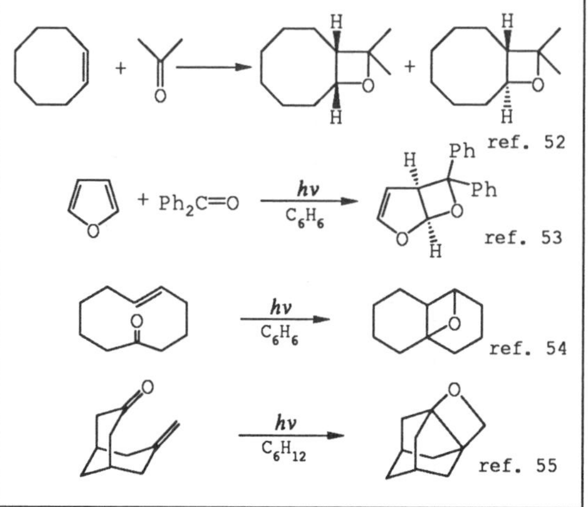

8.26

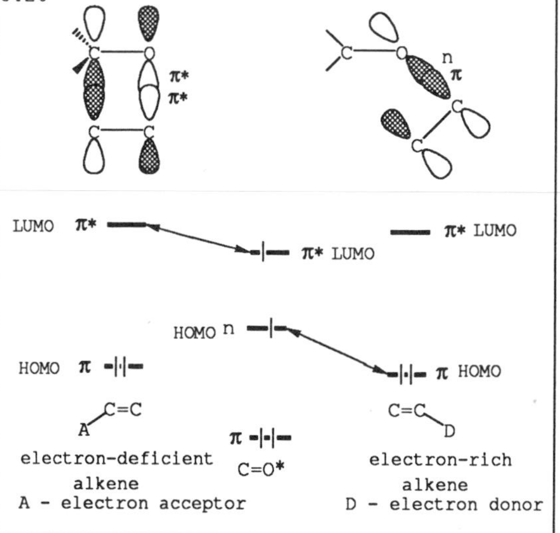

8.27

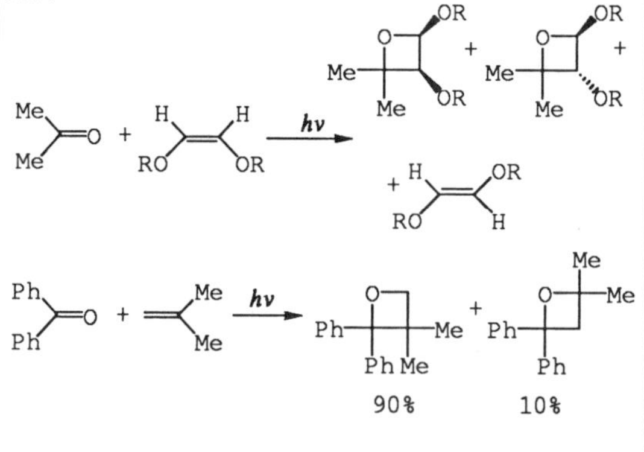

8.25. The *Paterno–Büchi reaction*,[50(a)] i.e., the photocycloaddition of an n,π^* carbonyl compound to an alkene in its ground state, yields an oxetane (oxacyclobutane) from either the S_1 or the T_1 state.[51] However, if the E_T of the alkene is less than that of the carbonyl compound, energy transfer to the alkene will occur (see 5.18). The immediate precursors of oxetanes are biradicals whose existence has been confirmed by picosecond spectroscopy.[50(b)] With unsymmetrical alkenes the more stable biradical intermediates are usually formed. For more details of *intramolecular oxetane formation, see 9.4.

8.26. Two frontier orbital mechanisms explain the formation of oxetanes: (1) LUMO—LUMO interaction, in which the half-occupied π^* carbonyl orbital interacts with the unoccupied π^* MO of an electron-deficient alkene in the plane of the molecule and a C,O-biradical is formed; (2) HOMO—HOMO interaction, in which the half-occupied n orbital (HOMO) of the carbonyl O atom interacts with the π orbital (HOMO) of an electron-rich alkene in a direction perpendicular to the molecular plane and a C,C-biradical is formed. The n,π^* and π,π^* interactions involve an electrophilic or a nucleophilic attack, respectively, on the π or π^* MO of the alkene.[3(a), 56]

8.27. Loss of stereospecificity in the formation of an oxetane arises from biradicals with sufficiently long lifetimes to permit rotation about the C—C bond of the alkene. The singlet biradical has a *tight geometry* (more hindered rotation), whereas the triplet biradical has a *loose geometry*.[57] In a more stereospecific reaction, tight geometry is preferred by electron-rich alkenes; loose geometry and nonstereospecificity is typical of electron-deficient alkenes. The cycloaddition of ketones to enol ethers is not regioselective and, although one of the isomers predominates, the selectivity is less than expected. This is explained by different degrees of participation of the S_1 and T_1 states.[58]

8.28

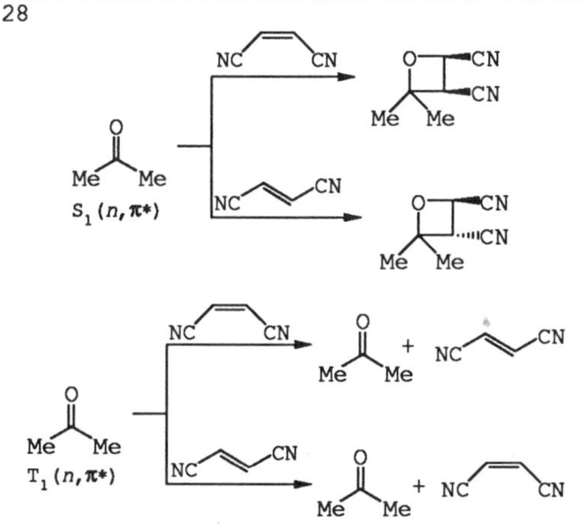

8.29

8.30

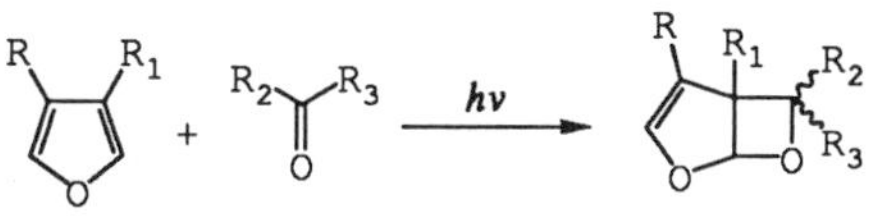

	yield
R=R$_1$=R$_2$=H; R$_3$=Et	80%
R=R$_1$=R$_2$=H; R$_3$=Ph	88%
R=R$_1$=R$_2$=H; R$_3$=C$_8$H$_{17}$	~100%
R=R$_1$=H; R$_2$=R$_3$=Ph	94%
R=R$_1$=Me; R$_2$=H R$_3$=PhCH$_2$OCH$_2$CH$_2$	63%

8.28. The course of the photocycloaddition of electron-deficient alkenes to ketones follows certain rules. While oxetanes are formed only from the S$_1$(n,π*) state, the T$_1$(n,π*) state stereospecifically sensitizes the *cis-trans* isomerization of electron-deficient alkenes and does not lead to oxetanes. This difference in the reactivities of the S$_1$(n,π*) and T$_1$(n,π*) excited states of a ketone toward electron-deficient alkenes, e.g., dicyanoethylenes, can be used as a chemical "titration" of the electronically excited states from the thermal decomposition of 1,2-dioxetanes (see 13.35).[59,60] The photocycloaddition of substituted acrylonitriles to ketones in the S$_1$(n,π*) state leads to oxetane formation, while [2+2] dimerization of the acrylonitrile occurs from the T$_1$(n,π*) state.[61]

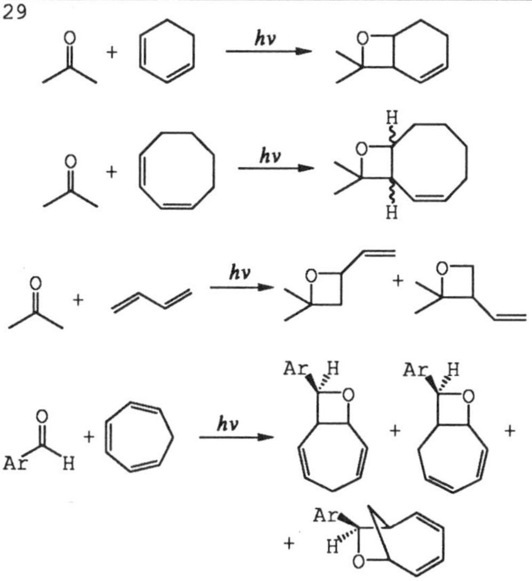

8.29. The E$_T$ of dienes is usually less than that of carbonyl compounds. However, the formation of oxetanes competes successfully with excitation energy transfer because dienes quench the T$_1$(n,π*) state, while the formation of oxetanes occurs from the S$_1$(n,π*) state of the carbonyl compounds.[62] Accordingly, vinyloxetane formation proceeds stereospecifically with acyclic dienes.[63] Conjugated dienes undergo addition more readily than simple alkenes. Although the E$_T$ of trienes is even less than that of dienes, cycloheptatriene forms oxetanes, as well as a [6+2] cycloaddition product.[64]

8.30. The photocycloaddition of aldehydes to furan yields *exo*-substituted photoproducts with a high degree of regio- and stereoselectivity.[65] The reaction involves the attack of the electron-deficient carbonyl O atom on the nucleophilic side (the 2 or 5 position) of the furan ring. The products can be viewed as Z enolates.[66] The photoaddition is synthetically significant, especially in the synthesis of natural products.

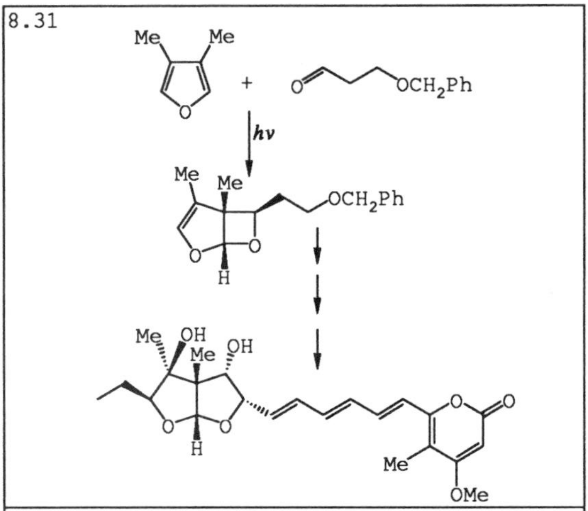

8.31

8.32

8.33

8.31. The significance of oxetanes from the aldehyde-furan photocycloaddition is illustrated by the synthesis of a number of natural products,[51,67] e.g., avenaciolide, crassin acetate, isolobophytolide, etc. The synthesis of the mycotoxin *asteltoxin* shown was achieved by a 16-step reaction sequence starting with the irradiation of 3-benzyloxpropanal in the presence of 3,4-dimethylfuran to form the *exo*-substituted photoadduct.[68]

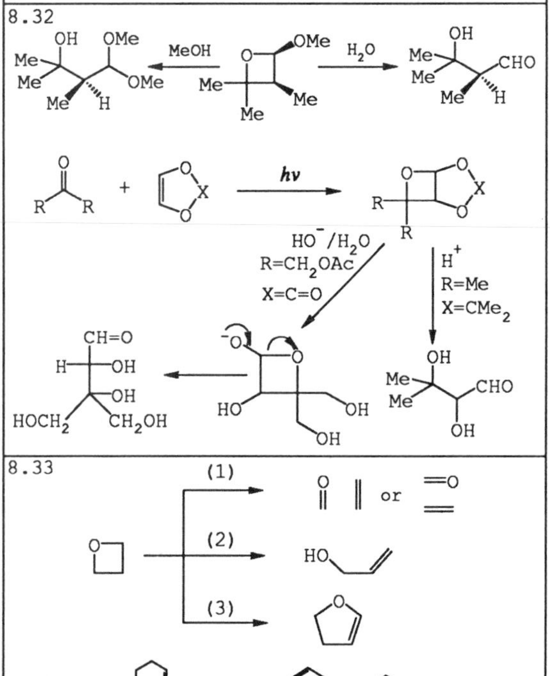

8.32. Alkoxyoxetanes, particularly the 2-isomers, are important because of their reactivity toward nucleophiles to form acetals, 1,3-diol derivatives, and β-hydroxyaldehydes (aldol). The alkoxyoxetane in the formation of a β-hydroxyaldehyde can be viewed as a protected aldol; the overall reaction is a stereoselective photoaldol reaction.[69] Vinylene carbonate (VC) is a useful reagent in photocycloadditions to carbonyl compounds, indicated by the synthesis of the branched-chain sugar, D,L-apiose.[70] When 1,3-diacetoxyacetone is photolyzed in the presence of VC, an oxetane is formed. Treatment of this product with base causes deacetylation and ring opening via a hemiacetal. Similarly, photocycloaddition of acetone to 2,2-dimethyldioxole after treatment with acid gives an overall 65% yield of a glyceraldehyde derivative.[71]

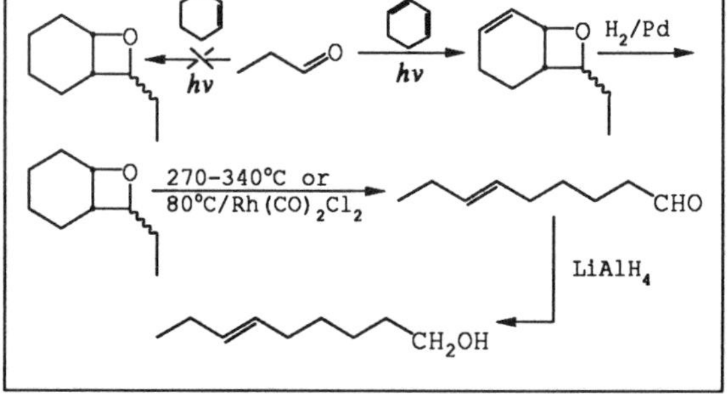

8.33. Oxetanes represent important intermediates in the synthesis of organic compounds. The significant reactions of oxetanes include: oxetane cracking (part of the carbonyl-alkene metathesis reaction), ring-chain isomerization, and ring expansion. These transformations depend on the controlled release of the oxetane ring strain and provide useful functional groups for further modification.[51] The carbonyl-alkene metathesis was employed in the synthesis of the *pheromone* of the Mediterranean fruit fly.[72]

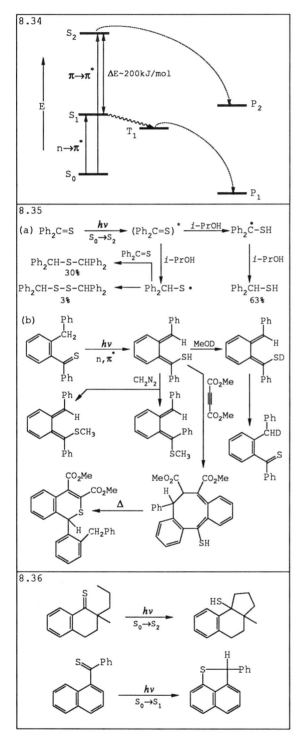

8.34. Thiocarbonyl compounds (thiones) exhibit a long wavelength absorption band, e.g., $Ph_2C = S$, $\lambda_{max} = 609$ nm ($\epsilon = 180$) in cyclohexane, attributed to an $n \rightarrow \pi^*$ transition that is completely separate from the $\pi \rightarrow \pi^*$ transition [$\lambda_{max} = 315$ nm ($\epsilon = 15,500$)]. The large energy gap between the S_1 and S_2 states and small $S_2 - S_1$ vibration overlap permit the selective formation of one of these states by irradiation with light of appropriate energy. With lower energy irradiation, the initially formed S_1 state converts by ISC to the reactive T_1 state; higher energy light produces the S_2 state.[73,74]

8.35. Irradiation of thiones in the presence of suitable H-donors gives, like carbonyl compounds, reduction products either intermolecularly (a) or intramolecularly (b). However, the intermediate thioketyl radical does not dimerize to the corresponding thio-analog of pinacol, but is further reduced to benzhydrylthiol. The simultaneously formed thiyl radical reacts with a second thione molecule to produce a sulfide and, to a lesser extent, dimerizes to the corresponding disulfide.[75] Intramolecular H-abstraction by diaryl thiones with an activated hydrogen atom (benzyl, but not methyl) leads to a photothioenol that can be trapped either by diazomethane or a suitable dienophile.[76] H–D exchange[77] also provides evidence of its formation.

8.36. Intramolecular H-abstraction leading to five-membered rings occurs with alkyl-aryl thiones containing a $\delta-CH$ bond. If the δ position is blocked, e.g., by oxygen, γ or ϵ insertion takes place with the formation of a cyclobutane or cyclohexane, respectively.[78] Another examle is the reductive cyclization of aromatic thiones having a vacant peri-position. Although with most thiones excitation to the S_1 state is followed by ISC to T_1, in this reaction the S_1 state is assumed to be the reactive state.[79] Alternant, but not nonalternant, hydrocarbons undergo reductive cyclizations.

8.37

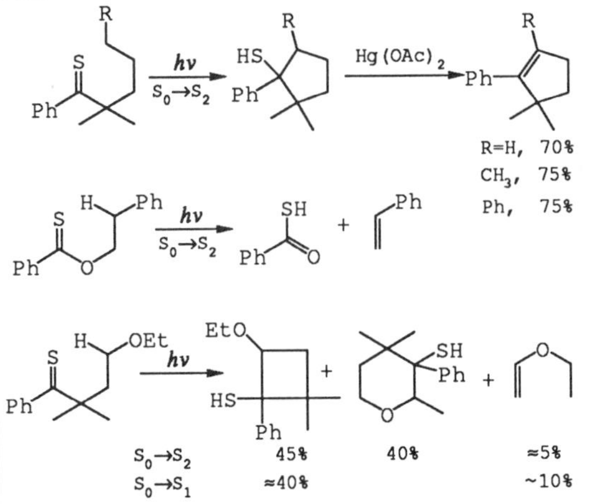

R=H, 70%
CH$_3$, 75%
Ph, 75%

S$_0$→S$_2$ 45%   40%   ≈5%
S$_0$→S$_1$ ≈40%   ~10%

8.38

λ>550nm

λ=366nm

(a) - dilute solution
(b) - conc. solution

8.39

X = Ph; CN

T$_1$(n,π*)

S$_2$(π,π*)

T$_1$(n,π*)

T$_1$(n,π*)

S$_2$(π,π*)

X = OEt; CN

8.37. As shown in 8.36, irradiation of alkyl-aryl thiones with no activated γ—CH bond results in cyclization, preferentially with the formation of cyclopentanethiols. These thiols are readily converted into cyclopentenes.[80] In the presence of an activated γ—CH bond, the Norrish Type II process takes place with the intermediate formation of a 1,4-biradical that subsequently undergoes either cyclization[80] and/ or β-cleavage.[81] In the last example shown, the oxygen atom activates both the γ- and ε—CH bond and a six-membered ring is also formed; this process does not occur if the reactive state is S$_1$.

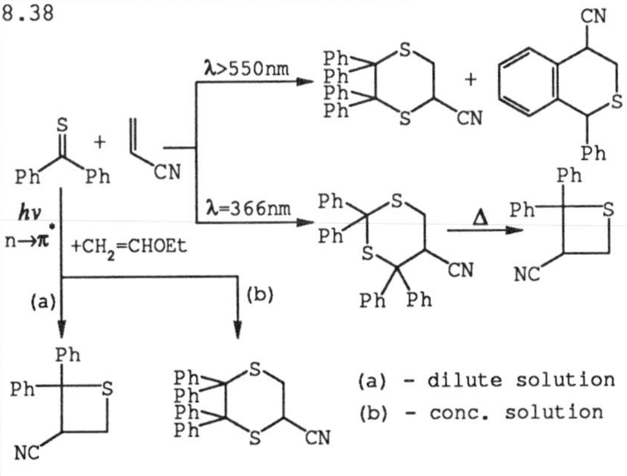

8.38. The photocycloaddition of alkenes to thiones is strongly wavelength dependent and thietanes, 1,3- or 1,4-dithianes, as well as 3-thiatetrahydronaphthalenes, are the obvious products. With longer wavelength irradiation the originally formed S$_1$ state undergoes ISC to the reactive T$_1$ state. The surprising ability of the S$_2$ state to undergo bimolecular reactions becomes evident when the large S$_2$-S$_1$ energy gap is considered.[82] There is also a different composition of products, depending on the electronic structure of the alkene (electron rich vs. electron poor).[83]

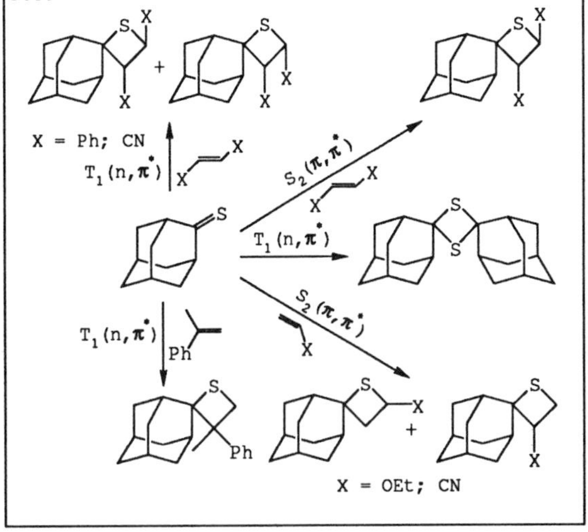

8.39. Of the photocycloadditions of alkenes to alicyclic thiones, the best known are the photocycloadditions to adamantanethione. If the adamantanethione is excited to its S$_1$ state, the T$_1$ state is the reactive state and the reaction apparently involves 1,4-biradical intermediates. Thus, it is a nonconcerted, i.e., a nonstereospecific, but regiospecific process.[84] On the other hand, if the reactive state is S$_2$, the photocycloaddition is a concerted stereospecific process.[85] Long wavelength irradiation of a cyclohexane solution of adamantanethione leads to the formation of its dimer, whereas short wavelength irradiation yields products of reductive alkylation.[75]

REFERENCES

(1) Jaffé, H. H. and Orchin, M. *Theory and Application of Ultraviolet Spectroscopy.* Wiley: New York, 1962.

(2) Turro, N. J. *J. Chem. Ed.* **1966**, *43*, 13.

(3) For reviews, see (a) Turro, N. J., Dalton, J. C., Dawes, K., Farrington, G., Hautala, R., Morton, D., Niemczyk, M., and Schore, N. *Acc. Chem. Res.* **1972**, *5*, 92. (b) Dalton, J. C. and Turro, N. J. *Ann. Rev. Phys. Chem.* **1970**, *21*, 499.

(4)(a) For the MNDO calculations of the potential surfaces for photochemical α-cleavage, see Reinsch, M., Höweler, U., and Klessinger, M. *Angew Chem., Int. Ed. Engl.* **1987**, *26*, 238. (b) For the *ab-initio* study of the α-cleavage of acetaldehyde, see Yadav, J. S. and Goddard, J. D. *J. Chem. Phys.* **1986**, *84*, 2682.

(5)(a) The selectivity of the α-cleavage of unsymmetrical ketones decreases at lower wavelengths and the quantum yield in fluid solutions is lowered due to the competing Norrish Type II reaction. See Noyers, W. A., Jr., Porter, G. B., and Jolley, J. E. *Chem. Rev.* **1956**, *56*, 49. (b) For the α-cleavage of acylgermanes and the subsequent reaction of the radicals formed with alkenes, see Kiyooka, S., Kaneko, Y., Matsue, H., Hamada, M., and Fujiyama, R. *J. Org. Chem.* **1990**, *55*, 5562.

(6)(a) Tert.-butyl ketone: Yang, N. C. and Feit, E. D. *J. Am. Chem. Soc.* **1968**, *90*, 504. (b) Dibenzyl ketone: Engel, P. S. *J. Am. Chem. Soc.* **1970**, *92*, 6074. (c) For the photolysis of dibenzyl ketone on zeolites, see Turro, N. J. and Zhang, Z. *Tetrahedron Lett.* **1987**, *28*, 5637. (d) Diphenylindanone: Quinkert, G., Optiz, K., Wiersdorff, W. W., and Finke, M. *Liebigs Ann. Chem.* **1966**, *963*, 44.

(7)(a) Robins, W. K. and Eastman, R. H. *J. Am. Chem. Soc.* **1970**, *92*, 6076. (b) Quinkert, G., Opitz, K., Wiersdorff, W. W., and Weinlich, J. *Tetrahedron Lett.* **1963**, 1863.

(8) For e.s.r. spectroscopy of radicals, see Paul, H. and Fischer, H. *Helv. Chim. Acta.* **1973**, *56*, 1575.

(9)(a) Turro, N. J., Cox, G. S., and Paczkowski, M. A. *Topics Curr. Chem.* **1985**, *129*, 57. (b) Turro, N. J. and Kraeutler, B. *J. Am. Chem. Soc.* **1978**, *100*, 7432.

(10)(a) Meier, G., Pfriem, S., Schäfer, U., Malsch, K.-D., and Matusch, R. *Chem. Ber.* **1981**, *114*, 3965. (b) Lei, X.-G., Doubleday, C. E., Zimmt, M. B., and Turro, N. J. *J. Am. Chem. Soc.* **1986**, *108*, 2444.

(11) Hemminger, J. C., Rusbult, C. F., and Lee, E. K. C. *J. Am. Chem. Soc.*, **1971**, *93*, 1867.

(12)(a) Eastman, R. H., Starr, J. E., Martin, R. S., and Sakata, M. K. *J. Org. Chem.*, **1963**, *28*, 2162. (b) Cooke, R. S. and Lyon, G. D. *J. Am. Chem. Soc.* **1971**, *93*, 3840.

(13) Weinshenker, N. M. and Greene, F. D. *J. Am. Chem. Soc.* **1968**, *90*, 506.

(14) For a review of the photochemistry of cyclic ketones, see (a) Weiss, D. S. *Org. Photochem.* **1981**, *5*, 347. (b) Heinrich, P. In *Houben-Weyl: Methoden der organischen Chemie.* Thieme: Stuttgart, 1975, Vol. 4/5b, p. 734. (c) Chapman, O. L. and Weiss, D. S. *Org. Photochem.* **1973**, *3*, 283. (d) Quinkert, G. *Angew. Chem., Int. Ed. Engl.* **1965**, *4*, 211. (e) For the effect of temperature on product distribution in the photolysis of cycloalkanones, see Doyle, J. D. *J. Chem. Soc., Perkin Trans. 1.* **1972**, 683. (f) For a SINDO/1 study of photoisomerization and fragmentation of cyclopentanone, see Müller-Remmers, P. L., Mishra, P. C., and Jug, K. *J. Am. Chem. Soc.*, **1984**, *106*, 2538.

(15) Morton, D. R. and Turro, N. J. *J. Am. Chem. Soc.*, **1973**, *95*, 3947.

(16) For review, see Yates, P. and Loutfy, R. O. *Acc. Chem. Res.* **1975**, *8*, 209.

(17)(a) Quinkert, G., Wegemund, B., Homburg, F., and Cimbollek, G. *Chem. Ber.* **1964**, *97*, 958. (b) For a review, see Gilbert, A. In *Photochemistry in Organic Synthesis.* Coyle, J. D., ed. The Royal Society of Chemistry: Burlington House, London, 1986, p. 80.

(18) Padwa, A., Carter, S. P., Nimmesgern, H., and Stull, P. D. *J. Am. Chem. Soc.* **1988**, *110*, 2894.

(19) Quinkert, G., Wegemund, B., and Blanke, E. *Tetrahedron Lett.*, **1962**, 22.

(20) For a review, see (a) Yates, P. *J. Photochem.* **1976**, *5*, 91. (b) Stohrer, W.-D., Jacobs, P., Kaiser, K. H., Wiech, G., and Quinkert, G. *Topics Curr. Chem.* **1974**, *46*, 181.

(21) Morton, D. R., Lee-Ruff, E., Southam, R. M., and Turro, N. J. *J. Am. Chem. Soc.* **1970**, *92*, 4649.

(22) Pirrung, M. C., Chang, V. K., and DeAmicis, C. V. *J. Am. Chem. Soc.*, **1989**, *111*, 5824, and references therein.

(23)(a) Quinkert, G., Wiech, G., and Stohrer, W.-D. *Angew. Chem., Int. Ed. Engl.* **1974**, *13*, 199, 200. (b) Turro, N. J. and McDaniel, D. M. *J. Am. Chem. Soc.* **1970**, *92*, 5727.

(24) Miller, R. D., Golitz, P., Janssen, J., and Lemmens, J. *J. Am. Chem. Soc.* **1984**, *106*, 7277.

(25) For a review, see Newton, R. F. In *Photochemistry in Organic Synthesis*. Coyle, J. D., ed. The Royal Society of Chemistry: Burlington House, London, 1986, p. 39.

(26) Pirrung, M. C. and DeAmicis, C. V. *Tetrahedron Lett.* **1988**, *29*, 159.

(27) For reviews of the synthetic application of photoreductions, see (a) Horspool, W. M. In *Photochemistry in Organic Synthesis*. Coyle, J. D., ed. The Royal Society of Chemistry: Burlington House, London, 1986, p. 61. (b) Cowan, D. O. and Drisko, R. L. *Elements of Organic Chemistry*, Plenum: New York, 1976, p. 75. (c) For a review of photochemical formation of 1,2-ethanediols, see Schönberg, A. *Preparative Organic Chemistry*. Springer Verlag: New York, 1968, p. 203.

(28)(a) For a general review of the photoreduction mechanism, see Wagner, P. J. *Topics Curr. Chem.* **1976**, *66*, 1. (b) For the reduction of carbonyl compounds by electron transfer, see Cohen, S. G., Parola, A., and Parsons, G. H., Jr., *Chem. Rev.* **1973**, *73*, 141.

(29)(a) Hammond, G. S., Moore, W. M., and Foss, R. D. *J. Am. Chem. Soc.* **1961**, *83*, 2798. (b) For the photochemistry of ketyl radicals, see Netto-Ferreira, J. C., Murphy, W. C., Redmond, R. W., and Scaiano, J. C. *J. Am. Chem. Soc.* **1990**, *112*, 4472, and references therein.

(30) For the photoreduction of π,π^* and n,π^* triplet carbonyls by amines, see Khan, J. and Cohen, S. G. *J. Org. Chem.* **1991**, *56*, 938.

(31)(a) Blank, B., Henne, A., Laroff, G. P., and Fischer, H. *Pure Appl. Chem.* **1975**, *41*, 475. (b) For a review of the photogeneration of unstable enols, see Capon, B. In *Chemistry of Enols*. Rappoport, Z., ed. Wiley: Chichester, United Kingdom, 1990, p. 307. (b) For a review of the photochemistry of enols, see Weedon, A. C. In *Chemistry of Enols*. Rappoport, Z., ed. Wiley: Chichester, United Kingdom, 1990, p. 591.

(32)(a) Keeffe, J. R., Kresge, A. J., and Schepp, N. P. *J. Am. Chem. Soc.* **1990**, *112*, 4862. (b) For a review, see Kresge, A. J. *Acc. Chem. Res.* **1990**, *23*, 43.

(33)(a) For the effect of α-silyl substitution on enol chemistry, see Kresge, A. J. and Tobin, J. B. *J. Am. Chem. Soc.* **1990**, *112*, 2805. (b) For the reversible ketene $\leftrightarrows$ ethynol photoisomerization, see Hochstrasser, R. and Wirz, J. *Angew. Chem., Int. Ed. Engl.* **1990**, *29*, 411.

(34) For 7-methylenebicyclo[3.3.2]nonan-3-one, see Mori, T., Yang, K. H., Kimoto, K., and Nozaki, H. *Tetrahedron Lett.* **1970**, 2419.

(35) For reviews, see (a) Wagner, J. P. In *Rearrangements in Ground and Excited States*. deMayo, P., ed. Academic Press: New York, 1980. (b) Wagner, P. J. *Acc. Chem. Res.* **1971**, *4*, 168. (c) Wagner, P. J. *Acc. Chem. Res.* **1983**, *16*, 461.

(36) For recent studies in this field, see (a) Aoyama, H., Arata, Y., and Omote, Y. *J. Org. Chem.* **1987**, *52*, 4639, and references therein. (b) Wagner, P. J. and Nahm, K. *J. Am. Chem. Soc.* **1987**, *109*, 4404, 6528. (c) Nuñez, A. and Weiss, R. G. *J. Am. Chem. Soc.* **1987**, *109*, 6215. (d) For modification of the photochemistry by cyclodextrin, see Redy, D. and Ramamurthy, V. *J. Org. Chem.* **1987**, *52*, 3953. (e) Macrocyclic diketones in solid state, see Lewis, T. J., Rettig, S. J., Scheffer, J. R., Trotter, J., and Wireko, F. *J. Am. Chem. Soc.* **1990**, *112*, 3679.

(37)(a) For a theoretical study of hydrogen transfer via six- and seven-membered transition states, see Dorigo, A. E. and Houk, K. N. *J. Am. Chem. Soc.* **1987**, *109*, 2195. (b) Dorigo, A. E., McCarrick, M. A., Loncharich, R. J., and Houk, K. N. *J. Am. Chem. Soc.* **1990**, *112*, 7508. (c) For a MINDO/3 study, see Dewar, M. J. S. and Doubleday, C., Jr. *J. Am. Chem. Soc.* **1978**, *100*, 4935. (d) For a molecular mechanics study of the Norrish Type II reaction of cyclodecanone, see Sauers, R. R. and Huang, S. Y. *Tetrahedron Lett.* **1990**, *31*, 5709.

(38) Yang, N. C., Elliott, S. P., and Kim, B. *J. Am. Chem. Soc.* **1969**, *91*, 7551.

(39)(a) Turro, N. J. and Weiss, D. S. *J. Am. Chem. Soc.* **1968**, *90*, 2185. (b) Lewis, F. D., Johnson, R. W., and Ruden, R. A. *J. Am. Chem. Soc.* **1972**, *94*, 4292. (c) Wagner, P. J., Kelso, P. A., Kemppainen, A. E., and Zepp, R. G. *J. Am. Chem. Soc.* **1972**, *94*, 7500.

(40)(a) For a review of the reaction of 1,4-biradicals, see Scaiano, J. C. *Acc. Chem. Res.* **1982**, *15*, 252. (b) For a review of the dynamics of flexible triplet biradicals, see Doubleday, C., Jr., Turro, N. J., and Wang, J.-F. *Acc. Chem. Res.* **1989**, *22*, 199. (c) For a CIDNP study, see Kaptein, R., de Kanter, F. J. J., and Rist, G. H. *J. Chem. Soc., Chem. Commun.* **1981**, 499. (d) For reaction path analysis of H-abstraction by the formaldehyde triplet state, see Severange, D., Pandey, B., and Morrison, H. *J. Am. Chem. Soc.* **1987**, *109*, 3231. (e) For structural effects in triplet biradical lifetimes, see Caldwell, R. A., Majima, T., and Pac, C. *J. Am. Chem. Soc.* **1982**, *104*, 650. (f) For an excited-chemistry of a 1,5-biradical, see Scaiano, J. C. and Wagner, P. J. *J. Am. Chem. Soc.* **1984**, *104*, 4626. (g) For decay of triplet biradicals, see Zimmt, M. B., Doubleday, C., Jr., and Turro, N. J. *J. Am. Chem. Soc.* **1986**, *108*, 3618. (h) For a study of photocyclization of 2-alkoxyphenylketones, see Wagner,

P. J., Meador, M. A., and Park, B.-S. *J. Am. Chem. Soc.* **1990,** *112,* 5199.

(41)(a) Yang, N. C. and Elliott, S. P. *J. Am. Chem. Soc.* **1969,** *91,* 7550.   (b) Casey, C. P. and Boggs, R. A. *J. Am. Chem. Soc.* **1972,** *94,* 6457.

(42) Sugimura, T. and Paquette, L. A. *J. Am. Chem. Soc.* **1987,** *109,* 3017.

(43) For examples of δ-hydrogen abstraction, see (a) Wagner, P. J. *Acc. Chem. Res.* **1989,** *22,* 83 (a review).   (b) Paquette, L. A., Ternansky, R. J., Balogh, D. W., and Kentgen, G. *J. Am. Chem. Soc.* **1983,** *105,* 5446 (dodecahedrane synthesis).   (c) Descotes, G. *Bull. Soc. Chim. Belg.* **1982,** *91,* 973.   (d) Bernasconi, G., Cottier, L., Descotes, G., Praly, J. P., Remy, G., Gernier-Loustalot, M. F., and Metras, F. *Carbohydr. Res.* **1983,** *115,* 106.

(44) Kraus, G. A. and Chen, L. *J. Am. Chem. Soc.* **1990,** *112,* 3464, and references therein.

(45) Cossy, J., Belotti, D., and Pete, J. P. *Tetrahedron Lett.* **1987,** *28,* 4545, 4547.

(46)(a) Winnick, M. A., Lee, C. K., Basu, S., and Saunders, D. S. *J. Am. Chem. Soc.* **1974,** *96,* 6182.   (b) Winnick, M. A. *Acc. Chem. Res.* **1977,** *10,* 173.   (c) Wagner, P. J. *Acc. Chem. Res.* **1983,** *16,* 461.

(47) For reviews, see   (a) Weedon, A. C. In *Chemistry of Enols.* Rappoport, Z., ed. Wiley: Chichester, United Kingdom, 1990, p. 591.   (b) Wilson, R. M. *Org. Photochem.* **1985,** *5,* Chapter 5.   (c) Sammes, P. G. *Tetrahedron.* **1976,** *32,* 405.

(48) For steric congestion effect, see Wagner, P. J. and Zhou, B. *J. Am. Chem. Soc.* **1988,** *110,* 611.

(49) For laser-jet photochemistry, see (a) Wilson, R. M., Hannemann, K., Peters, K., and Peters, E.-M. *J. Am. Chem. Soc.* **1987,** *109,* 4741.   (b) Wilson, R. M., Hannemann, K., Heineman, W. R., and Kirchhoff, J. R. *J. Am. Chem. Soc.* **1987,** *109,* 4743.

(50)(a) Büchi, G., Inman, C. G., and Lipinski, E. S. *J. Am. Chem. Soc.* **1954,** *76,* 4327.   (b) Freilich, S. C. and Peters, K. S. *J. Am. Chem. Soc.* **1985,** *107,* 3819.

(51) For reviews, see   (a) Ninomiya, I. and Naito, T. *Photochemical Synthesis.* Academic Press: San Diego, 1989, p. 138.   (b) Carless, H. A. J. In *Photochemistry in Organic Synthesis.* Coyle, J. D., ed. The Royal Society of Chemistry: Burlington House, London, 1986, p. 95.   (c) Carless, H. A. J. In *Synthetic Organic Photochemistry.* Horspool, W. M., ed. Plenum: New York, 1984, p. 425.   (d) Jones, G., II. *Org. Photochem.* **1981,** *5,* 1.   (e) Baldwin, S. W. *Org. Photochem.* **1981,** *5,* 123.

(52) For cyclooctene/acetone, see Shima, K., Sakai, Y., and Sakurai, H. *Bull. Chem. Soc. Japan.* **1971,** *44,* 215.

(53) For furan/benzophenone, see Schenck, G. O., Hartmann, W., and Steinmetz, R. *Chem. Ber.* **1963,** *96,* 498.

(54) For 5-cyclodecenone, see Lange, G. L. and Bosch, M. *Tetrahedron Lett.* **1971,** 315.

(55) Mori, T., Yang, K. H., Kimoto, K., and Nozaki, H. *Tetrahedron Lett.* **1970,** 2419.

(56)(a) For the possible role of the free radical ions in the oxetane formation with electron-rich alkenes, see Mattay, J., Gersdorf, J., Leismann, H., and Steenken, S. *Angew. Chem., Int. Ed. Engl.* **1984,** *23,* 249. See also Mattay, J. *Angew. Chem., Int. Ed. Engl.* **1987,** *26,* 825.

(57) Michl, J. *Mol. Photochem.* **1972,** *4,* 243, 257, 287.

(58) Turro, N. J. and Wriede, P. A. *J. Am. Chem. Soc.* **1970,** *92,* 320.

(59) Dalton, J. C., Wriede, P. A., and Turro, N. J. *J. Am. Chem. Soc.* **1970,** *92,* 1318.

(60) For reviews, see   (a) Turro, N. J. and Ramamurthy, V. In *Rearrangements in Ground and Excited States.* deMayo, P., ed. Academic Press: New York, 1980, Vol. 3.   (b) Turro, N. J., Lechtken, P., Schore, N. E., Schuster, G., Steinmetzer, H.-C., and Yekta, A. *Acc. Chem. Res.* **1974,** *7,* 97.

(61) Barltrop, J. A. and Carless, H. A. J. *J. Am. Chem. Soc.* **1972,** *94,* 1951, and references therein.

(62)(a) With cyclohexadiene, dimerization occurs in most cases. For more details, see Saltiel, J., Coates, R. M., and Dauben, W. G. *J. Am. Chem. Soc.* **1966,** *88,* 2745. Valentine, D., Turro, N. J., and Hammond, G. S. *J. Am. Chem. Soc.* **1964,** *86,* 5202. Turro, N. J., Wriede, P. A., and Dalton, J. C. *J. Am. Chem. Soc.* **1968,** *90,* 3274.   (b) For cyclooctadiene, see Shima, K., Sakai, Y., and Sakurai, H. *Bull. Chem. Soc. Japan.* **1971,** *44,* 215.

(63) For butadiene, see Barltrop, J. A. and Carless, H. A. J. *J. Am. Chem. Soc.* **1972,** *94,* 8761.

(64) Yang, N. C. and Chiang, W. *J. Am. Chem. Soc.* **1977,** *99,* 3163.

(65)(a) Toki, S., Shima, K., and Sakurai, H. *Bull. Chem. Soc. Japan.* **1965,** *38,* 760.   (b) Shima, K. and Sakurai, H. *Bull. Chem. Soc. Japan.* **1966,** *39,* 1806.

(66) Schreiber, S. L., Hoveyda, A. H., and Wu, H.-J. *J. Am. Chem. Soc.* **1983,** *195,* 660.

(67)(a) For chiral induction in oxetane formation, see Buschmann, H., Scharf, H.-D., Hoffmann, N., Plath, M. W., and Runsink, J. *J. Am. Chem. Soc.* **1989**, *111*, 5367, and references therein.   (b) Schreiber, S. L. *Science.* **1985**, *227*, 857.   (c) Oppolzer, W. *Acc. Chem. Res.* **1982**, *15*, 135.

(68) Schreiber, S. L. and Satake, K. *J. Am. Chem. Soc.* **1984**, *106*, 4186.

(69) Schroeter, S. H. *J. Org. Chem.* **1969**, *34*, 1188.

(70) Araki, Y., Nagasawa, J., and Ishido, Y. *J. Chem. Soc., Perkin Trans. 1.* **1981**, 12.

(71) Scharf, H.-D. and Mattay, J. *Tetrahedron Lett.* **1976**, 3509.

(72) Jones, G., Acquandre, M. A., and Carmody, M. A. *J. Chem. Soc., Chem. Commun.* **1975**, 206.

(73) For reviews of thiocarbonyls, see   (a) Ramamurthy, V. *Org. Photochem.* **1985**, *7*, 231.   (b) de Mayo, P. *Acc. Chem. Res.* **1976**, *9*, 52.   (c) Steer, R. P. and Ramamurthy, V. *Acc. Chem. Res.* **1988**, *21*, 380.   (d) Coyle, J. D. *Tetrahedron.* **1985**, *41*, 5393.

(74) For the electronic spectra of thiones, see (a) Steer, R. P. *Rev. Chem. Intermed.* **1981**, *4*, 1. (b) Mahaney, M. and Huber, J. R. *J. Mol. Spectrosc.* **1981**, *87*, 438.   (c) Blackwell, D. S. L., Liao, C. C., Loutfy, R. O., de Mayo, P., and Paszyc, S. *Mol. Photochem.* **1972**, *4*, 171.

(75)(a) Ohno, A. and Kito, N. *Bull. Chem. Soc. Japan.* **1973**, *46*, 2487.   (b) Ohno, A. and Kito, N. *Int. J. Sulfur Chem.* **1971**, *1*, 26.   (c) For the mechanism of adamantanethione photoreduction, see Law, K. Y. and de Mayo, P. *J. Am. Chem. Soc.* **1977**, *99*, 5813; **1979**, *101*, 3251.

(76)(a) Kito, N. and Ohno, A. *J. Chem. Soc., Chem. Commun.* **1971**, 1338.   (b) Kito, N. and Ohno, A. *Int. J. Sulfur Chem.* **1973**, *8*, 427.

(77) Kito, N. and Ohno, A. *J. Chem. Soc., Chem. Commun.* **1971**, 1338.

(78)(a) de Mayo, P. and Suau, R. *J. Am. Chem. Soc.* **1974**, *96*, 6807.   (b) Ho, K. W. and deMayo, P. *J. Am. Chem. Soc.* **1979**, *101*, 5725.   (c) Basu, S., Couture, A., Ho, K. W., Hoshino, M., de Mayo, P., and Suau, R. *Can. J. Chem.* **1981**, *59*, 246.

(79) Cox, A., Kemp, D. R., Lapouyade, R., de Mayo, P., Joussot-Dubien, J., and Bonneau, R. *Can. J. Chem.* **1975**, *53*, 2386.

(80) deMayo, P. and Suau, R. *J. Am. Chem. Soc.* **1974**, *96*, 6807.

(81) Barton, D. H. R., Bolton, M., Magnus, P. D., Marathe, K. G., Poulton, G. A., and West, P. J. *J. Chem. Soc., Perkin Trans. 1.* **1973**, 1574.

(82) de Mayo, P. and Shizuka, H. *J. Am. Chem. Soc.* **1973**, *95*, 3942.

(83)(a) Ohno, A. *Int. J. Sulfur Chem. B.* **1971**, *6*, 183. (b) Tsunchihashi, G., Yamauchi, M., and Fukuyama, M. *Tetrahedron Lett.* **1967**, 1971.   (c) Liao, C. C. and de Mayo, P. *J. Chem. Soc., Chem. Commun.* **1971**, 1525.

(84) Lawrence, A. H., Liao, C. C., de Mayo, P., and Ramamurthy, V. *J. Am. Chem. Soc.* **1976**, *98*, 2219.

(85)(a) Lawrence, A. H., Liao, C. C., de Mayo, P., and Ramamurthy, V. *J. Am. Chem. Soc.* **1976**, *98*, 3572.   (b) Rajee, R. and Ramamurthy, V. *Tetrahedron Lett.* **1978**, 3463.

Chapter 9

PHOTOCHEMISTRY OF ENONES, DIENONES, AND QUINONES

PHOTOCHEMISTRY OF ENONES, DIENONES, AND QUINONES

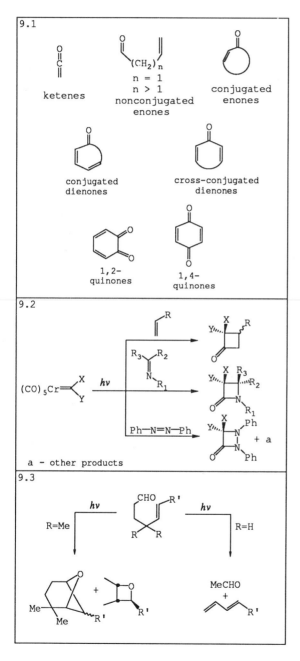

9.1

ketenes

(CH₂)ₙ
n = 1
n > 1
nonconjugated
enones

conjugated
enones

conjugated
dienones

cross-conjugated
dienones

1,2-
quinones

1,4-
quinones

9.2

(CO)₅Cr=〈X/Y   hv

a - other products

9.3

hv   R=Me

CHO R'
R   R

hv   R=H

MeCHO
+

Me
Me   R'

+   R'

9.1. This chapter is concerned with several classes of compounds that result from the combination of a carbonyl group and a C=C bond. Their reactivity is not only a function of electronic configuration and multiplicity of the excited state, but it also depends on the relative positions of the carbonyl group and the C=C double bond. The simplest combination of C=O and C=C is an *enone*. If one C atom is shared by the double bond and the carbonyl group, i.e., C=C=O, the double bonds are cumulated and the molecule is a ketene. Ketenes are synthetically valuable for the preparation of cyclobutanones by stereospecific [2+2] cycloaddition reactions with alkenes. However, their extreme reactivity and the lack of viable methods for the synthesis of suitably substituted ketenes present serious obstacles to their full synthetic exploitation.

9.2. Irradiation (visible light, ambient temperature) of heteroatom-stabilized chromium carbene complexes produces species that react as ketenes and facilitates the study of other synthetically interesting ketene reactions. By altering the substituents on the carbene and in the π system, this method has become increasingly useful for the synthesis of organic compounds under mild conditions. These reactions usually proceed in good yield with a high degree of stereo- and regioselectivity.[1,2]

9.3. If the two chromophores in nonconjugated enones are sufficiently separated, products result from the photochemistry of the individual chromophores. A Norrish Type II fragmentation occurs in the photolysis of the unsaturated aldehyde shown if it has an allylic hydrogen (R=H). If R=CH₃, the intramolecular formation of two isomeric oxetanes occurs (see 8.25).[3]

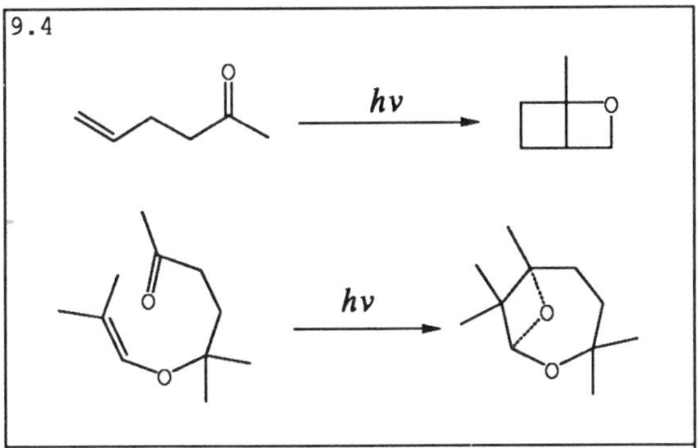

9.4. The *intra*molecular Paterno–Büchi reaction, first performed with 5-hexen-2-one,[4] leads to a bicyclic oxetane. This originally inefficient reaction has developed into a highly efficient and versatile photochemical method for the synthesis of a variety of compounds that are difficult or impossible to prepare by other methods.[5,6] Hundreds of examples of the intramolecular Paterno–Büchi reaction have been reported.[7] The efficiency of these reactions can be attributed to the rapid rate of interaction between the excited C=O group and the C=C bond. This combination of substrates allows the formation of one regioisomer. Thus, yields are high and there are usually no byproducts. This oxetane formation occurs not only with γ,δ-enones, but also with enones in which C=C is further separated from the carbonyl group (see also 9.6).[8]

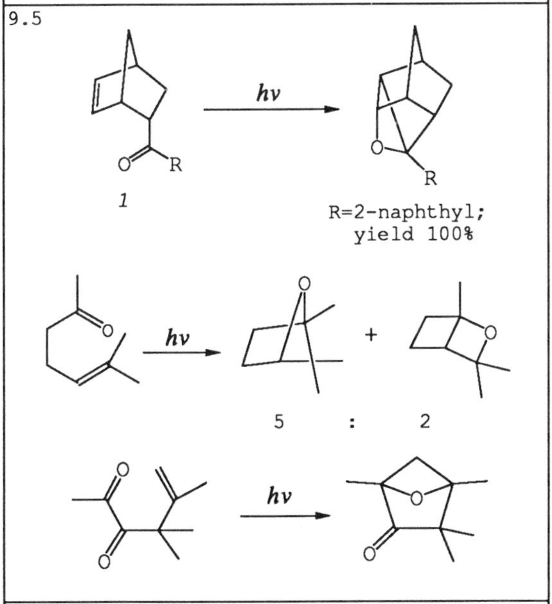

9.5. Because the T_1 state of 2-acenaphthone is π,π^*, an oxetane is not formed by an intermolecular reaction. In contrast, the 2-naphthyl ketone (*1*) gives an oxetane in quantitative yield.[9] Whenever possible, a five-membered ring by an $x[2+2]$ cycloaddition is favored over cyclobutane ring formation by a "normal" $[2+2]$ cycloaddition[10] (rule of five, see 6.2).

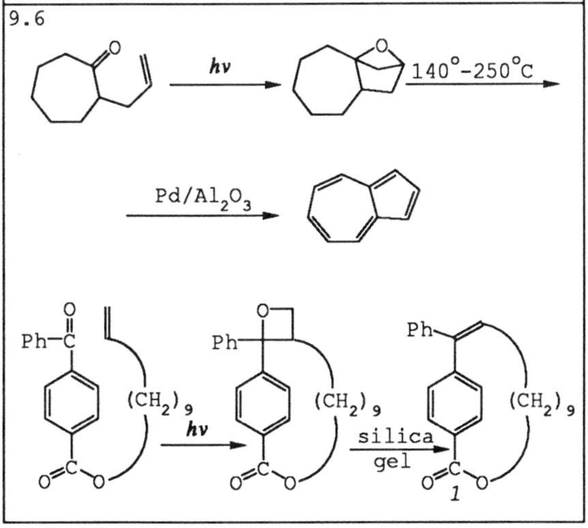

9.6. The carbonyl-alkene metathesis has been widely used with a variety of bicyclic oxetanes. For example, azulene has been synthesized via an oxetane by irradiation of 2-allylcycloheptanone and subsequent thermolysis and dehydration-dehydrogenation.[11] Another example is the synthesis of the large ring lactone *1*. Irradiation of the unsaturated ester of benzophenone-4-carboxylic acid results in an 83% yield of the bicyclic oxetane. Subsequent treatment of the oxetane with silica gel causes ring cleavage to *1* in 90% yield.[12]

9.7

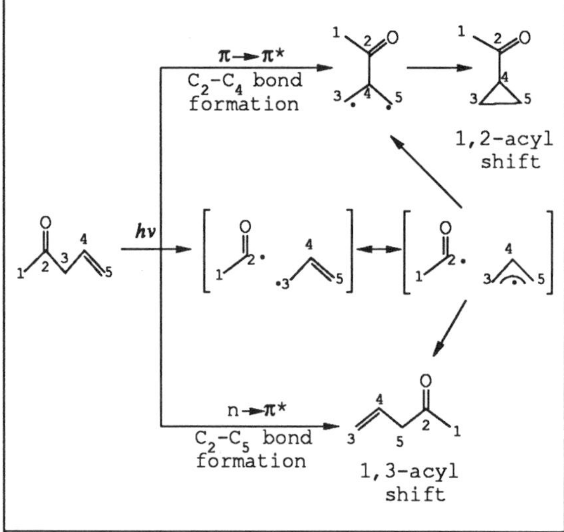

9.8

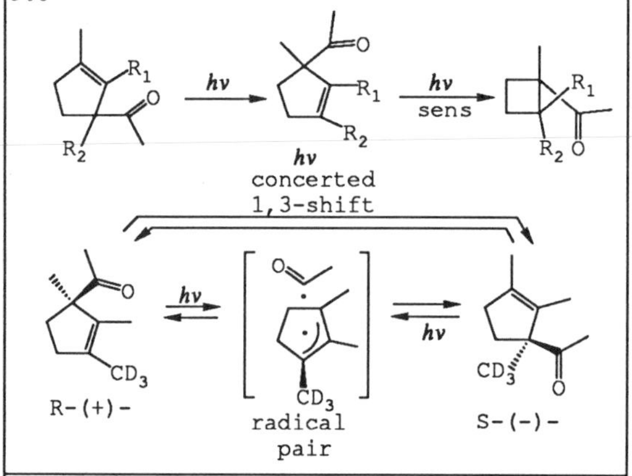

9.9

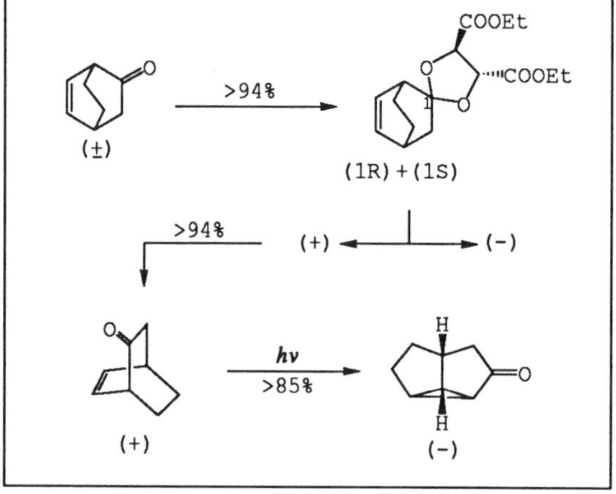

9.7. The most photochemically important nonconjugated enones are β,γ-unsaturated ketones.[13] Their irradiation induces sigmatropic reactions that are directed by the electronic configuration and multiplicity of the excited state. The reactions are 1,2-acyl shifts (oxadi-π-methane rearrangements), which occur from the lowest T_1 (π,π^*) state (the lowest T state for most β,γ-enones), and 1,3-acyl shifts, which occur upon direct irradiation from the S_1 or T_2 (n,π^*) state with the formation of an acyl-allyl radical pair. It has been suggested that, while at higher temperature the radical pair is formed predominantly from the S_1 state, at lower temperature the T_2 state is involved.[14] Accordingly, temperature-activated α-cleavage proceeds from the S_1 state.[15] Although both rearrangements involve biradical or radical pair intermediates, they are generally stereospecific.

9.8. 2-cyclopentenyl methyl ketones were the first examples of the rearrangement of β,γ-enones.[16] Photo-CIDNP and radical-trapping experiments[17] indicate that the 1,3-acyl shift is a radical process that proceeds predominantly via a caged radical pair, which also leads to α-cleavage products. The stereospecificity of this reaction was shown by the irradiation of enantiomers in which the R$-$(+) enantiomer led to the rearranged S$-$(-) isomer with 20% racemization.[14]

9.9. While the direct irradiation of the bicyclo[2.2.2]octenone produces a bicyclo[4.2.0]octenone as a result of a 1,3-acyl shift, the triplet-sensitized reaction yields the tricyclo-[3.3.0.0^{2,8}]octan-2-one, an oxadi-π-methane rearrangement product. This reaction is the key step in the synthesis of [5$-$5] fused ring systems. The stereospecificity of the rearrangement provides the opportunity to produce a pure diastereomer of tricyclooctanone by the irradiation of an optically active bicyclooctenone. This method is important in the synthesis of natural products.[18,19]

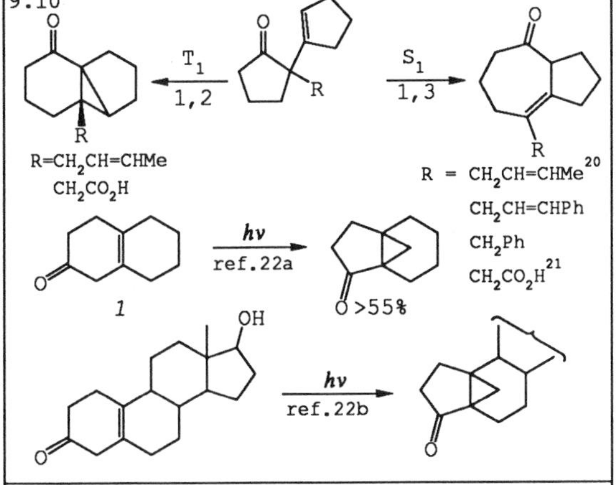

9.10. Cyclic β,γ-unsaturated ketones undergo both 1,2- and 1,3-acyl shifts, usually with ring contraction. A 1,2 shift leads to a cyclopropane derivative, while a 1,3 shift can sometimes result in ring expansion. The different course of the photolysis of the homologous enones *1* and *2* (see 9.11) is remarkable. Since enone *1* has a much more rigid structure than enone *2*, the planes of the carbonyl group and the C=C bond in *1* are separated. As a result, α-cleavage in the S_1 state does not compete efficiently with intersystem crossing to T_1, from which α-cleavage occurs with the formation of a triplet biradical. A subsequent 1,2-acyl shift occurs.

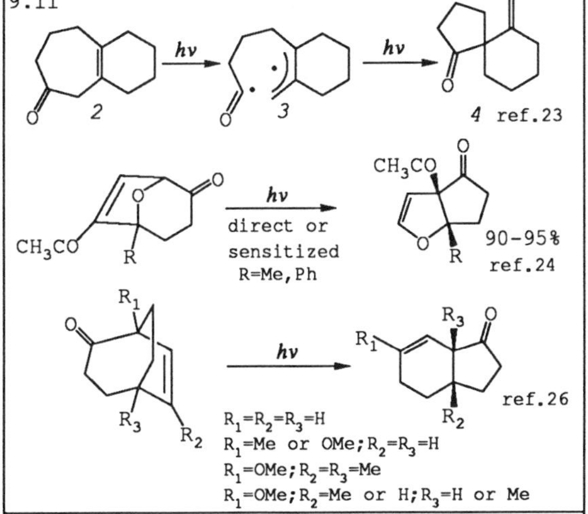

9.11. Overlap of the carbonyl group and the double bond in enone *2* sufficiently increases the efficiency of the α-cleavage so that it competes successfully with intersystem crossing from the S_1 state. The resulting biradical *3* generates product *4* in good yield.[23] The 1,3-acyl shift in cyclic β,γ-enones is a useful method for altering ring size and connectivity. For example, the rearrangement of bicyclo[3.2.2]non-6-en-2-one to bicyclo[4.3.0]-non-2-en-9-one has been employed as a key step in the synthesis of natural products.[24,25]

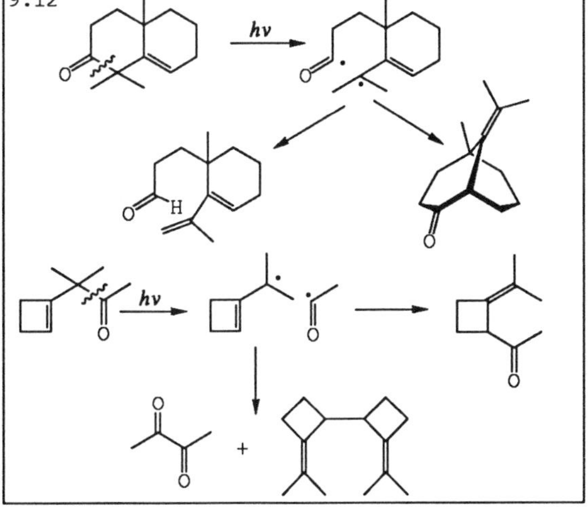

9.12. The 1,3-acyl shift competes with other reactions of acyl-allyl radical pairs, particularly with enones that are disubstituted between the C=O and C=C groups. Apparently, intersystem crossing is sufficiently fast to compete with recombination of the radicals. Intramolecular radical pairs undergo either a hydrogen atom shift to the acyl radical with the formation of an aldehyde or a 1,3-acyl shift.[27] Intermolecular radical pairs by diffusion from the solvent cage result in the recombination of identical radicals.[28] Products of a Norrish Type II reaction,[29] in addition to oxetanes, are also formed.[30]

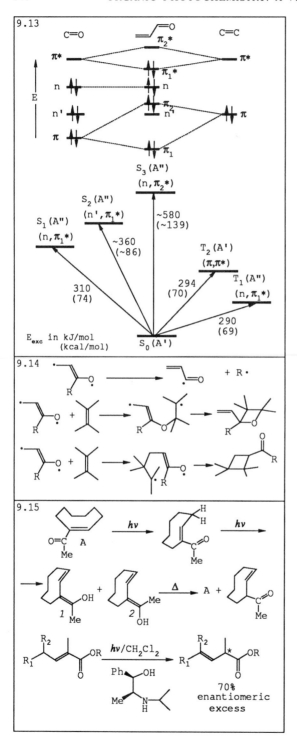

9.13. Simple conjugated enones exhibit a weak absorption band above ~ 280 nm ($n \to \pi^*$ transition) and a strong band above ~ 200 nm ($\pi \to \pi^*$ transition).[31] The excited states of conjugated enones and dienones have characteristics of alkenes and carbonyl compounds. While the levels of the n orbitals, because of their different symmetry, depend very slightly on the interaction of the two chromophores, the π and π^* orbitals of both chromophores are split. Thus, a number of n,π^* singlet states result. T_1 also has an n,π^* configuration.[32] However, since the π,π^* singlet-triplet splitting is large, the second lowest excited state is the triplet π,π^* state.

9.14. The valence-bond structures of a conjugated enone in an n,π^* or π,π^* state are similar to those of a saturated carbonyl compound in the same electronic state. Generally, the photochemistry of conjugated enones proceeds via an n,π^* or π,π^* T_1 state. Some fundamental photoreactions of enones, e.g., α-cleavage, dimerization, and oxetane formation, can be predicted by the biradical intermediates, shown for ease of visualization.[33,34]

9.15. UV irradiation of conjugated carbonyl compounds with one H atom in the γ position results in deconjugation to β,γ-unsaturated ketones via a dienol intermediate. Irradiation of 1-acetylcyclooctene in acetonitrile at 350 nm gives an 80% yield of a mixture of *1* and *2* in a ratio of 5:1. These enols can be isomerized thermally to a mixture of the starting 1-acetylcyclooctene and the "deconjugated" 3-acetyl-cyclooctene.[35] The photoenolization occurs after the *cis-trans* isomerization of the ring since abstraction of a γH atom is only possible from the *cis* isomer. If the starting enone has an α-alkyl substituent, the dienol intermediate is prochiral and an enantioselective protonation occurs in an aprotic solvent in the presence of a chiral aminoalcohol.[36,37]

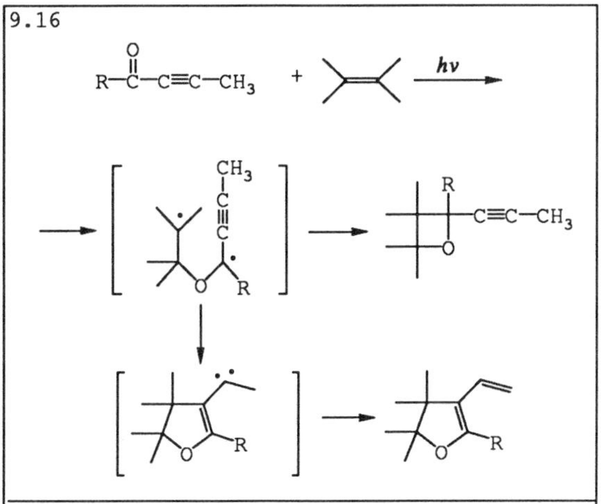

9.16. Upon irradiation in the presence of alkenes, α,β-acetylenic ketones undergo both [3+2] and [2+2] cycloaddition reactions. The [3+2] cycloaddition proceeds through a biradical intermediate that cyclizes to a carbene. The product resulting from an intramolecular rearrangement of the carbene is a vinyldihydrofuran derivative and, with some substrates, the biradical recombines to produce an oxetane.[38]

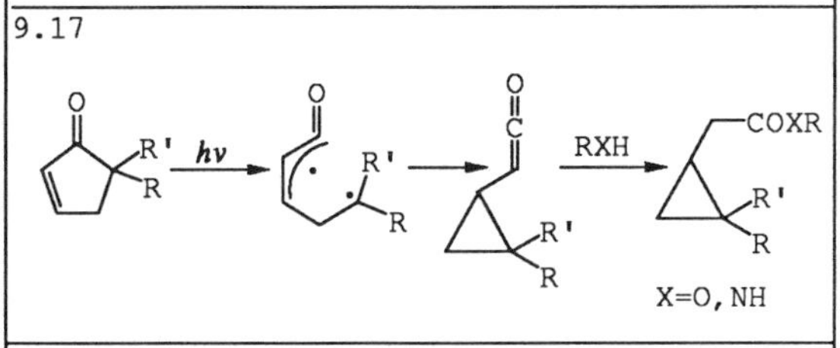

9.17. α-cleavage of conjugated enones is frequently enhanced by radical-stabilizing substituents in position 5. While dimerization competes efficiently with α cleavage in cyclopentenone, the predominant reaction of the S_1 state of 5,5-disubstituted cyclopentenones, particularly in dilute solution, is α-cleavage.[39]

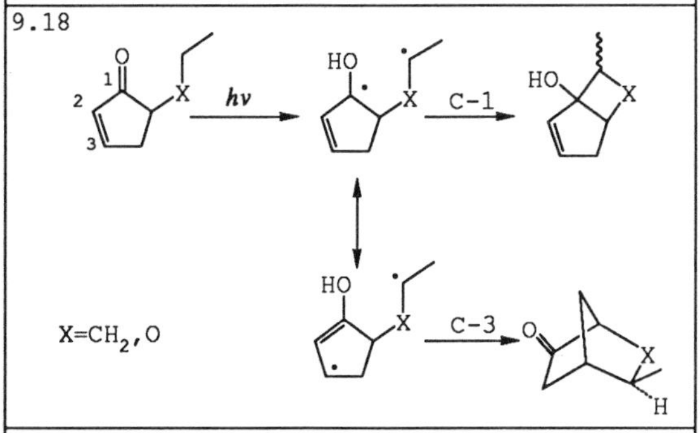

9.18. Intramolecular abstraction of a hydrogen atom is analogous to the primary photochemical step in the Norrish Type II fragmentation, but without cleavage of the β bond. Instead, the biradical is stabilized by the formation of a bond with either the C-1 carbon atom to yield cyclobutanols or the C-3 carbon atom to give cyclohexanones (cyclic enones yield bicyclic ketones).[39,40]

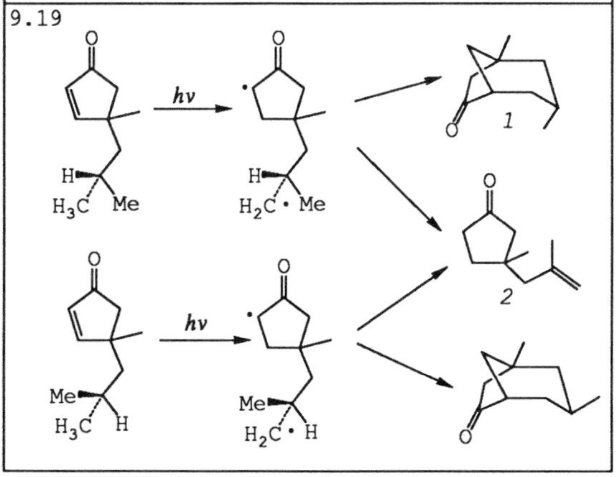

9.19. A study of the geometric requirements of this reaction shows that a diastereomeric biradical participates almost exclusively (90%) and produces *1*.[41] The preferential formation of the biradical cannot be explained by steric effects since mechanical models show that the two conformers leading to the different biradicals are very similar. The result is consistent with the suggestion that intramolecular hydrogen atom transfer to the β carbon atom of an enone proceeds through a $^3(\pi,\pi^*)$ excited state, which is pyramidal at the β-C atom. Since the H atom transfer to the O atom of the enone takes place in the $^3(n,\pi^*)$ state, which is thought to be planar at the β atom, a higher yield of product *2* is expected.[42]

9.20

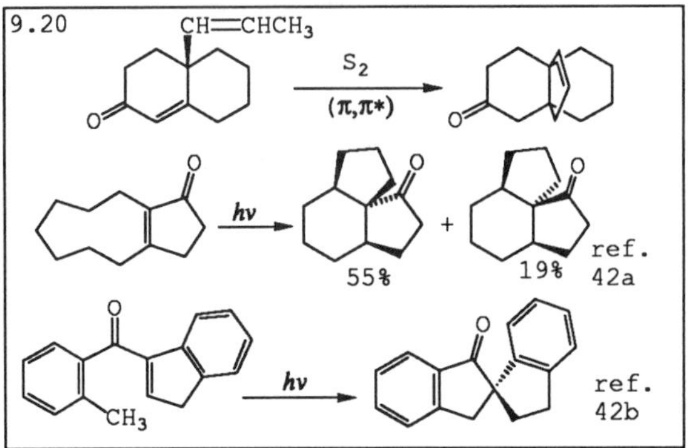

ref. 42a

55%   19%

ref. 42b

9.21

39%

MeOH

51%

9.22

l = liquid; *s* = solid

9.20. This intramolecular cyclization, viewed as a $[2\sigma + 2\pi]$ cycloaddition of a C—H bond to the C=C bond of the enone, is an efficient process (Φ = 0.29) that occurs in the S_2 state (λ = 254 nm). In the S_1 state (λ = 300 nm) *cis-trans* isomerization of the side chain takes place. This cyclization represents a procedure for the preparation of angularly fused tricyclic compounds by irradiation of bicyclic conjugated enones.[43] Other examples of $[2\sigma + 2\pi]$ cycloaddition reactions of conjugated enones result in spirocyclic compounds.[44] The reactions apparently proceed by a H-atom abstraction by the β-carbon atom of the enone in the $T(n,\pi^*)$ state.

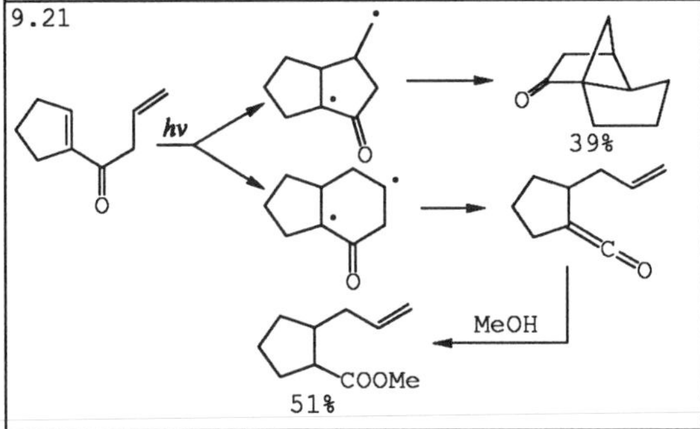

9.21. Irradiation of 1,5-hexadien-3-ones causes intramolecular photocyclization, leading to the formation of a 1,5-biradical in accord with the "rule of five" (see 6.2). However, the reaction is less regioselective than that of 1,5-dienes. Alkyl substitution in position 5 and the incorporation of a conjugated double bond into a five- or a six-membered ring favors 1,6-cyclization.[45]

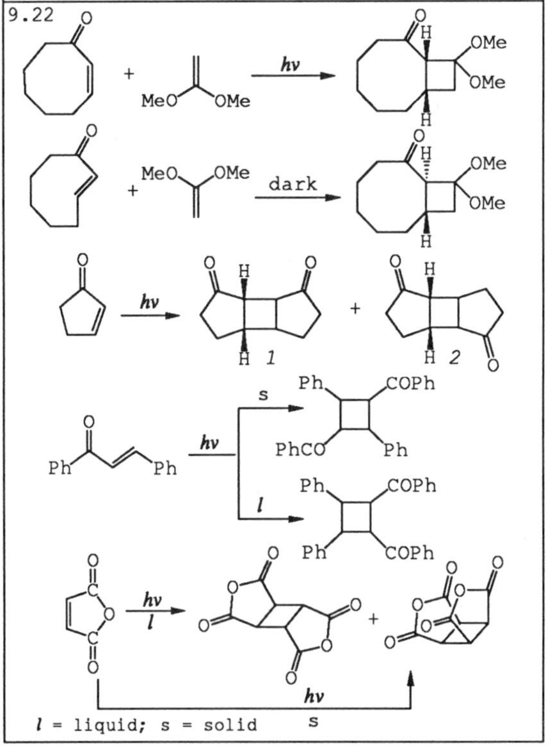

9.22. Photodimerization or photocycloaddition reactions of conjugated ketones, especially cyclic enones, with the formation of a cyclobutane ring are extremely facile and preparatively important.[46] The irradiation of cyclopentenone[47] or cyclohexenone[48] produces a mixture of two *anti* dimers (head-to-head and head-to-tail) in a ratio that is a function of both the concentration and the solvent (see 15.20 and 15.21). The reaction proceeds through a $^3(\pi,\pi^*)$ state via a 1,4-biradical. However, *cis-trans* isomerization is preferred in higher cycloenones. The resulting *trans*-cycloenones do undergo dimerization, but in a thermal process. The regio- and stereoselectivity of the dimerization reaction also depend on the reaction medium.[50]

9.23

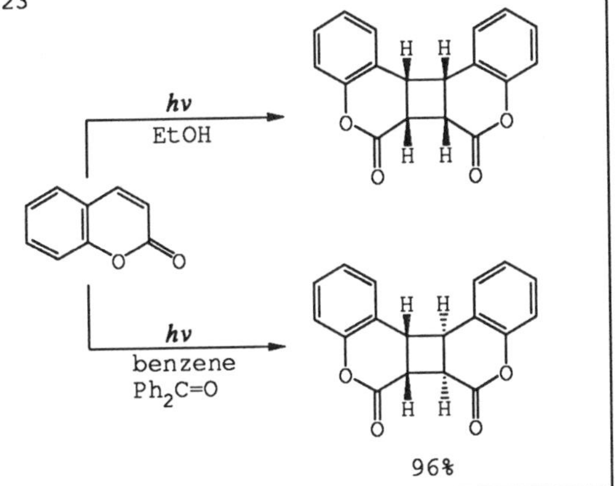

96%

9.24

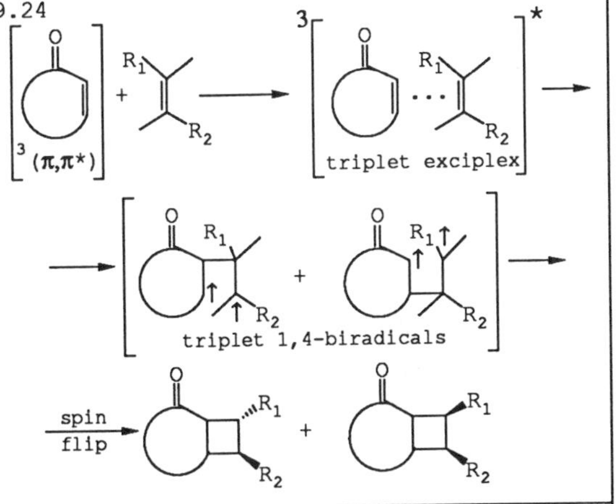

9.25

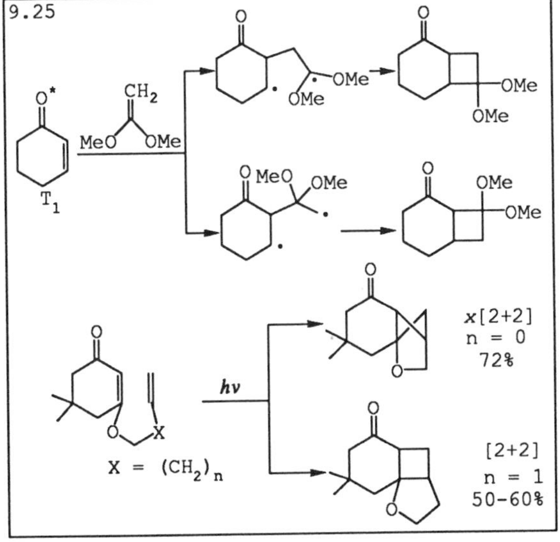

9.23. The photodimerization of coumarin is regioselective. Direct irradiation in ethanol gives the *syn* dimer, while dimerization does not occur in less polar solvents, such as benzene. Direct irradiation of a benzene solution induces self-quenching of the S_1 state. However, sensitization with benzophenone leads to a high yield of the *anti* dimer. This reaction involves the T_1 state of coumarin in both polar and nonpolar solvents.[51,52] Because of enhanced excimer formation, the dimerization proceeds at a remarkably increased rate in an aqueous medium compared with other solvents.[53]

9.24. The conjugated enone/alkene [2+2] photocycloaddition (photoannelation) reaction is one of the most synthetically useful of all photoreactions of enones,[54] particularly in the synthesis of cage compounds[55] and natural products.[56] The double bond of an enone, which undergoes cycloaddition to an alkene, is part of a six- (or less) membered ring.[57] Exceptions to this are oxygen β-substituted cyclooctenones.[58] The initial excitation is probably S_1 (n,π^*), followed by intersystem crossing to the T_1 (n,π^*) or the twisted (π,π^*) state, which is apparently the reactive state.[59] The cycloaddition proceeds through a short-lived exciplex[60] that collapses to 1,4-biradical(s) by bond formation at atom *2* and/or *3* of the enone. Since the stereochemical integrity of the ethylene fragment is lost, both *cis*- and *trans*-alkenes give mixtures of all possible stereoisomers of the cycloadduct.

9.25. The regiochemistry of the intermolecular cycloaddition is poor, except for highly electron-rich alkenes (which react more readily than electron-deficient alkenes).[60] This contrasts with the intramolecular pathway in which the C=C bond is separated from the enone by 2–4 atoms. In general, the five-membered ring formation is favored in the initial radical addition of the excited state to the alkene (rule of five, see 6.2). If a five-membered ring cannot form, a six-membered ring is then preferred.[61]

9.26

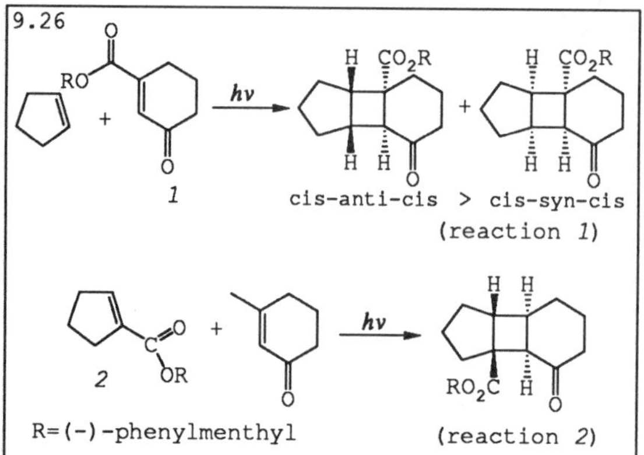

cis-anti-cis > cis-syn-cis
(reaction 1)

R=(-)-phenylmenthyl   (reaction 2)

9.27

9.28

9.26. Asymmetric inductions are obtained by mixed [2+2] cyclobutane-forming photoadditions when various chiral substituents are attached to the enone moiety *1* (reaction *1*) or the alkene moiety *2* (reaction 2). Inductions of 79% and 57%, respectively, were observed in studies with the chiral (−)-phenylmenthyl group in toluene. The polarity of the solvent and the irradiation temperature have a significant effect on the amount of asymmetric induction.[62]

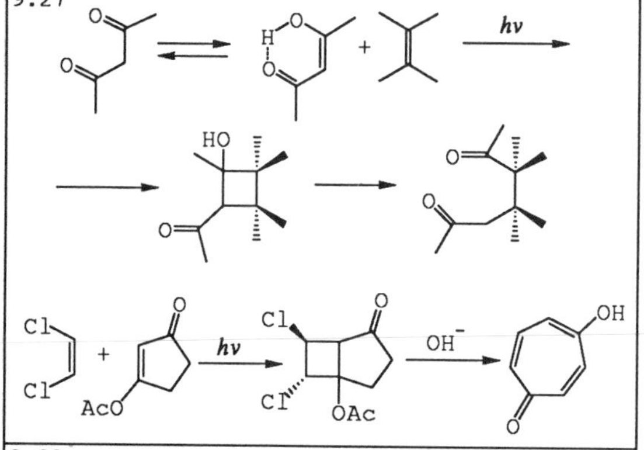

9.27. The synthetic significance of the intermolecular enone-alkene photocycloaddition was extended by irradiating β-diketones in the presence of alkenes to produce 1,5-diketones (deMayo reaction). The enol of the 1,3-diketone, which is restricted in a six-membered ring by an intramolecular hydrogen bond, is an intermediate. Photoaddition of an alkene to this enol gives a β-hydroxy ketone, which undergoes retroaldolization to the 1,5-diketone. With cyclic β-dicarbonyl compounds or their enol acetates (acylates), the resulting bicyclic β-hydroxyketones easily undergo the retroaldol reaction with concomitant expansion of the ring by two C atoms.[63] One of the first applications of this method was the synthesis of γ-tropolone.[64,65]

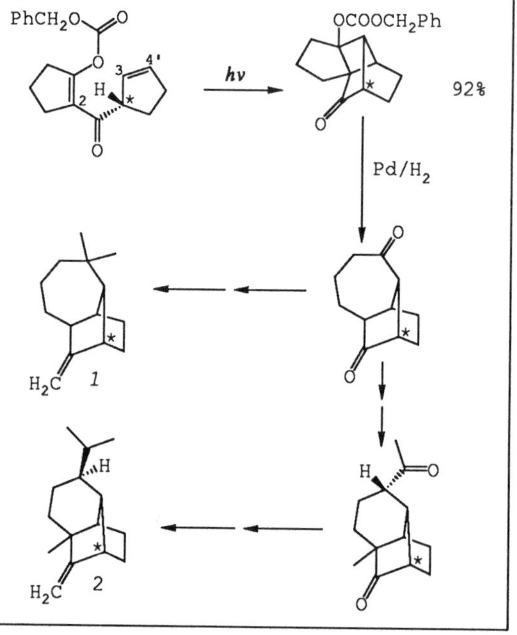

9.28. One disadvantage of this bimolecular reaction is the inability to predict its regiochemistry. On the other hand, the potential of a *regioselective annelative two-carbon ring expansion* from cyclic 1,3-diketones, or their enol derivatives having alkenyl chains, has been studied recently. The first example of one of these intramolecular pathways is the reaction leading to the facile enantioselective synthesis of (+)-longifolene (*1*)[66] and (+)-sativene (*2*).[67] The key step is an efficient and regioselective photochemical addition leading exclusively to the bonding of C−2 to C−4'.

9.29

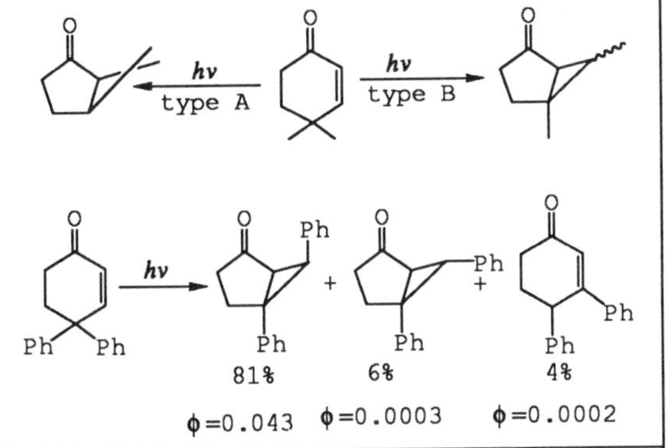

$\phi=0.043$ $\phi=0.0003$ $\phi=0.0002$

9.30

^1B $^1n,\pi*$ $^1\pi,\pi*$ Z

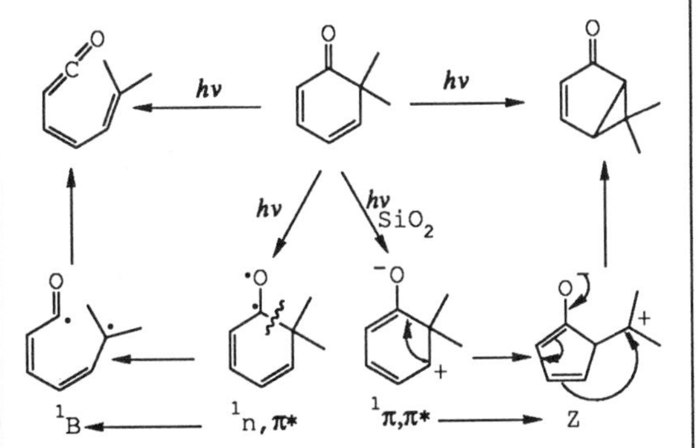

9.31

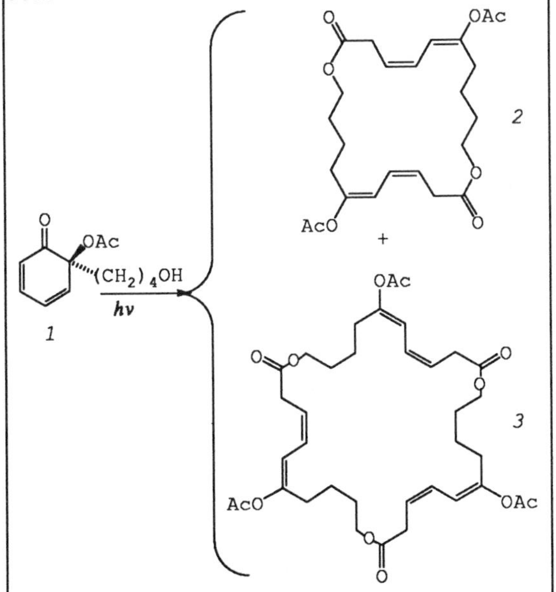

9.29. When bimolecular reactions (cycloaddition and dimerization) are suppressed by the use of dilute solutions, the products of intramolecular rearrangement, namely, derivatives of bicyclo[3.1.0]hexan-2-one, are obtained by irradiation of 4,4-disubstituted 2-cyclohexen-1-ones. In 4,4-dialkylcyclohexenones, the C atom of the cyclohexene ring migrates (type A);[68] in 4,4-diaryl derivatives, the aryl group migrates (type B). The irradiation of 4,4-diphenyl-2-cyclohexen-1-one at 300–340 nm ($n \rightarrow \pi*$ band) results predominantly in the formation of the *trans*- and *cis*-5,6-diphenylbicyclo-[3.1.0]hexan-2-ones in a ratio of 140 : 1 ($\Phi_1 : \Phi_2$). The preferential formation of the *trans* isomer is attributed to a concerted mechanism governed by orbital overlap.[69]

9.30. The rearrangement of linear conjugated 2,4- and cross-conjugated 2,5-cyclohexadien-1-ones has received considerable attention.[70] 2,4-cyclohexadien-1-ones undergo two different photoreactions from the S_1 state. In the S_1 (n,$\pi*$) state α cleavage takes place yielding a ketene via a singlet biradical, ^1B. The addition of a protic nucleophile results in the corresponding derivative of hexadienoic acid. In the S_1 ($\pi,\pi*$) state, e.g., by irradiation in trifluoroethanol or in a slurry of silica gel in cyclohexane, a [1,2] sigmatropic shift occurs via a dipolar (zwitterionic) intermediate, Z.[71]

9.31. Photolytic opening of the 2,4-cyclohexadien-1-one ring with the formation of a ketene has been used in the synthesis of macrocyclic mono-, di-, and trilactones.[72] Irradiation of *1* at 340 nm in the presence of the nonprotic nucleophile 1,4-diaza-bicyclo[2.2.2] octane, which activates the ketene toward weak nucleophiles, leads to a 30% yield of the 22-membered ring dilactone *2* and a 6% yield of the 33-membered trilactone *3*. This strategy was also used in the synthesis of the lichen macrolide (+)-aspicilin.[73]

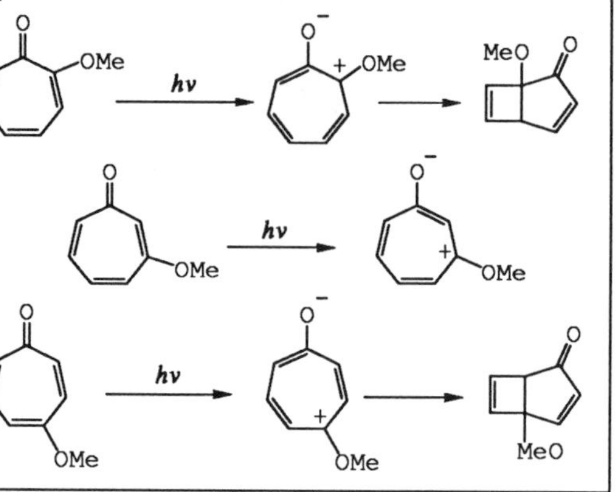

9.32. Unusual photochemical valence isomerizations have been observed for tropolone methyl ethers in which a charge transfer mechanism is assumed. β-tropolone methyl ether should not undergo photochemical valence isomerizations according to this mechanism; this is confirmed by experiment.[74,75] Similar isomerizations were observed from the irradiation of the alkaloid colchicine. Enantioselective intramolecular photoreactions of tropolone alkyl ethers in a chiral crystalline inclusion complex result in optically active 1-alkoxybicyclodienones.[76]

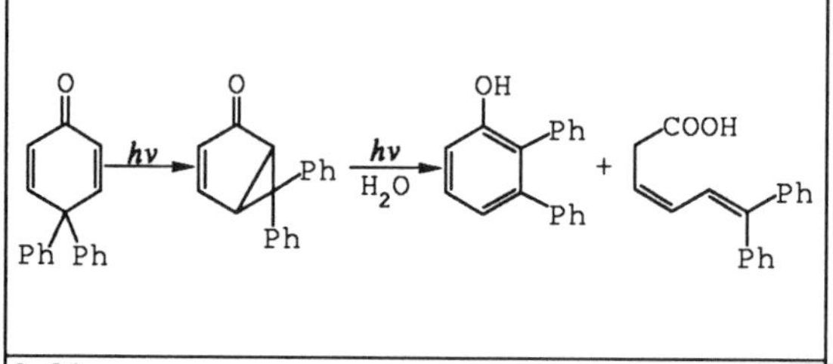

9.33. In the type A rearrangement of cross-conjugated 4,4-disubstituted 2,5-cyclohdexadienones, the first photochemical step occurs with a high quantum yield ($\Phi = 0.85$); the efficiency of the second step is substantially lower ($\Phi = 0.013$). The formation of bicyclo[3.1.0]hex-3-en-2-one is formally analogous to a 1,2-acyl shift and occurs from the T_1 (n,π^*) state. The unsaturated acid arises from the S_1 state of the bicyclic enone and the phenol is produced from its T_1 state.[77]

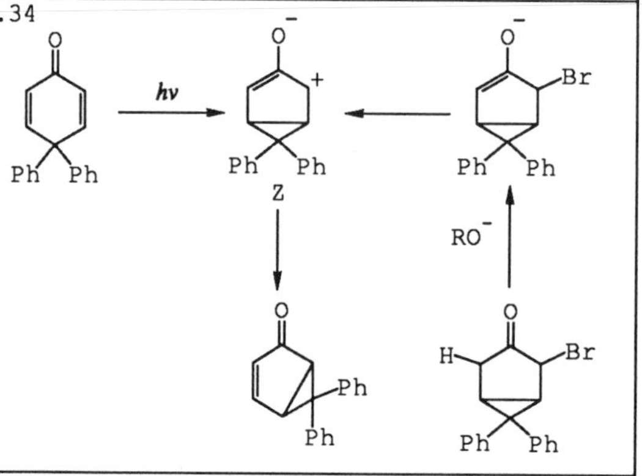

9.34. The mechanism proposed for the rearrangment of 2,5-cyclohexadienones involves: (1) excitation to the $^1(n,\pi^*)$ state; (2) intersystem crossing to the triplet state; (3) C(3)−C(5) rebonding with the formation of the triplet oxyallyl biradical; (4) subsequent electronic transition to the ground state of the dipolar species Z, a key intermediate; and (5) [1,4] sigmatropic shift to the final bicyclic ketone. Confirmation of the participation of the zwitterionic intermediate in the rearrangement is provided by the thermal preparation of the photoproduct by the reaction of the 2-bromoketone with base.[78,79]

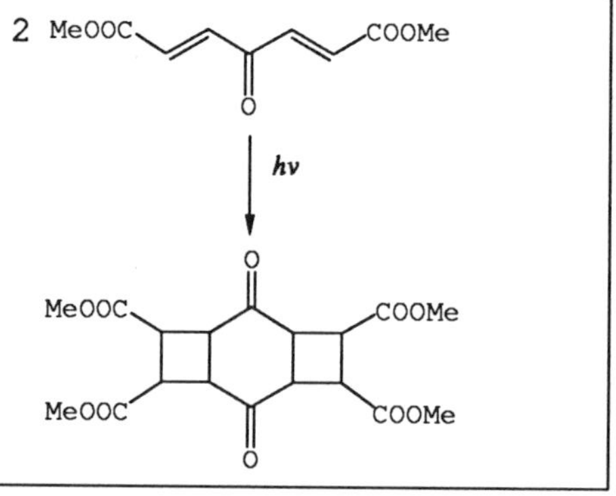

9.35. Interestingly, the irradiation of a linearly conjugated dienone gives a tricyclic dimer as a result of a double [2+2] cycloaddition reaction.[80]

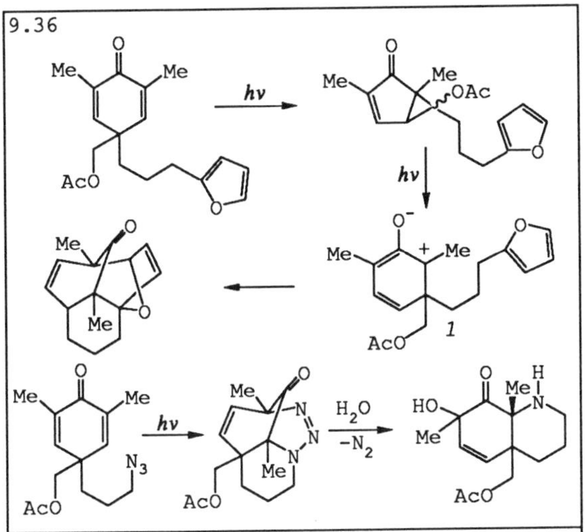

9.36. Intramolecular cycloadditions from the oxyallyl zwitterion *1*, produced by consecutive photorearrangements of 4,4-disubstituted 2,5-cyclohexadien-1-ones, are important in the preparation of carbocyclic and heterocyclic ring systems.[81]

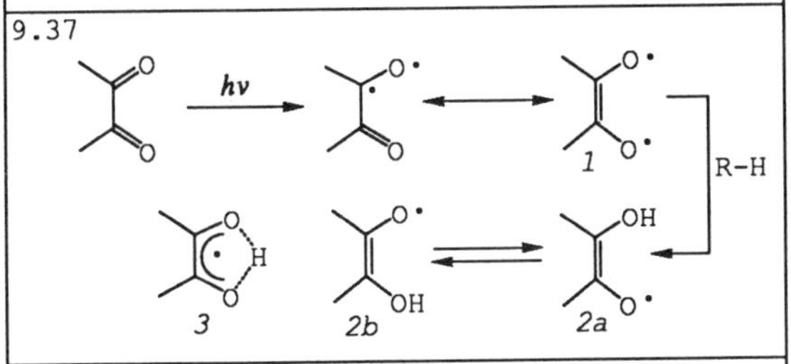

9.37. o-quinones, such as phenanthrenequinone (PQ) and 1,2-naphthoquinone (1,2-NQ) have properties similar to α-diketones, especially their VIS absorption (λ > 450 nm). Upon irradiation they undergo efficient intersystem crossing to produce long-lived triplets that undergo photoreactions typical of excited carbonyl groups. For example, biradical *1* reacts with suitable hydrogen atom donors to form the semidione radical *2* (confirmed by ESR and flash photolysis). Two tautomeric forms of the semidione radical (*2a* and *2b*) should be formed from unsymmetrical o-quinones. An ESR study has shown, however, that either the semidione radical has the symmetrical structure *3* or a rapid equilibrium exists between the tautomeric radicals.[82]

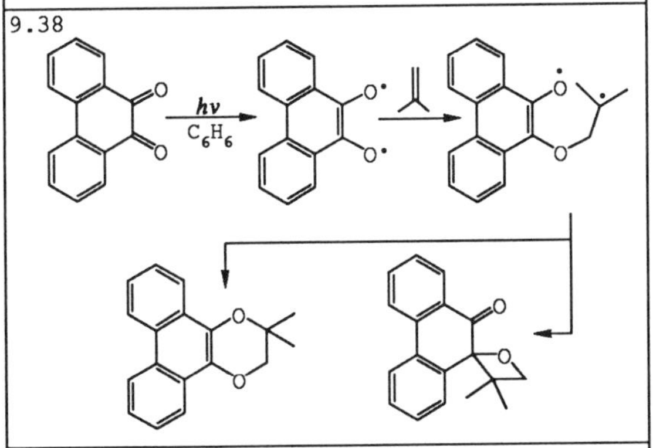

9.38. The reactive state of PQ is T_1 (n,π*), which should participate in either carbonyl or aromatic type excited state processes. The most common photoreactions of o-quinones in the presence of alkenes are [2+2] cycloadditions giving spirooxetanes and [4+2] cycloadditions yielding derivatives of dioxene (Schönberg adducts). Both products are formed in a two-step nonstereospecific process. Small amounts of reductive photoadducts demonstrate the H atom's abstracting ability of the T(n,π*) state.[83]

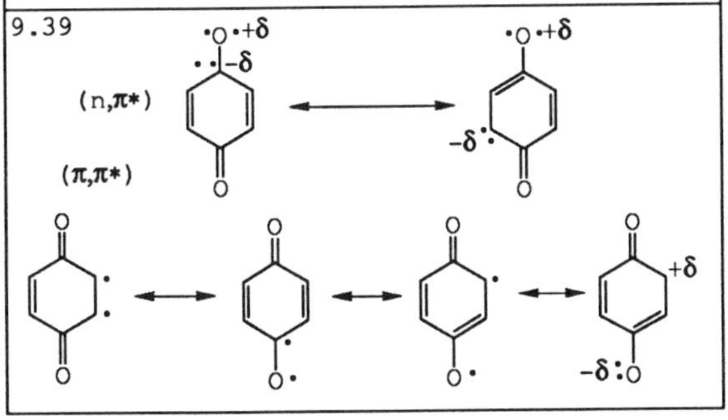

9.39. The electronic structure of *p*-quinones in the excited state is very similar to that of conjugated enones. For ease of visualization the photoreactions of *p*-quinones can be explained by valence structures of n,π* and π,π* states. While the n,π* states participate in the photoreduction and oxetane formation, the π,π* state is responsible for photodimerization and photocycloaddition reactions.

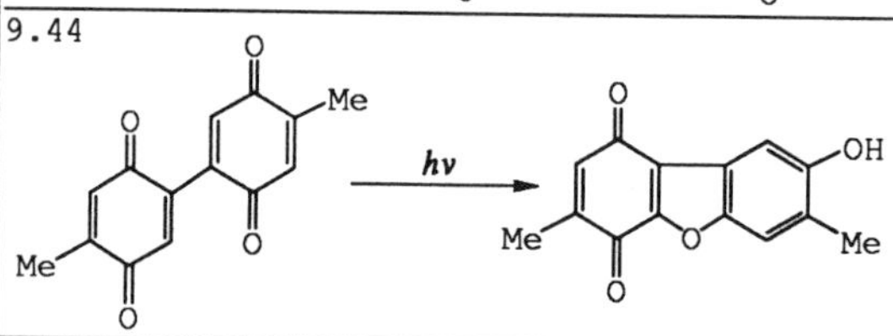

9.40. Most photoreactions of p-quinones can be demonstrated by 1,4-naphthoquinone (1,4-NQ). Remarkably, benzofuran yields both an oxetane and a cycloaddition product from the C=C bond of 1,4-NQ.[84]

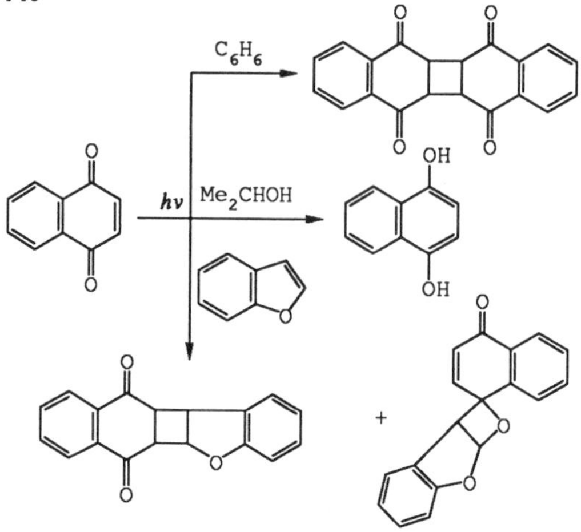

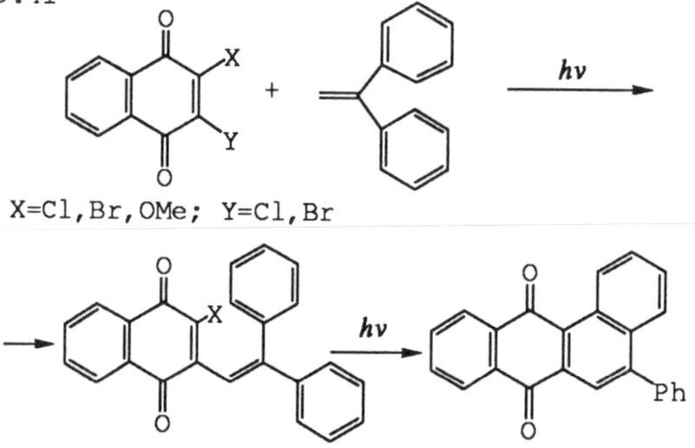

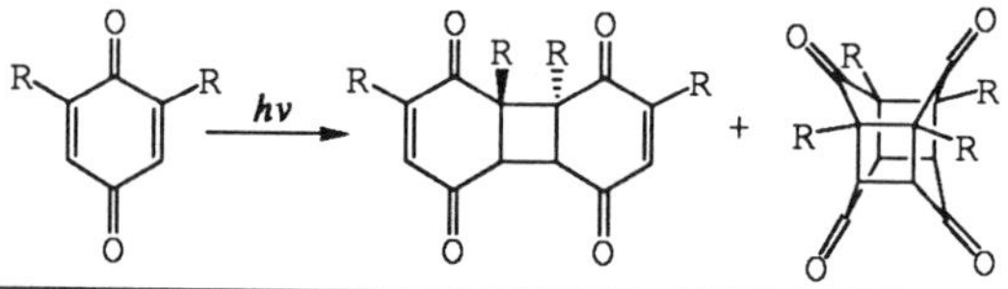

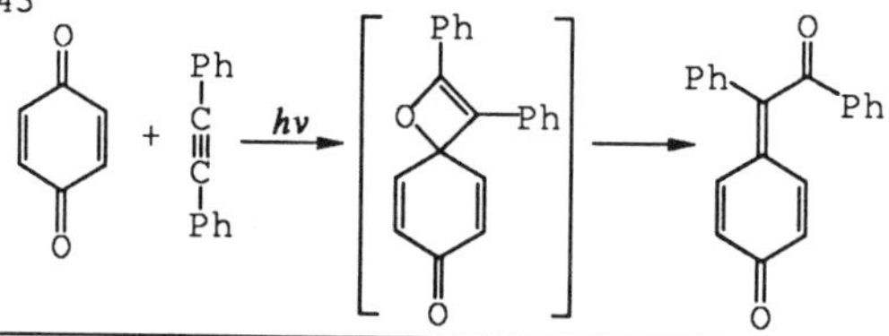

9.41. Photochemical reactions of halo-1,4-NQs with 1,1-diphenylethylenes involve substitution of a halogen atom by the ethylenic group (ethylene adduct), followed by photocyclization and the formation of 5-phenyl-benz[a]anthracene-7,12-diones. This reaction was shown to proceed via photoinduced electron transfer and provides a general method for the synthesis of polyaromatic compounds.[85]

9.42. The photodimerization reactions of p-quinones yield both *cis-anti-cis* and *cis-syn-cis* isomers of cyclobutanes. The latter are photochemically unstable and undergo further intramolecular [2+2] cycloaddition reactions to produce tetracyclic cage tetraketones. These compounds reconvert thermally to benzoquinones, but with the substituents in different positions.[86]

9.43. While alkenes form oxetanes and other products with 1,4-quinones, irradiation of alkynes yields unstable oxetenes, which spontaneously undergo rearrangement and subsequent ring opening. A spirocyclic oxetene is presumed to be an intermediate in the photocycloaddition of 1,4-benzoquinone to diphenylacetylene in benzene solution. The product of the rearrangement is a quinonemethide derivative.[87]

9.44. 2,3' - bis - p - benzoquinones undergo both thermal and photochemical cyclization with the formation of furan derivatives. The thermal cyclization proceeds with very low yields; the photoreaction occurs only in protic solvents (alcohols, aqueous THF, acetic acid) and has not been observed in benzene, toluene, carbon tetrachloride, etc.[88]

REFERENCES

(1)(a) For the original paper, see Hegedus, L. S., McGuire, M. A., Schultze, L. M., Yijun, C., and Anderson, O. P. *J. Am. Chem. Soc.* **1984,** *106*, 2680. For recent papers, see (b) Hegedus, L. S., Imwinkelried, R., Alarid-Sargent, M., Dvorak, D., and Satoh, Y. *J. Am. Chem. Soc.* **1990,** *112*, 1109. (c) Söderberg, B. C., Hegedus, L. S., and Sierra, M. A. *J. Am. Chem. Soc.* **1990,** *112*, 4364. For a review, see (d) Hegedus, L. S. In *New Aspects of Organic Chemistry, Proc. Int. Kyoto Conf. 4th. 1988,* Yoshida, Z., Shiba, T., and Ohshiro, Y., eds. Kodansha: Tokyo, Japan, 1989, p. 39.

(2) For a preparative procedure, see Hegedus, L. S., McGuire, M. A., and Schultze, L. M. *Org. Synth.* **1987,** *65*, 140.

(3) Kossanyi, J., Guiard, B., and Furth, B. *Bull. Soc. Chim. France.* **1974,** 305.

(4) Srinivasan, R. *J. Am. Chem. Soc.* **1960,** *82*, 775.

(5)(a) Meinwald, J. and Chapman, O. L. *J. Am. Chem. Soc.* **1968,** *90*, 3218. (b) Mazzocchi, P. H., Klinger, L., Edwards, M., Wilson, P., and Shook, D. *Tetrahedron Lett.* **1983,** *24*, 143.

(6) For the employment in the synthesis of tricyclo-ketones, see (a) Gleiter, R. and Kissler, B. *Tetrahedron Lett.* **1987,** *27*, 6157. (b) Nakazaki, M., Naemura, K., and Kondo, Y. *J. Org. Chem.* **1976,** *41*, 1229. (c) Sauers, R. R., Rousseau, A. D., and Byrne, B. *J. Am. Chem. Soc.* **1975,** *97*, 4947. (d) For the synthesis of azulenes, see Jost, P., Chaquin, P., and Kossanyi, J. *Tetrahedron Lett.* **1980,** *21*, 465. (e) For the synthesis of medium size ethers and acetals, see Carless, H. A. J., Beanland, J., and Mwesigye-Kibende, S. *Tetrahedron Lett.* **1987,** *28*, 5933.

(7) For a review, see (a) Carless, H. A. J. In *Synthetic Organic Photochemistry*, Horspool, W. M., ed. Plenum: New York, 1984, p. 425. (b) Jones G., II. *Org. Photochem.* **1981,** *5*, 1. (c) Meier, H. In *Methoden der organischen Chemie Houben-Weyl*, Thieme: Stuttgart, 1975, Vol. 4/5b, p. 838.

(8)(a) Fried, J., Kittisopukil, S., and Hallinan, E. A. *Tetrahedron Lett.* **1984,** *25*, 4329. (b) Carless, H. A. J. and Fekarurhobo, G. K. *J. Chem. Soc., Chem. Commun.* **1984,** 667.

(9)(a) Sauers, R. R. and Rousseau, A. D. *J. Am. Chem. Soc.* **1972,** *94*, 1776. (b) For the preparative procedure for the intramolecular oxetane formation from 5-acetylnorbornene, see Sauers, R. R., Schinski, W., and Sickles, B. *Org. Photochem. Synth.* **1971,** *1*, 76.

(10) Yang, N. C., Nussim, M., and Coulson, D. R. *Tetrahedron Lett.* **1965,** 1525.

(11) Jost, P., Chaquin, P., and Kossanyi, J. *Tetrahedron Lett.* **1980,** 465.

(12) Bichan, D. and Winnik, M. *Tetrahedron Lett.* **1974,** 3857.

(13) For reviews, see (a) Schuster, D. I. In *Rearrangements in Ground and Excited States.* deMayo, P., ed. Academic Press: New York, 1980, Vol. 3, p. 167. (b) Houk, K. N. *Chem. Ber.* **1976,** *76*, 1.

(14) Sadler, D. E., Wendler, J., Olbrich, G., and Schaffner, K. *J. Am. Chem. Soc.* **1984,** *106*, 2064, and references therein.

(15)(a) Baggiolini, E., Hamlow, H. P., and Schaffner, K. *J. Am. Chem. Soc.* **1970,** *92*, 4906. (b) For the stabilizing effect of the Me_3Si group on the decarbonylation of β,γ-enals, see Hwu, J. R. and Furth, P. S. *J. Am. Chem. Soc.* **1989,** *111*, 8834.

(16) See references 2a–2e in Ref. 14.

(17) See reference 2f in Ref. 14.

(18)(a) Ritterskamp, P., Demuth, M., and Schaffner, K. *J. Org. Chem.* **1984,** *49*, 1155. (b) Demuth, M., Chandrasenkhar, S., and Schaffner, K. *J. Am. Chem. Soc.* **1984,** *106*, 1092. (c) For a review, see Demuth, M. and Schaffner, K. *Angew. Chem., Int. Ed. Engl.* **1982,** *21*, 820. (d) Givens, R. S., Oettle, W. F., Coffin, R. L., and Carlson, R. G. *J. Am. Chem. Soc.* **1971,** *93*, 3957.

(19) For further examples of bicyclooctenone rearrangements employed in the synthesis of natural products, see (a) Uyehara, T., Osanai, K., Sugimoto, M., Suzuki, I., and Yamamoto, Y. *J. Am. Chem. Soc.* **1989,** *111*, 7264 (*gymnomitrol*), and references therein. (b) Demuth, M., Ritterskamp, P., and Schaffner, K. *Helv. Chim. Acta.* **1984,** *67*, 2023 (*coriolin*). (c) Yates, P. and Stevens, K. E. *Tetrahedron.* **1981,** *37*, 4401.

(20) Koppes, M. J. C. and Cerfontain, H. *Recl. Trav. Chim. Pays-Bas.* **1985,** *104*, 272.

(21)(a) van der Veen, R. H. and Cerfontain, H. *Tetrahedron.* **1985,** *41*, 585, and references therein. (b) For an example of a stereospecific version of this 1,3-acyl shift, see Coffin, R. L., Givens, R. S., and Carlson R. G. *J. Am. Chem. Soc.* **1974,** *96*, 7554.

(22)(a) Williams, J. R. and Ziffer, H. *Tetrahedron.* **1968,** *24*, 6725. (b) For the preparative procedure, see Williams, J. R. and Ziffer, H. *Org. Photochem. Synth.* **1976,** *2*, 65.

(23)(a) Williams, J. R. and Sarkisian, G. M. *Chem. Commun.* **1971**, 1564. (b) Williams, J. R. and Sarkisian, G. M. *J. Org. Chem.* **1980**, *45*, 5088.

(24) Padwa, A., Zhi, L., and Fryxell, G. E. *J. Org. Chem.* **1991**, *56*, 1077.

(25) Uyehara, T., Furuta, T., Kabasawa, Y., Yamada, J., Kato, T., and Yamamoto, Y. *J. Org. Chem.* **1989**, *53*, 3669 (ptilacaulin).

(26)(a) Uyehara, T., Kabasawa, Y., Kato, T., and Furuta, T. *Tetrahedron Lett.* **1985**, *26*, 2343. (b) Uyehara, T., Kabasawa, Y., Furuta, T., and Kato, T. *Bull. Chem. Soc. Japan.* **1986**, *59*, 539.

(27)(a) Paquette, L. A. and Meehan, G. V. *J. Org. Chem.* **1969**, *34*, 450. (b) Furutachi, N. and Hayashi, H. *Tetrahedron Lett.* **1972**, 1061.

(28) Turro, N. J., Lechtken, P., Schuster, G., Orell, J., and Steinmetz, H.-J. *J. Am. Chem. Soc.* **1974**, *96*, 1627.

(29) Kiefer, E. F. and Carlson, D. A. *Tetrahedron Lett.* **1967**, 1617.

(30) Schexnayder, M. A. and Engel, P. S. *J. Am. Chem. Soc.* **1975**, *97*, 4825.

(31)(a) Meier, H. In *Methoden der organischen Chemie Houben-Weyl*, Thieme: Stuttgart, 1975, Vol. 4/5b, p. 898. (b) Trecker, D. J. *Org. Photochem.* **1969**, *2*, 71.

(32) McCullough, J. J., Ohorodnyk, H., and Santry, D. P. *Chem. Commun.* **1969**, 570.

(33) Turro, N. J. In *Technique of Organic Chemistry.* Leermakers, P. A. and Weissberger, A., eds. Wiley–Interscience: New York, 1969, Vol. 14, p. 233.

(34)(a) Weedon, A. In *Synthetic Organic Photochemistry.* Horspool, W. M., ed. Plenum: New York, 1984. (b) Baldwin, S. W. *Org. Photochem.* **1981**, *5*, 123.

(35)(a) Noyori, R., Inoue, H., and Kato, M. *J. Am. Chem. Soc.* **1970**, *92*, 6699. (b) The quantum yield of the photochemical deconjugation of acyclic 3-alkyl conjugated ketones increases in the presence of a weak base. See Ricard, R., Sauvage, P., Wan, C. S. K., Weedon, A. C., and Wong, D. F. *J. Org. Chem.* **1986**, *51*, 62, and references therein. (c) For the acid-catalyzed photochemical deconjugation of 3-alkyl-2-cyclohexenones, see Rudolph, A. and Weedon, A. C. *J. Am. Chem. Soc.* **1989**, *111*, 8756.

(36)(a) Piva, O., Henin, F., Muzart, J., and Pete, J.-P. *Tetrahedron Lett.* **1987**, *28*, 4825. (b) Henin, F., Mortezaei, R., Muzart, J., and Pete, J.-P. *Tetrahedron Lett.* **1985**, *26*, 4945.

(37) For diastereoselective deconjugation of α,β-unsaturated esters, see (a) Mortezaei, R., Awandi, D., Henin, F., Muzart, J., and Pete, J.-P. *J. Am. Chem. Soc.* **1988**, *110*, 4824, and references therein. (b) Piva, O., Mortezaei, R., Henin, F., Muzart, J., and Pete, J.-P. *J. Am. Chem. Soc.* **1990**, *112*, 9263. (c) Piva, O. and Pete, J.-P. *Tetrahedron Lett.* **1990**, *31*, 5157.

(38)(a) Agosta, W. C., Caldwell, R. A., Jay, J., Johnston, L. J., Venepalli, B. R., Scaiano, J. C., Singh, M., and Wolff, S. *J. Am. Chem. Soc.* **1987**, *109*, 3050, and references therein. (b) Guérin, B. and Johnston, L. J. *J. Org. Chem.* **1989**, *54*, 3176, and references therein.

(39) Agosta, W. C. and Smith, A. B. *J. Am. Chem. Soc.* **1971**, *93*, 5513.

(40) For the use of the intramolecular H-abstraction in the synthesis of natural products, see Paquette, L. A., Pansegrau, P. D., Wiedeman, P. E., and Springer, J. P. *J. Org. Chem.* **1988**, *53*, 1461.

(41)(a) Noyori, R., Inoue, H., and Kato, M. *J. Am. Chem. Soc.* **1970**, *92*, 6699. (b) Wolff, S., Schreiber, W. L., Smith, A. B., III, and Agosta, W. C. *J. Am. Chem. Soc.* **1972**, *94*, 7797.

(42)(a) Ariel, S., Askari, S., Scheffer, J. R., and Trotter, J. *J. Org. Chem.* **1989**, *54*, 4324, and references therein. (b) However, *ab initio* studies conclude that, for acrolein, the β-carbon atom remains planar in the $^3(\pi,\pi^*)$ state.

(43)(a) Nobs, F., Burger, U., and Schaffner, K. *Helv. Chim. Acta.* **1977**, *60*, 1607. (b) For the transformation of hydroxychalcones to flavanoids by photoinduced single-electron transfer, see Pandey, G., Krishna, A., and Kumaraswamy, G. *Tetrahedron Lett.* **1987**, *28*, 4615.

(44)(a) Tobe, Y., Iseki, T., Kakiuchi, K., and Odaira, Y. *Tetrahedron Lett.* **1984**, *25*, 3895. (b) Mehta, G. and Subrahmanyam, D. *Tetrahedron Lett.* **1987**, *28*, 479. (c) For an intermolecular photoaddition of a C−H bond to a conjugated enone catalyzed by alkylcobaloximes, see Kijima, M., Miyamori, K., and Sato, T. *J. Org. Chem.* **1987**, *52*, 706.

(45)(a) For a review of synthetic applications of intramolecular enone-alkene photocycloadditions, see Crimmins, M. T. *Chem. Rev.* **1988**, *88*, 1453. (b) Becker, D. and Haddad, N. *Org. Photochem.* **1989**, *10*, 1. (c) For the preparative photocyclization of 1,5-hexadien-3-one, see Bond, F. T., Jones, H. L., and Scerbo, L. *Org. Photochem. Synth.* **1971**, *1*, 33.

(46) For examples of preparative photodimerizations of conjugated enones, see (a) Trecker, D. J., Gris-

wold, A. A., and Chapman, O. L. *Org. Photochem. Synth.* **1971**, *1*, 62 (isophorone dimers). (b) Cargill, R. L. *Org. Synth.* **1984**, *62*, 118 ([2+2] cycloaddition of ethylene to 3-methylcyclohexenone).

(47)(a) Wagner, P. J. and Buchech, D. J. *J. Am. Chem. Soc.* **1969**, *91*, 5090. (b) Ruhlen, J. L. and Leermakers, P. A. *J. Am. Chem. Soc.* **1967**, *89*, 4944. (c) Eaton, P. E. and Hurt, W. S. *J. Am. Chem. Soc.* **1966**, *88*, 5038.

(48)(a) Lam, E. Y. Y., Valentine, E., and Hammond, G. S. *J. Am. Chem. Soc.* **1967**, *89*, 3482. (b) For a recent contribution to the photochemistry of 4-pyranone, see West, F. G., Fischer, P. V., and Willoughby, C. A. *J. Org. Chem.* **1990**, *55*, 5936.

(49)(a) 2-cyclooctenone: Eaton, P. E. and Lin, K. *J. Am. Chem. Soc.* **1964**, *86*, 2087. (b) 2-cycloheptenone: Bonneau, R., de Violet, P. F., and Joussot-Dubien, J. *Nouveau J. Chem.* **1977**, *1*, 31, and references therein.

(50) See Chapter 14, Ref. 5(b), p. 312 and Chapter 15, Ref. 31(a)–(c).

(51)(a) Hammond, G. S., Stout, C. A., and Lamola, A. A. *J. Am. Chem. Soc.* **1964**, *86*, 3103. (b) Krauch, C. H., Farid, S., and Schenck, G. O. *Chem. Ber.* **1966**, *99*, 625. (c) Schenck, G. O., von Wilucki, I., and Krauch, C. H. *Chem. Ber.* **1962**, *95*, 1409.

(52)(a) For a preparative example of cyclomerization of 7,7'-pentamethylene-dioxycoumarin, see Loos, H., Leenders, L., and DeSchryver, F. C. *Org. Photochem. Synth.* **1976**, *2*, 41. (b) For the preparation of photodimerization of *trans*-cinnamic acid to α-truxillic acid, see Farnum, D. G. and Mostashari, A. J. *Org. Photochem. Synth.* **1971**, *1*, 103.

(53)(a) Ito, Y., Kajita, T., Kunimoto, K., and Matsuura, T. *J. Org. Chem.* **1989**, *54*, 587, and references therein. (b) For Lewis acid catalysis of coumarin photodimerization, see Lewis, F. D. and Barancyk, S. V. *J. Am. Chem. Soc.* **1989**, *111*, 8653.

(54) For reviews, see (a) Kossanyi, J. *Pure Appl. Chem.* **1979**, *51*, 181. (b) Lenz, G. *Rev. Chem. Intermed.* **1981**, *4*, 369. (c) Baldwin, S. W. *Org. Photochem.* **1981**, *5*, 123. (d) Oppolzer, W. *Acc. Chem. Res.* **1982**, *15*, 135. (e) Margaretha, P. *Top. Curr. Chem.* **1982**, *103*, 1. (f) Weedon, A. C. In *Synthetic Organic Photochemistry*. Horspool, W. M., ed. Plenum: New York, 1984, p. 61. (g) Wong, H. N. C., Lau, K.-L., and Tam, K.-F. *Top. Curr. Chem.* **1986**, *133*, 83. (h) Wender, P. A. In *Photochemistry in Organic Synthesis.* Coyle, J. D., ed. The Royal Society of Chemistry; London, 1986, p. 163.

(i) Horspool, W. M. In *Photochemistry in Organic Synthesis.* Coyle, J. D., ed. The Royal Society of Chemistry; London, 1986, p. 210. (j) Ninomiya, I. and Naito, T. *Photochemical Synthesis.* Academic Press: San Diego, 1989, p. 58.

(55)(a) Mehta, G., Padma, S., Osawa, E., Barbiric, D. A., and Mochizuki, Y. *Tetrahedron Lett.* **1987**, *28*, 1295. (b) Marchand, A. P., Annapurna, P., Reddy, S. P., Watson, W. H., and Nagl, A. *J. Org. Chem.* **1989**, *54*, 187.

(56)(a) Paquette, L. A., Lin, H.-S., Gun, B. P., and Coghlan, M. J. *J. Am. Chem. Soc.* **1988**, *110*, 5818. (b) Winkler, J. D., Muller, C. L., and Scott, R. D. *J. Am. Chem. Soc.* **1988**, *110*, 4831. (c) Crimmins, M. T. and Gould, L. D. *J. Am. Chem. Soc.* **1987**, *109*, 6199. (d) Bernassan, J.-M., Bouillot, A., Fétizon, M., Hanna, I., and Maia, E. R. *J. Org. Chem.* **1987**, *52*, 1993, and references therein.

(57)(a) deMayo, P., Nicholson, A. A., and Tchir, M. F. *Can. J. Chem.* **1969**, *47*, 711. (b) For the photocycloaddition of alkynes, see Smith, A. B., III and Boschelli, D. *J. Org. Chem.* **1983**, *48*, 1217. (c) Wang, T.-Z. and Paquette, L. A. *J. Org. Chem.* **1986**, *51*, 5232. (d) For the cycloalkene ring-size-dependent reversal of regioselectivity in photocycloadditions of cyclohexenones, see Lange, G. L., Organ, M. G., and Lee, M. *Tetrahedron Lett.* **1990**, *31*, 4689.

(58) Pirrung, M. C. and Webster, N. J. G. *J. Org. Chem.* **1987**, *52*, 3603, and references therein.

(59) For some cyclohexenones a twisted $^3(\pi, \pi^*)$ state is suggested to be the reactive species. See (a) Schuster, D. I., Bonneau, R., Dunn, D. A., Rao, J. M., and Joussiet-Dubieu, J. *J. Am. Chem. Soc.* **1984**, *106*, 2706. (b) Mintas, M. and Schuster, D. I. *J. Am. Chem. Soc.* **1988**, *110*, 2305.

(60)(a) Corey, E. J., Bass, J. D., LaMathieu, R., and Mitra, R. B. *J. Am. Chem. Soc.* **1964**, *86*, 5570. (b) Schuster, D. I., Heibel, G. E., Brown, P. B., Turro, N. J., and Kumar, C. V. *J. Am. Chem. Soc.* **1988**, *110*, 8261. (c) For mechanistic alternatives, see Schuster, D. I., Brown, P. B., Capponi, L. J., Rhodes, C. A., Scaiano, J. C., Tucker, P. C., and Weir, D. *J. Am. Chem. Soc.* **1987**, *109*, 2533.

(61)(a) Tamura, Y., Ishibashi, H., Hirai, M., Kita, Y., and Ikeda, M. *J. Org. Chem.* **1975**, *40*, 2702. (b) Schell, F. M., Cook, P. M., Hawkinson, S. W., Cassidy, R. E., and Thiessen, W. E. *J. Org. Chem.* **1979**, *44*, 1380. (c) For the preparative photocyclization of 1.5-hexadien-3-one, see Bond, F. T., Jones, H. L., and Scerbo, S. *Org. Photochem. Synth.* **1971**, *1*, 33.

(62)(a) Lange, G. L., Decicco, C., and Lee, M. *Tetrahedron Lett.* **1987**, *28*, 2833. (b) For the utilization of these cycloaddition reactions in a metathetical approach to medium-ring compounds, see Wender, P. A. and Manly, C. J. *J. Am. Chem. Soc.* **1990**, *112*, 8579, and references therein.

(63) For a review, see Ref. 54(i), p. 70.

(64) Challand, B. D., Hikino, H., Kornis, G., Lange, G., and deMayo, P. *J. Org. Chem.* **1969**, *34*, 794.

(65) For additional applications of the deMayo reaction in natural product syntheses, see (a) Nakane, M., Gollman, H., Hutchinson, L. R., and Knutson, P. L. *J. Org. Chem.* **1980**, *45*, 2536 (swerside). (b) Baldwin, S. W. and Crimmins, M. T. *J. Am. Chem. Soc.* **1982**, *104*, 1132 (sarracenin). (c) Winkler, J. D. and Hershberger, P. M. *J. Am. Chem. Soc.* **1989**, *111*, 4852 [(−)-perhydromistrionicotoxin].

(66) Oppolzer, W. and Godel, T. *J. Am. Chem. Soc.* **1978**, *100*, 2583.

(67) Oppolzer, W. *Pure Appl. Chem.* **1981**, *53*, 1181.

(68)(a) Zimmerman, H. E. *Pure Appl. Chem.* **1964**, *9*, 493. (b) Chapman, O. L. and Weiss, D. S. *Org. Photochem.* **1973**, *3*, 197. (c) Schuster, D. I., Greenberg, M. M., Nuñez, I. M., and Tucker, P. C. *J. Org. Chem.* **1983**, *48*, 2615, and references therein.

(69)(a) Zimmerman, H. E. and Hancock, K. G. *J. Am. Chem. Soc.* **1968**, *90*, 3749. (b) Zimmerman, H. E. *Tetrahedron.* **1974**, *30*, 1617. (c) For the most recent study, see Zimmerman, H. E. and Clair, D. S. *J. Org. Chem.* **1989**, *54*, 2125, and references therein.

(70)(a) Quinkert, G., Kleiner, E., Freitag, B.-J., Glenneberg, J., Billhardt, U.-M., Cech, F., Schmieder, K. R., Schudok, C., Steinmetzer, H.-C., Bats, J. W., Zimmermann, G., Dürner, G., Rehm, D., and Paulus, E. F. *Helv. Chim. Acta.* **1986**, *69*, 469. (b) Quinkert, G. *Angew. Chem., Int. Ed. Engl.* **1975**, *14*, 790. (c) Quinkert, G. *Pure Appl. Chem.* **1973**, *33*, 285.

(71) For the mechanism of this photocleavage, see references 22–29 in Ref. 68(a). (b) For diene ketenes from σ-quinol acetates, see Ref. 70(a).

(72) Quinkert, G., Fischer, G., Billhardt, U.-M., Glenneberg, J., Hertz, U., Dürner, G., Paulus, E. G., and Bats, J. W. *Angew. Chem., Int. Ed. Engl.* **1984**, *23*, 440.

(73)(a) Quinkert, G., Heim, N., Glenneberg, J., Billhardt, U.-M., Autze, V., Bats, J. W., and Dürner, G. *Angew. Chem., Int. Ed. Engl.* **1987**, *26*, 362. (b) For the synthesis of the saffron pigment *dimethylcrocetin*, see Quinkert, G., Schmieder, K. R., Dürner, G., Hache, K., Stegk, A., and Barton, D. H. R. *Chem. Ber.* **1977**, *110*, 3582.

(74) Pasto, D. J. *Org. Photochem.* **1967**, *1*, 155.

(75)(a) For photodimerization products of tropone, see Feldman, K. S., Come, J. H., Kosmider, B. J., Smith, P. M., Rotella, D. P., and Wu, M.-J., *J. Org. Chem.* **1989**, *54*, 592, reference 18b. (b) For an *intra*molecular photocycloaddition of an alkynyl tropone, see Feldman, K. S., Come, J. H., Fegley, G. J., Smith, B. D., and Parvez, M. *Tetrahedron Lett.* **1987**, *28*, 607. (c) For an *intra*molecular photocycloaddition of an alkenyl tropone, see Feldman, K. S., Wu, M.-J., and Rotella, D. P. *J. Am. Chem. Soc.* **1990**, *112*, 8490. (d) For the role of zwitterionic intermediates in the photochemistry of 4-pyranones, see West, F. G., Fischer, P. V., and Willoughby, C. A. *J. Org. Chem.* **1990**, *55*, 5936.

(76)(a) Kaftory, M., Yagi, M., Tanaka, K., and Toda, F. *J. Org. Chem.* **1988**, *53*, 4391. (b) For eight-membered ring dienones and trienones, see Toda, F., Tanaka, K., and Oda, M. *Tetrahedron Lett.* **1988**, *29*, 653.

(77) For reviews, see (a) Schaffner, K. and Demuth, M. In *Rearrangements in Ground and Excited States*, deMayo, P., ed. Academic Press: New York, 1980, Vol. 3, p. 281. (b) Schuster, D. I. *Acc. Chem. Res.* **1978**, *11*, 65. For the preparation of lumisantonin, see Schuster, D. I. and Fabian, A. C. *Org. Photochem. Synth.* **1971**, *1*, 65.

(78)(a) Zimmerman, H. E., Crumrine, D. S., Döpp, D., and Huyfer, P. S. *J. Am. Chem. Soc.* **1969**, *91*, 434. (b) Pirrung, M. C. and Nunn, D. S. *Tetrahedron Lett.* **1988**, *29*, 163.

(79)(a) Zimmerman, H. E. and Lynch, D. C. *J. Am. Chem. Soc.* **1985**, *107*, 7745. (b) Schultz, A. G., Myong, S. O., and Puig, S. *Tetrahedron Lett.* **1984**, *25*, 1011. (c) Schuster, D. I. and Liu, K.-C. *Tetrahedron.* **1981**, *37*, 3329.

(80)(a) Rowland, N. E., Sondheimer, F., Bullock, G. A., LeGoff, E., and Grohmann, K. *Tetrahedron Lett.* **1970**, 4769. (b) Corse, J., Finkle, B. J., and Lundin, R. E. *Tetrahedron Lett.* **1961**, 1.

(81) Schultz, A. G., Geiss, W., and Kullnig, R. K. *J. Org. Chem.* **1989**, *54*, 3158, and references therein.

(82) For a review of the photochemistry of (a) α-diketones: Rubin, M. B. *Top. Curr. Chem.* **1985**, *129*, 1. (b) quinones: Bruce, J. M. In *The Chemistry of Quinoid Compounds*, Patai, S., ed. Wiley: New York, 1974, Vol. 1, Part 1, Chapter 9.

(c) Maruyama, K. In *The Chemistry of Quinoid Compounds*, Patai, S., ed. Wiley: New York, 1974, Vol. 2, Part 1, Chapter 13. (d) For a reinvestigation of the photochemistry of 2-alkoxy-1,4-NQ, see Kraus, G. A., Shi, J., and Reynolds, D. *Synth. Commun.* **1990,** *20*, 1837.

(83)(a) Becker, R. S., Natarajan, L. U., Lenoble, C., and Harvey, R. G. *J. Am. Chem. Soc.* **1988,** *110*, 7163, and references therein. (b) Ci, X., daSilva, R. S., Nicodeti, D., and Whitten, E. G. *J. Am. Chem. Soc.* **1989,** *111*, 1337.

(84)(a) Rennert, J., Japar, S., and Guttman, M. *J. Am. Chem. Soc.* **1968,** *90*, 464. (b) Krauch, C. H. and Farid, S. *Tetrahedron Lett.* **1966,** 4783. (c) Schulte-Frohlinde, D. and Sonntag, C. V. *Z. Phys. Chem.* **1965,** *44*, 314. (d) For photochemical reactions of isoprenoid 1,4-quinones, see Iwamoto, H. *J. Org. Chem.* **1988,** *53*, 1507, and references therein.

(85) Maruyama, K. and Imahori, H. *J. Org. Chem.* **1989,** *54*, 2692.

(86)(a) Cookson, R. C., Cox, D. A., and Hudec, J. *J. Chem. Soc.* **1962,** 1717. (b) Bryce-Smith, D. and Gilbert, A. *J. Chem. Soc.* **1964,** 2428.

(87) Bryce-Smith, D., Fray, G. I., and Gilbert, A. *Tetrahedron Lett.* **1964,** 2137.

(88) Shand, A. J. and Thomson, R. H. *Tetrahedron.* **1963,** *19*, 1919.

Chapter 10
PHOTOCHEMISTRY OF CARBOXYLIC ACID DERIVATIVES

PHOTOCHEMISTRY OF CARBOXYLIC ACID DERIVATIVES

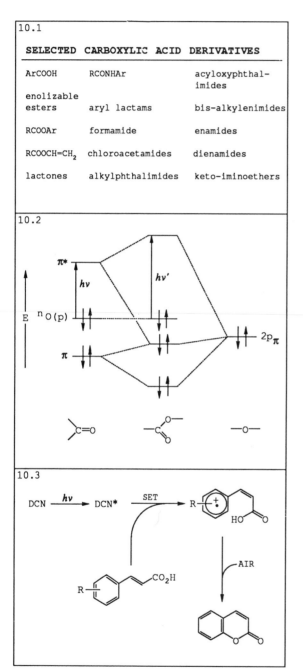

10.1

SELECTED CARBOXYLIC ACID DERIVATIVES

ArCOOH	RCONHAr	acyloxyphthal-imides
enolizable esters	aryl lactams	bis-alkylenimides
RCOOAr	formamide	enamides
RCOOCH=CH₂	chloroacetamides	dienamides
lactones	alkylphthalimides	keto-iminoethers

10.1. Functional derivatives of carboxylic acids in which the carbonyl group is conjugated with a heteroatom represent a diverse class of compounds that differ in their chemical and physical properties, including UV-VIS spectroscopic properties. In addition to the perturbed carbonyl chromophore, other chromophores may be present that can also affect the course of primary photochemical processes. In the following sections the photochemical reactivity of selected representatives of this group of compounds will be discussed.

10.2. The interaction of a fully occupied $2p_\pi$ atomic orbital (unshared electron pair) of the oxygen atom with the π molecular orbital of the carbonyl group causes an increase in the energy of the π, particularly the π^* MOs. The n,π^* excitation energy in this group of compounds is higher than that of simple carbonyl compounds. Acids, esters, lactones, and lactams exhibit absorption maxima of low intensity ($\epsilon <$ 100 Lmol^{-1} cm^{-1}) in the short wavelength region ($\lambda_{max} < 220$ nm) and undergo a hypsochromic shift in polar solvents (see 2.36).[1] Thiolesters and N-alkylated imides absorb at longer wavelengths.[2] Aliphatic carboxylic acids, esters, and amides possess an S_1 state of n,π^* and a T_1 state of π,π^* configuration, while their aromatic analogs have a π,π^* configuration in both the S_1 and T_1 states.[3]

10.3. The irradiation of substituted cinnamic acids in the presence of an electron-transfer sensitizer, such as 1,4-dicyanonaphthalene (DCN), produces an aromatic radical-cation by a single-electron-transfer (SET) process. Subsequent intramolecular nucleophilic attack by the OH group produces, after oxidation by air dissolved in the solvent, a coumarin derivative.[4,5]

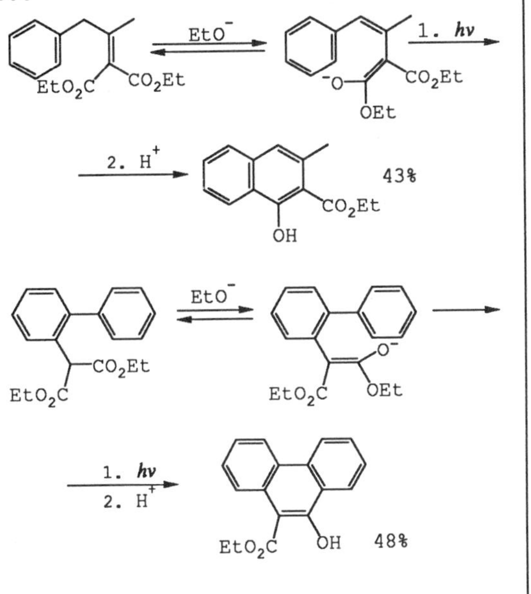

10.4. Enolizable esters, such as sodium enolates of arylalkylidenemalonates, arylmalonates, or the corresponding cyanoacetates, which include a conjugated *cis*-triene system, undergo a facile photochemical conrotatory ring closure (intramolecular *photoacylation*). Upon protonation of the resulting cyclohexadienolate a satisfactory yield of the final product is obtained. Unlike photocyclization-oxidation, this method is not limited to the synthesis of phenanthrene or higher aromatics (see 6.31 through 6.34), but is also used for the synthesis of naphthalene derivatives.[6]

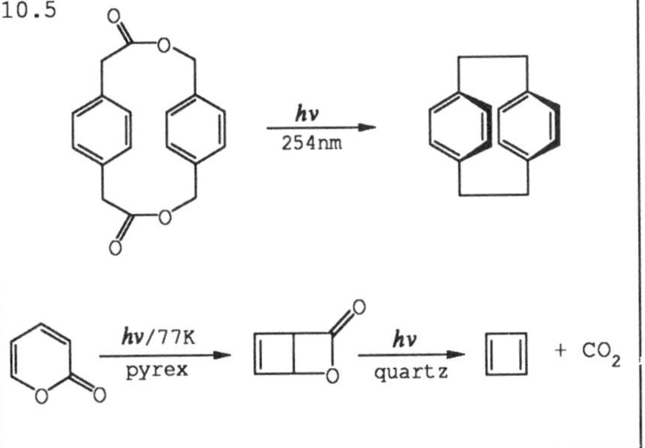

10.5. One of the most important photochemical reactions of esters and lactones is the photoextrusion of CO_2 (*photodecarboxylation*), which is particularly efficient in the presence of radical-stabilizing chromophores (phenyl, furanyl, etc.) in an appropriate position in the ester.[7] The reaction is initiated from the S_1 state with homolytic cleavage of the C—O bond and the simultaneous or stepwise formation of a radical pair in a solvent cage. Recombination of the radicals occurs preferentially in the cage (in the photolysis of acyclic esters no statistical ratio of products is formed).[8] The photoextrusion of CO_2 in cyclic di- and tetraesters apparently proceeds stepwise to yield [2.2]*para*cyclophane.[9] This method was also used for the matrix isolation of cyclobutadiene.[10]

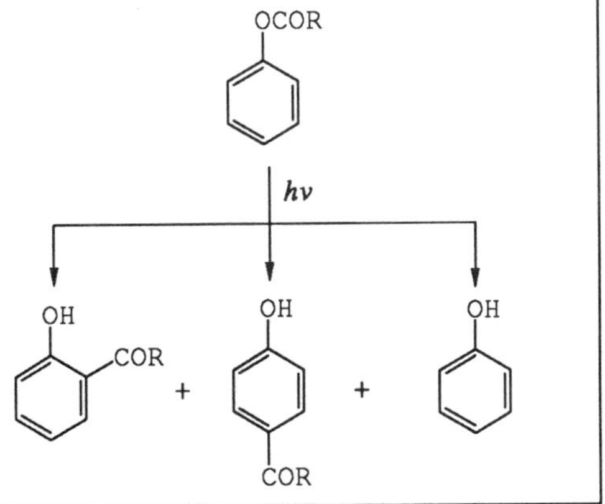

10.6. A common photoprocess of aromatic compounds is *lateral-nuclear rearrangement*. The most extensively studied reaction of this type is the *photo-Fries rearrangement* of O-aryl esters which, by a formal 1,3- or 1,5-acyl shift, produces a mixture of *ortho* and *para* acylphenols and a minor amount of phenol.[11] The photo-Fries rearrangement depends on the light absorption by the aryloxy moiety and occurs intramolecularly from the S_1 (π,π^*) state. The quantum yield increases with decreasing solvent viscosity.[12]

10.7

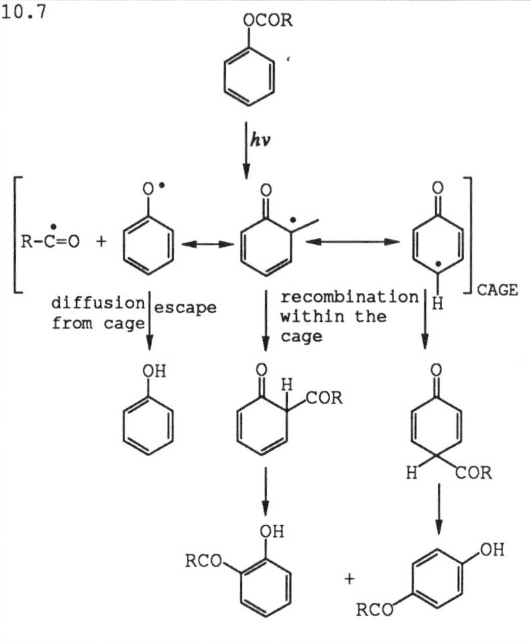

10.8

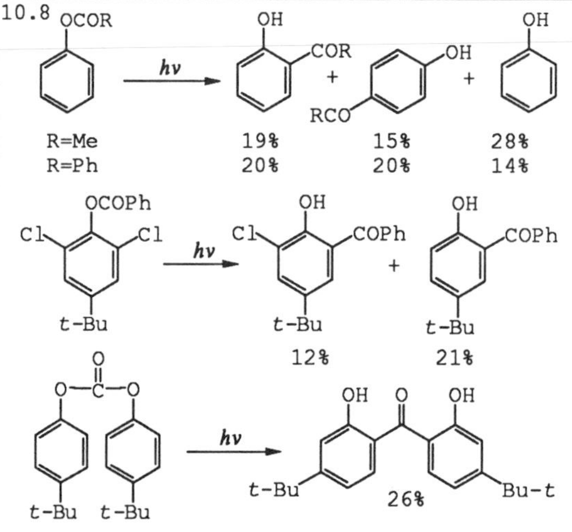

10.9

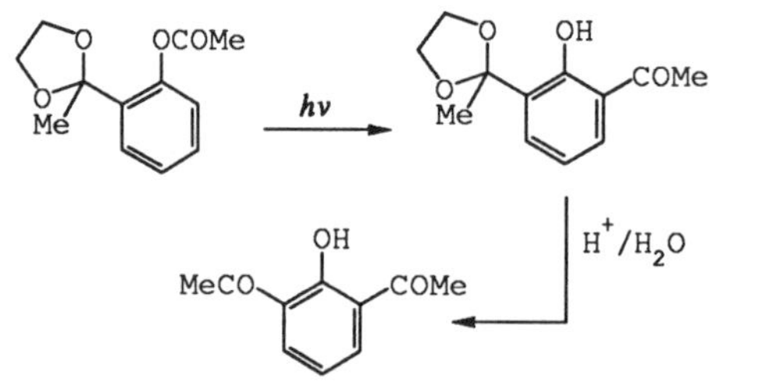

10.7. This generally accepted mechanism assumes a primary cleavage of the C—O bond with the formation of a singlet radical pair captured in a solvent cage.[13] Recombination of the radicals yields two products resulting from *ortho* and *para* rearrangement (in the gas phase no rearrangement products are formed).[14] Flash photolysis experiments have confirmed the existence of a phenoxy radical and a cyclohexadienone intermediate.[15] The diffusion of phenoxy radicals out of the cage leads to the formation of phenol by abstraction of a hydrogen atom from the solvent.

10.8. Polar solvents (t-butyl alcohol) favor the rearrangement, whereas nonpolar solvents (cyclohexane) support the formation of a phenol.[16] At high temperature the *ortho* isomer usually predominates, while at low temperature more *para* isomer is formed.[17] Electron-donating groups usually increase (acceptors decrease) the quantum yield.[18] A double rearrangement has been reported with diaryl carbonates[19] and terephthalates.[20] The photo-Fries rearrangement is synthetically significant in spite of its low yields.[21] It does not usually exhibit elimination or rearrangement of the alkyl (particularly t-butyl) groups as in its thermal counterpart. However, the displacement of a chlorine atom or a methoxy group by an acyl group or a hydrogen atom has been observed.[22,23]

10.9. Another advantage of the photochemical version of the Fries rearrangement reaction is its ability to be used with esters that are either inert[24,25] or that decompose in the presence of Lewis acids, as illustrated by the rearrangement of aryl esters containing an acetal group.[26] The rearrangement of aryl esters of unsaturated acids, e.g., cinnamic acid, leads to chalcones that cyclize to flavanones[27] (usually 4-chromanones) in low yield. Irradiation in the presence of a base leads directly to the cyclized product in a high yield.[28]

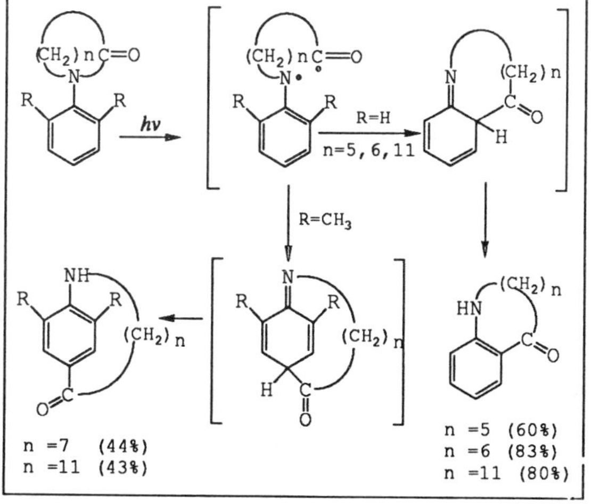

10.10. The rearrangement of vinyl esters of carboxylic acids, which upon irradiation form β-diketones (1,3-acyl shift), is analogous to the photo-Fries rearrangement reaction.[29] Sufficient enolization of the resulting diketone is a necessary condition for the success of this reaction since the enol acts as an efficient internal filter of radiation and as a quencher. This prevents subsequent photochemical reactions, such as decarbonylation, but at the same time substantially limits the degree of conversion. This rearrangement, which also occurs with dienol acetates (1,5-acyl shift), has been utilized in the transformation of natural products, particularly steroids.[29(b)]

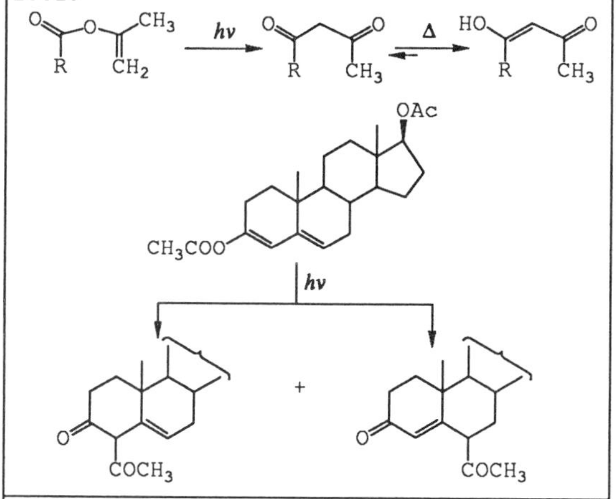

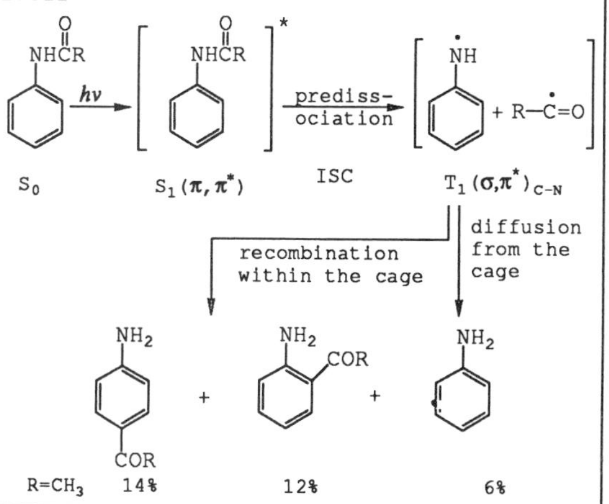

10.11. N-arylamides undergo a photo-Fries type rearrangement both in solution and in the solid state[30] by a mechanism very similar to that of esters. The reaction arises from the lowest singlet state and the experimental results are consistent with a mechanism involving predissociation by ISC to form a radical pair in a solvent cage.[31] Unlike aryl esters, the yields are significantly reduced in polar solvents. The ratio of the two isomeric products is close to 1.[32] The rearrangement of acylcarbazoles has been reported;[33] the reaction with aroylcarbazoles is wavelength dependent.[34]

10.12. Irradiation of N-phenyl- and O-arylcarbamates produces a mixture of *ortho* and *para* aminobenzoates and hydroxybenzamides,[35,36] respectively, while N-phenyllactams yield either macrocyclic aminoketones[37] or paracyclophane derivatives, when the hydrogen atoms in the *ortho* positions are substituted, e.g., by methyl groups.[38,39] Lateral-nuclear rearrangements involving other classes of compounds [allyl-aryl ethers (*photo-Claisen reaction*),[40] diaryl ethers,[41] and N-benzylanilines[42]] have limited synthetic value.

10.13

$$R—CH=CH_2 + HCONH_2 \xrightarrow{h\nu} RCH_2CH_2CONH_2$$

Alkene	Product	Yield[a]
		(%)
1-hexene	heptanamide	50
1-octene	nonanamide	61
1-decene	undecanamide	67
cyclohexene	cyclohexane-carboxamide	65
11-undecen-oic acid	dodecanediamide	90

[a]based on conversion of the alkene

10.14

10.15

10.13. The acetone photosensitized addition of formamide to terminal alkenes produces carboxamides in good yield.[43] The reaction utilizes a radical pathway (in the photolysis of formamide at 77K, the radical, $\cdot CONH_2$, was detected by ESR) and as a result, the products are usually those formed by anti-Markovnikov addition. Telomers and adducts of the photosensitizers are byproducts of this reaction. Internal alkenes yield a mixture of both possible 1:1 adducts. However, a triplet excitation energy transfer, followed by E-Z isomerization, can also occur. Aromatic hydrocarbons produce the corresponding carboxamides, however, with lower yields than alkenes.[44]

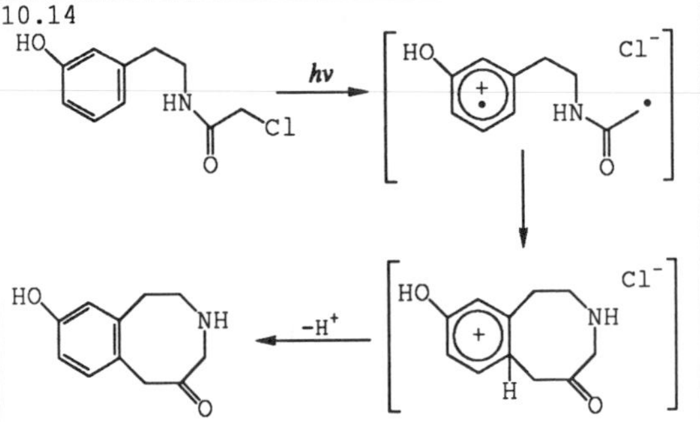

10.14. The photocyclization of ω-arylalkylamides of chloroacetic acid, which is initiated in protic solvents by a facile photochemically induced electron transfer, is synthetically interesting.[45] Usually the sequence starts with a chloroacetamide of a biogenic amine or its derivative, which undergoes a different cyclization reaction depending on the position and quality of the substituent in the phenyl ring. In the example shown, the product is a seven-membered ring lactam (benzazepinone).[46] In aprotic solvents, photocyclization via a biradical and reductive elimination of a chlorine atom are the major routes.[47]

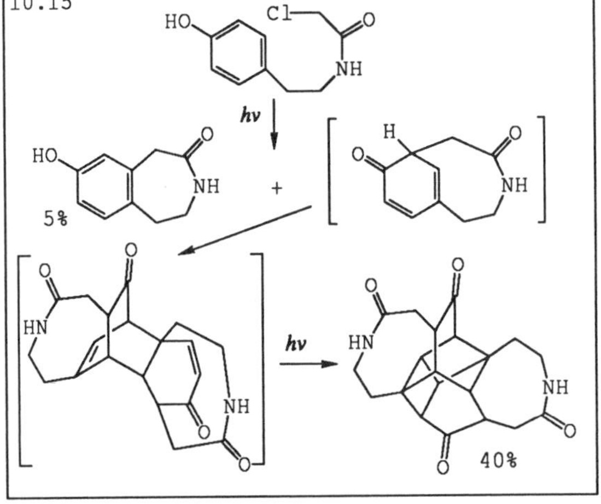

10.15. With a hydroxy group in the *para* position the yield of the corresponding benzazepinone is low. A cage compound from a thermal Diels-Alder "dimerization" of the keto-tautomer of the initially formed metacyclophane, followed by a photochemical [2+2] cycloaddition reaction, is the major product.[48] The alkyl chain between the aromatic ring and the amidic nitrogen can be extended. When it is composed of five methylene groups, a ten-membered lactam ring is formed, in addition to a product with a metacyclophane structure and other byproducts.[49]

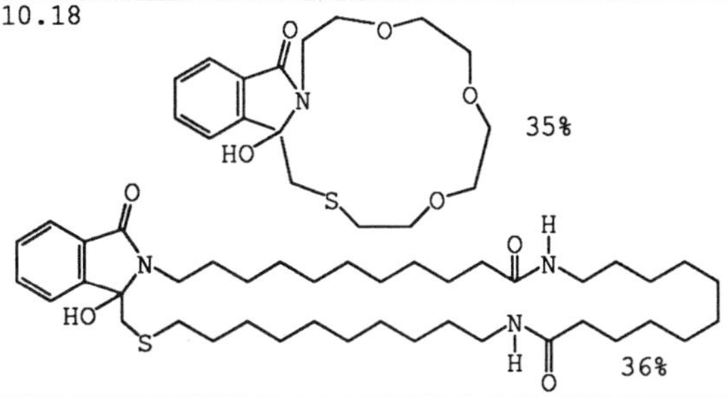

10.16. Of all cyclic amides, the photochemistry of N-alkylphthalimides has been the most extensively studied. N-alkylphthalimides undergo most of the photoreactions of simple amidic carbonyl compounds, as well as some additional specific reactions.[50] One of these is photocyclization involving an intramolecular abstraction of a δ-H atom by the carbonyl group of the phthalimide in its T_1 (π,π*) state. The reaction is solvent dependent; β-cleavage and H-transfer occur in acetone and acetonitrile, while only benzazepinedione formation was observed in ethanol.[51]

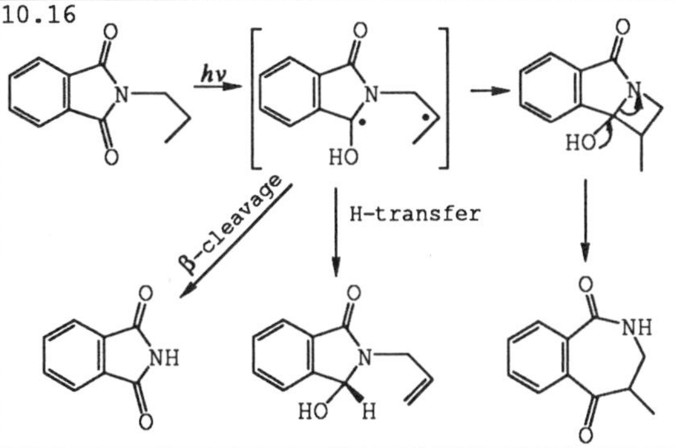

10.17. Insertion of a heteroatom (N, O, S) into the side chain of the N-alkylphthalimide leads to two significant changes: (1) in the position of the radical in the chain directed by the heteroatom; (2) in the mechanism of the photocyclization. It was suggested that the intermediate biradical is formed by an intramolecular electron and subsequent proton transfer either from the methyl or the methylene group adjacent to the heteroatom, particularly sulfur.[52,53] This mechanism is supported by the observation of a weak charge-transfer band with N-methylphthalimide and methyl butyl sulfide.[53] The method is particularly significant for N-methylthioalkylphthalimides since the sulfur atom increases the efficiency of the electron transfer to such an extent that the cyclic system is formed in quite high yield.[53,54]

10.18. The photocyclization of N-methylthioalkylphthalimides is preparatively important because of the remarkable efficiency in the formation of macrocyclic systems. As shown in the previous figure, up to 16-membered rings have been prepared with the parent system. The concept was extended to N-alkyl chains containing heteroatoms (O) and/or amide and ester groups, together with the methylthio group, and gives rise to a variety of macrocyclic systems in good yield.[53,55]

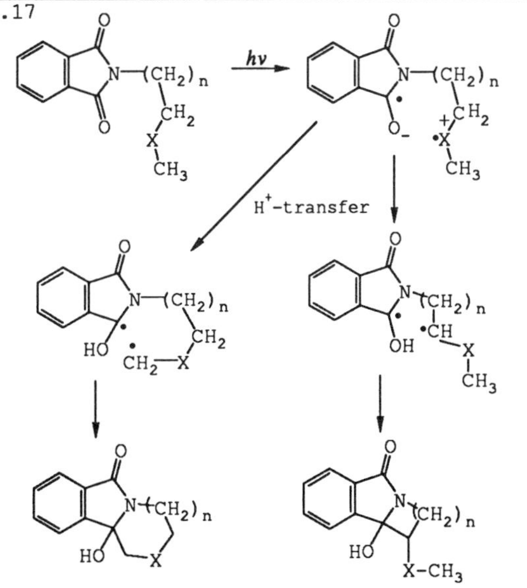

	X=NPh	X=S	X=S
n=4	15%	78%	6%
n=5	10%	58%	10%
n=9	8%	26%	4%
n=11		25%	4%

10.19

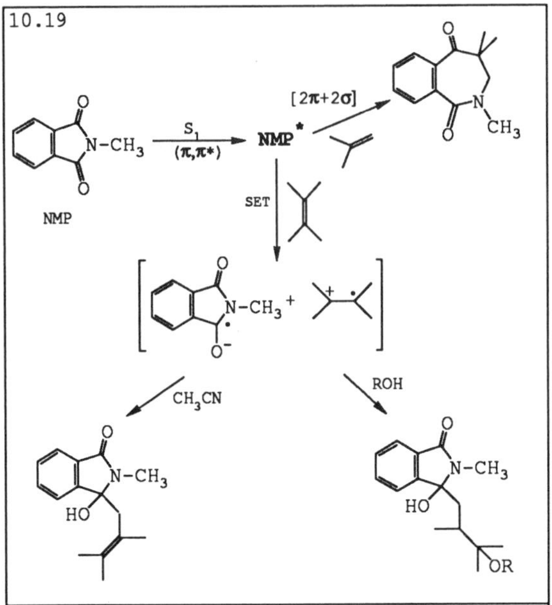

NMP

10.20

1

4

yield

n=2 68%
n=5 65%

2 41%

3 41%

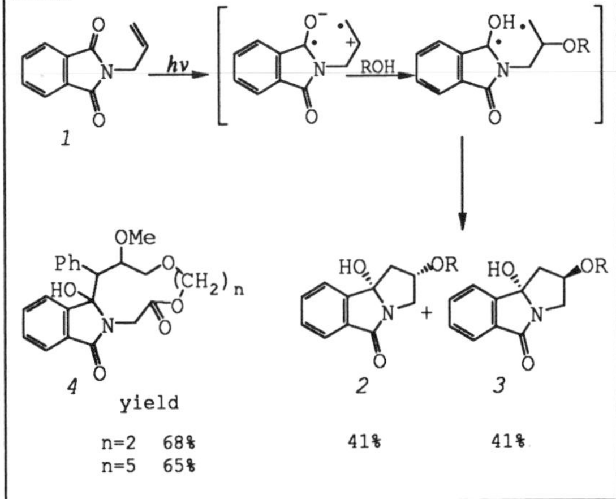

10.21

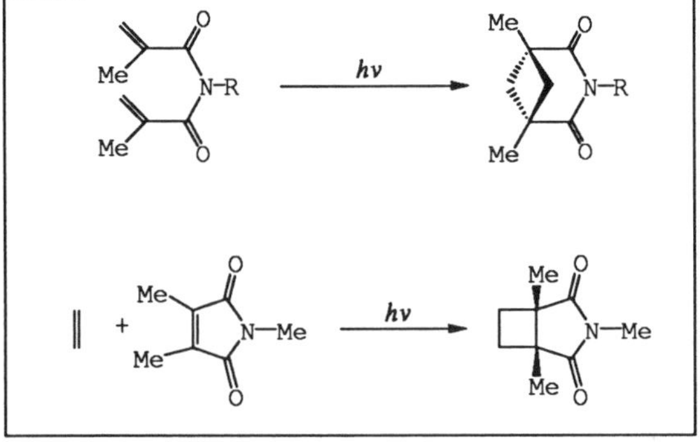

10.19. The photochemistry of N-methylphthalimide (NMP) in the presence of alkenes has been extensively studied.[50,56] Two competitive reactions were observed from the $S_1(\pi,\pi^*)$ state: (1) the regio- and stereocontrolled $[2\pi + 2\sigma]$ cycloaddition of the alkene to the C(O)—N σ-bond with the formation of dihydrobenzazepinediones; (2) addition reactions of alkenes (and alcohols) via a photostimulated single-electron transfer (SET).[57] The rate of SET is a function of the ionization potential (IP) of the alkene and occurs only with alkenes of IP < 9 eV. A small amount of the corresponding oxetane is also formed. Another minor product was reported recently with cyclohexene, i.e., its *para* cycloadduct to the benzene moiety of phthalimide.[58] The synthetic utility of the $[2\pi + 2\sigma]$ cycloaddition,[59] as well as the photochemistry of N-methyl dicarboximides of other aromatic compounds, have been reported.[56,60]

10.20. The intramolecular version of the above reaction is illustrated by the irradiation of N-allylphthalimide (*1*) in methanol to form cyclic products (*2*) and (*3*).[61] The mechanism involves an intramolecular electron transfer in the $S_1(\pi,\pi^*)$ state with the formation of a radical-ion pair, followed by an anti-Markovnikov addition of alcohol and recombination of the biradical. This surprisingly efficient method was utilized in the synthesis of macrocyclic compounds, e.g., (*4*).[62] However, with certain N-alkenylphthalimides the major product is an oxetane.[63]

10.21. It is significant that *bis*-methacryl-N-arylimide undergoes *crossbonding* by a formal $x[2+2]$ intramolecular photocycloaddition to form a product[64] with a different structure than that from the [2+2] cycloaddition of ethylene to N-aryldimethylmaleimide (a product of *parallel bonding*), which is a minor product of this photocyclization.

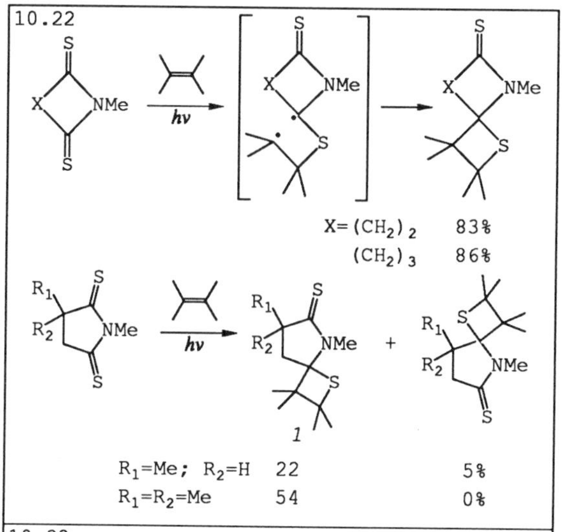

X= $(CH_2)_2$ 83%
$(CH_2)_3$ 86%

R_1=Me; R_2=H   22   5%
R_1=R_2=Me   54   0%

82%

R_1=R_2=H   42%
R_1=R_2=   45%

58%

35%

10.22. Increased attention has been given recently to the photochemistry of aliphatic and aromatic cyclic mono- and dithioamides, which differs significantly from that of their oxygen counterparts [aliphatic[65] or aromatic (see 10.16 through 10.20) cyclic imides]. In the same way that carbonyl photochemistry has been extended to the imide system in which the carbonyl chromophore is conjugated with an amide group, substitution of sulfur for oxygen leads to unique thioimide photochemistry. By analogy to the formation of simple thietanes by photocycloaddition of alkenes to thiocarbonyl compounds (see 8.38 and 8.39), thioimides yield *spirocyclic* thietanes by an *inter*molecular pathway.[66] Although two isomeric thietanes can be formed from unsymmetrical thioimides, generally isomer *1* is the major or even the sole product, probably because of steric hindrance.[67]

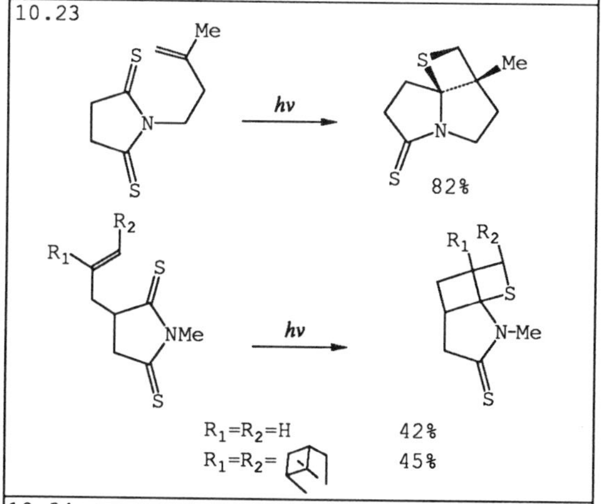

10.23. Analogous to thietane formation, biradicals are the commonly accepted intermediates in the cycloaddition of thioimides. The more stable biradicals are formed as a result of n → π* excitation, followed by ISC from the S_1 to the T_1 state. The *intramolecular* pathway of the photocycloaddition of the C=C bond of the alkene to the C=S group of the thioimide proceeds with the formation of tricyclic or polycyclic thietanes, which are sometimes highly strained.[68]

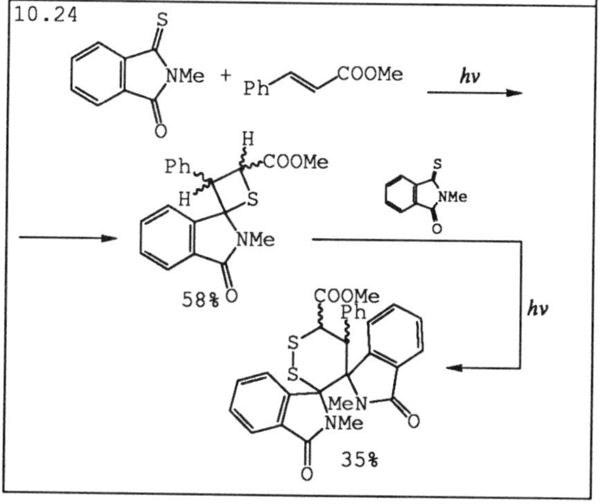

10.24. The difference between cyclic thioimides and their oxygen analogs can be demonstrated by the example shown here and in 10.19. Thietanes formed by irradiation of phthalthioimides in the presence of alkenes in which the C=C bond is conjugated with an aromatic ring are photochemically unstable and, after photolysis of the C—S bond, undergo reaction with a second molecule of phthalthioimide to form a 1,2-dithiane derivative.[69]

10.25

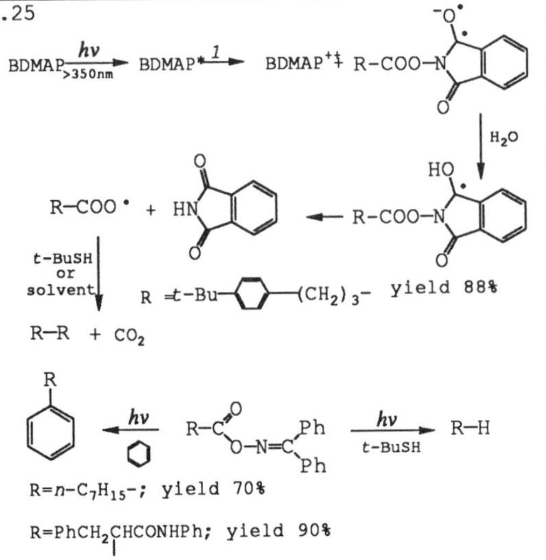

10.26

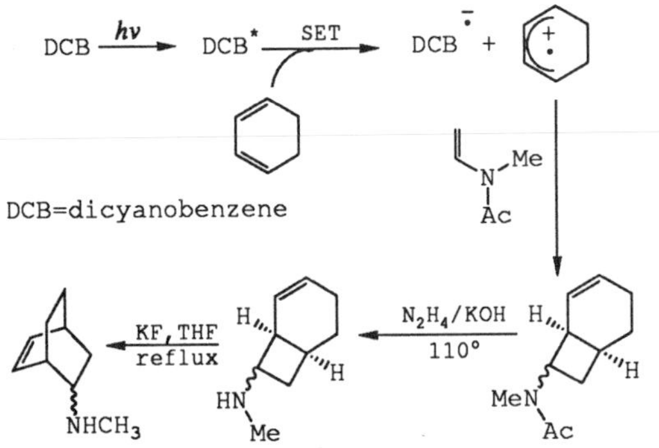

DCB = dicyanobenzene

10.27

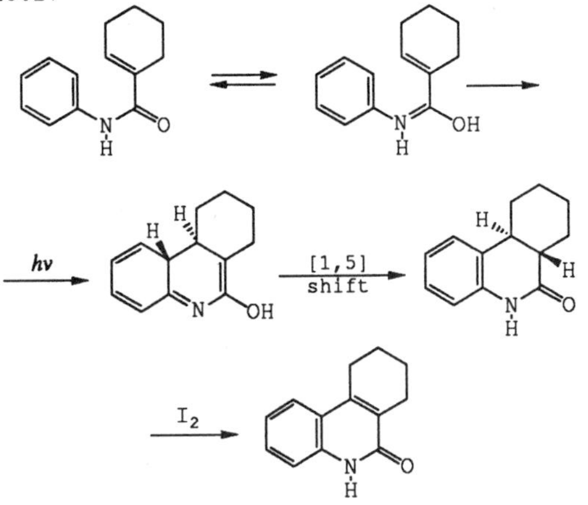

10.25. By irradiation of aqueous solutions of N-acyloxyphthalimides in the presence of an electron-transfer photosensitizer [1,6-*bis*(dimethylamino)pyrene, BDMAP] with visible light (λ = 350–450 nm), a smooth decarboxylation proceeds in high yield[70] via a photosensitized single-electron transfer. This reaction is an extension of the photodecarboxylation of oxime esters and, although it has several restrictions,[71] it can be used for the alkylation of benzene and pyridine.[72]

10.26. Because of the electron-rich dienophilic nature of enamides, their synthetic potential in the Diels–Alder reaction is hindered. However, the desired Diels–Alder adduct can be obtained indirectly in satisfactory yield by the radical-cation cyclobutanation of the enamide using photosensitized electron transfer, followed by an aminyl-anion assisted vinylcyclobutane rearrangement.[73]

10.27. Replacement of the central double bond in the hexatriene system by a carboxamide group, $-$CONH$-$ $\leftrightharpoons$ $-$C(OH)$=$N$-$, a new chromophore, a *dienamide*, is formed whose fundamental photochemical reaction is a conrotatory cyclization. There are three possible structural patterns of the dienamide, depending on whether either one or both double bonds are part of a (hetero)arene moiety. The photocyclization of dienamides has been extensively investigated and utilized in the synthesis of a wide variety of alkaloids.[74] The N-enacyl arylamines, in which one of the double bonds is part of the aromatic amine moiety, upon irradiation, form tetrahydroquinol-2-ones, which are starting substrates for the synthesis of berberine alkaloids. When the irradiation is carried out in the presence of iodine, 2-quinolones are obtained.[75] The mechanism consists of enolization followed by conrotatory ring closure and then a [1,5]suprafacial hydrogen shift.[76]

10.28. Another class of dienamides, N-aroyl derivatives of enamines, particularly N-cyclohexenylbenzamides in which one of the double bonds is part of the aroyl moiety, has been explored to a greater extent than the previous class, particularly for the synthesis of protoberberine alkaloids.[77] Their irradiation produces either derivatives of hexahydro- or tetrahydrophenanthridone depending on the reaction conditions.[78] With *ortho*-substituted derivatives the cycloaddition occurs in both positions with possible migration and/or elimination.[79]

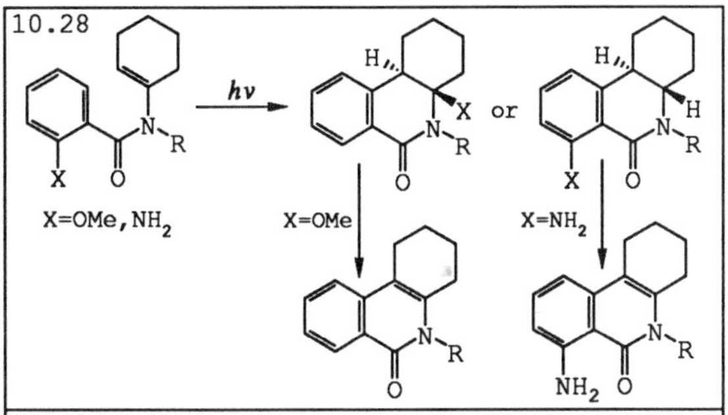

10.29. An important extension of nonoxidative and oxidative photocyclization processes of N-aroyl enamines is *reductive photocyclization*, based on irradiation of the substrate in the presence of a complex hydride.[80] This method has been used successfully in the total synthesis of a number of indole alkaloids,[81] e.g., yohimbine[82] and ergot alkaloids.[83]

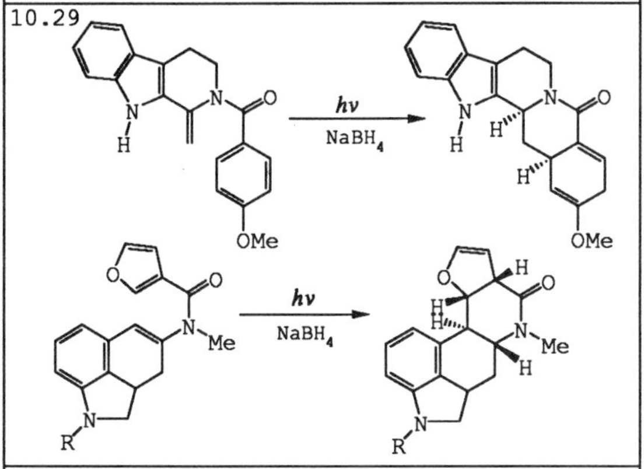

10.30. Irradiation of N-aroyl arylamines, e.g., benzanilides, in which both double bonds of the dienamide system are part of an arene moiety,[84] produces phenanthridone derivatives in the presence of iodine. A good leaving group (Br, I, MeO) in the *ortho* position is required for efficient photocyclization.[85,86]

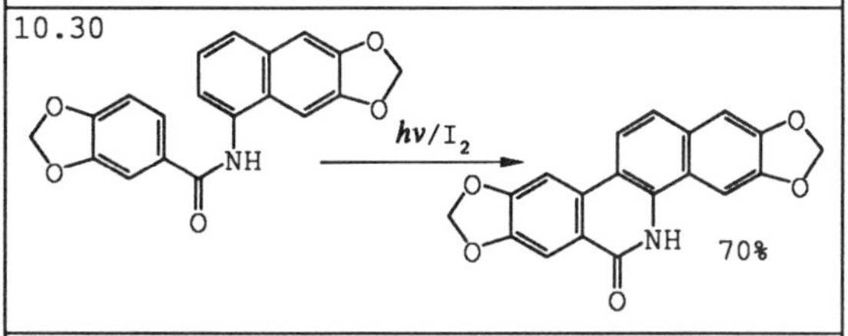

10.31. O-alkyl derivatives of the lactim form of dicarboxylic acid imides are ketoimino ethers. Their irradiation induces a rearrangement that leads to isocyanates of alkoxycarbocyclic compounds. The reaction occurs from the S_1 (n,π^*) state; it is highly stereospecific and initiated by an α-cleavage that produces a singlet σ,π-biradical followed by rapid ring closure. The cyclization rate must be several orders of magnitude greater than the rate of rotation about the C—C bond to explain the stereospecificity of the reaction.[87] This method is extremely useful in the synthesis of cyclopropane derivatives and bicyclo[n.1.0] systems.[88]

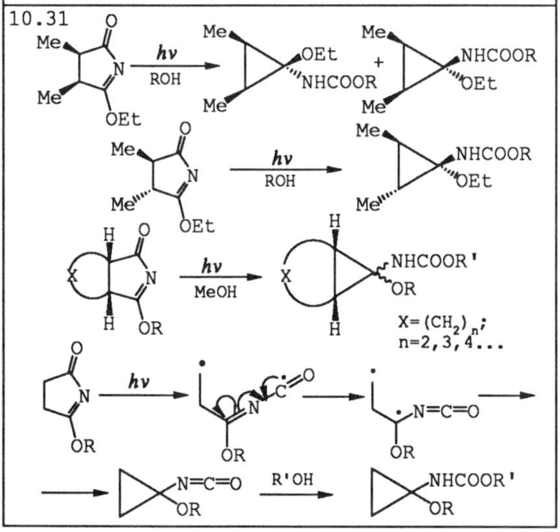

REFERENCES

(1) Lambert, J. B., Shurvell, H. G., Verbit, L., Cooks, R. G., and Stout, G. H. In *Organic Structural Analysis*. Macmillan: New York, 1976.

(2) Kanaoka, Y. and Hatanaka, Y. *J. Org. Chem.* **1976,** *41,* 400.

(3) Figure adapted from Coyle, J. D. *Chem. Rev.* **1978,** *78,* 97.

(4) Pandey, G., Krishna, A., and Rao, M. *Tetrahedron Lett.* **1986,** *27,* 4075.

(5) The method has been utilized in the synthesis of the 2,4-chromene system. See Pandey, G. and Krishna, A. *J. Org. Chem.* **1988,** *53,* 2364.

(6) Yang, N. C., Lin, L. C., Shani, A., and Yang, S. S. *J. Org. Chem.* **1969,** *34,* 1845.

(7) For a review, see Givens, R. S. *Org. Photochem.* **1981,** *5,* 309.

(8) Matuszewski, B., Givens, R. S., and Neywick, C. *J. Am. Chem. Soc.* **1973,** *95,* 595.

(9) Kaplan, M. L. and Truesdale, E. A. *Tetrahedron Lett.* **1976,** 3665.

(10)(a) Chapman, O. L., De La Cruz, D., Roth, R., and Pacansky, J. *J. Am. Chem. Soc.* **1973,** *95,* 1337. (b) Chapman, O. L., McIntosh, C. L., and Pacansky, J. *J. Am. Chem. Soc.* **1973,** *95,* 614.

(11)(a) For an excellent review of the literature up to 1969, see Bellus, D. *Adv. Photochem.* **1971,** *8,* 109. (b) For other reviews, see Pfau, M. and Julliard, M. *Bull. Soc. Chim. France.* **1977,** 785. For earlier reports on the photo-Fries rearrangement, see (c) Anderson, J. C. and Reese, C. B. *J. Chem. Soc.* **1963,** 1781. (d) Finnegan, R. A. and Mattice, J. J. *Tetrahedron.* **1965,** *21,* 1015.

(12) Sandner, M. R., Hedaya, E., and Trecker, D. J. *J. Am. Chem. Soc.* **1968,** *90,* 7249.

(13)(a) Adam, W., deSanabia, J. A., and Fischer, H. *J. Org. Chem.* **1973,** *38,* 2571. (b) For a short review of the mechanistic aspects of the photo-Fries reaction, see Shine, H. J. and Subotkowski, W. *J. Org. Chem.* **1987,** *52,* 3815, and references therein.

(14) Meyer, J. W. and Hammond, G. S. *J. Am. Chem. Soc.* **1972,** *94,* 2219.

(15) Kalmus, C. E. and Hercules, D. N. *J. Am. Chem. Soc.* **1974,** *96,* 449.

(16) Plank, D. A. *Tetrahedron Lett.* **1968,** 5423.

(17) For changes in regioselectivity (a) by complex formation, see 15.34; (b) in micelles: Singh, A. K. and Raghuraman, T. S. *Tetrahedron Lett.* **1985,** *26,* 4125.

(18) Slama, P., Bellus, D., and Hrdlovic, P. *Collect. Czech. Chem. Commun.* **1968,** *33,* 3752.

(19) Horspool, W. M. and Pauson, P. L. *J. Chem. Soc.* **1965,** 5162.

(20) Lappin, G. R. and Zannucci, J. S. *J. Org. Chem.* **1970,** *35,* 3679.

(21) For a photo-Fries rearrangement of a naphthyl ester as a key step in the synthesis of spinochrome A, see Farina, F., Martinez-Utrilla, R., and Paredes, M. C. *Tetrahedron.* **1982,** *38,* 1531.

(22) Kobsa, H. *J. Org. Chem.* **1962,** 2293, and references therein.

(23) Hageman, H. J. *Tetrahedron.* **1969,** *25,* 6015.

(24) Beugelmans, R. and LeGoff, M.-T. *Bull. Soc. Chim. France.* **1972,** 1106.

(25) LeGoff, M.-T. and Beugelmans, R. *Ibid.* **1972,** 1115.

(26) Garcia, H., Miranda, M. A., and Primo, J. *J. Chem. Res. (S).* **1985,** 100.

(27)(a) Obara, H., Takahashi, H., and Hirano, H. *Bull. Chem. Soc. Japan.* **1969,** *42,* 560. (b) Bhatia, V. K. and Kagan, J. *Chem. Ind.* **1970,** 1203. (c) Ramakrishnan, V. T. and Kagan, J. *J. Org. Chem.* **1970,** *35,* 2901.

(28) Primo, J., Tormo, R., and Mirando, M. A. *Heterocycles.* **1982,** *19,* 1819.

(29) For R=CH$_3$, see (a) Yogev, A., Gorodetsky, M., and Mazur, Y. *J. Am. Chem. Soc.* **1964,** *86,* 5208. (b) Gorodetsky, M. and Mazur, Y. *J. Am. Chem. Soc.* **1964,** *86,* 5213. For R = Ph, see Feldkimel-Gorodetsky, M. and Mazur, Y. *Tetrahedron Lett.* **1963,** 369. For cyclic enol esters, see Pappas, S. P., Alexander, J. E., Long, G. L., and Zehr, R. D. *J. Org. Chem.* **1972,** *37,* 1258.

(30) Shizuka, H. and Tanaka, I. *Bull. Chem. Soc. Japan.* **1969,** *42,* 909.

(31)(a) Shizuka, H. and Tanaka, I. *Bull. Chem. Soc. Japan.* **1968,** *41,* 2343. (b) Shizuka, H. *Bull. Chem. Soc. Japan.* **1969,** *42,* 57. (c) Parker, C. A. and Barnes, W. J. *Analyst.* **1957,** *82,* 606. (d) Stegemeyer, H. *Naturwiss.* **1966,** *53,* 582.

(32) Elad, D., Rao, D. V., and Stenberg, V. I. *J. Org. Chem.* **1965,** *30,* 3252.

(33) Shizuka H., Kato, M., Ochiai, T., Matsui, K., and Morita, T. *Bull. Chem. Soc. Japan.* **1970,** *43,* 67.

(34) Ghosh, S., Das, T. K., Datta, D. B., and Mehta, S. *Tetrahedron Lett.* **1987**, *28*, 4611.

(35) Trecker, D. J., Foote, R. S., and Osborn, C. L. *Chem. Commun.* **1968**, 1034.

(36) Bellus, D. and Schaffner, K. *Helv. Chim. Acta.* **1968**, *51*, 221.

(37) Fischer, M. *Chem. Ber.* **1969**, *102*, 342.

(38) Fischer, M. *Tetrahedron Lett.* **1969**, 2281.

(39) For another example of a lactone rearrangement, see Gandhi, R. P., Singh, M., Sachdeva, Y. P., and Mukherji, S. M. *Tetrahedron Lett.* **1973**, 661.

(40) Carroll, F. A. and Hammond, G. S. *Israel J. Chem.* **1972**, *10*, 613.

(41) Hageman, H. J. and Huysmans, W. G. B. *Rec. Trav. Chim.* **1972**, *91*, 528.

(42) Herweh, J. E. and Hoyle, C. E. *J. Org. Chem.* **1980**, *45*, 2195.

(43) Elad, D. and Rokach, J. *J. Org. Chem.* **1964**, *29*, 1885.

(44) Elad, D. *Tetrahedron Lett.* **1963**, 77.

(45) For a review, see Sundberg, R. J. *Org. Photochem.* **1983**, *6*, 121.

(46) Naruto, S., Yonemitsu, O., Kataoka, N., and Kimura, K. *J. Am. Chem. Soc.* **1971**, *93*, 4053.

(47)(a) Okuno, Y. and Yonemitsu, O. *Tetrahedron Lett.* **1974**, 1169. (b) Okuno, Y. and Yonemitsu, O. *Chem. Pharm. Bull.* **1975**, *23*, 1039.

(48) Iwakuma, T., Nakai, H., Yonemitsu, O., and Witkop, B. *J. Am. Chem. Soc.* **1974**, *96*, 2564.

(49) Hamada, T., Ohmori, M., and Yonemitsu, O. *Tetrahedron Lett.* **1977**, 1519.

(50) For exhaustive reviews, see (a) Kanaoka, Y. *Acc. Chem. Res.* **1978**, *11*, 407. (b) Mazzocchi, P. H. *Org. Photochem.* **1981**, *5*, 421. (c) Coyle, J. D. In *Synthetic Organic Photochemistry*, Horspool, W. M., ed. Plenum: New York, 1984, p. 259.

(51) Kanaoka, Y., Migita, Y., Koyama, K., Sato, Y., Nakai, H., and Mizoguchi, T. *Tetrahedron Lett.* **1973**, 1193.

(52) Machida, M., Kakechi, H., and Kanaoka, Y. *Heterocycles.* **1977**, *7*, 273.

(53) Sato, Y., Nakai, H., Mizoguchi, T., Hatanaka, Y., and Kanaoka, Y. *J. Am. Chem. Soc.* **1976**, *98*, 2349.

(54) Sato, Y., Nakai, H., Mizoguchi, T., and Kanaoka, Y. *Tetrahedron Lett.* **1976**, 1889.

(55) For the use of this method as a key step in the synthesis of the berberine alkaloid chilenene ring system, see King, C. R. and Ammon, H. L. *Tetrahedron Lett.* **1987**, *28*, 2473.

(56)(a) Kubo, Y., Togawa, S., Yamane, K., Takuwa, A., and Araki, T. *J. Org. Chem.* **1989**, *54*, 4929, and references therein. (b) Maruyama, K. and Kubo, Y. *J. Org. Chem.* **1985**, *50*, 1426.

(57)(a) Mazzochi, P. H., Wilson, P., Khachik, F., Klingler, L., and Minamikawa, S. *J. Org. Chem.* **1983**, *48*, 2891. (b) Mazzochi, P. H., Minamikawa, S., and Wilson, P. *J. Org. Chem.* **1985**, *50*, 2681. (c) Mazzochi, P. H. and Fritz, G. *J. Am. Chem. Soc.* **1986**, *108*, 5362.

(58) Schwack, W. *Tetrahedron Lett.* **1987**, *28*, 1869.

(59)(a) Maruyama, K., Ogana, T., Kubo, Y., and Araki, T. *J. Chem. Soc., Perkin Trans. 1.* **1985**, 2025. (b) Machida, M., Oda, K., and Kanaoka, Y. *Tetrahedron.* **1985**, *41*, 4995.

(60) Somich, C., Mazzocchi, P. H., and Ammon, H. L. *J. Org. Chem.* **1987**, *52*, 3614.

(61) Maruyama, K., Kubo, Y., Machida, M., Oda, K., Kanaoka, Y., and Fukuyama, K. *J. Org. Chem.* **1978**, *43*, 2303.

(62)(a) Maruyama, K. and Kubo, Y. *J. Am. Chem. Soc.* **1978**, *100*, 7772. (b) Maruyama, K., Kubo, Y., Machida, M., Oda, K., Kanaoka, Y., and Furuyama, K. *J. Org. Chem.* **1978**, *43*, 2303.

(63) Mazzocchi, P. H., Klingler, L., Edwards, M., Wilson, P., and Shook, D. *Tetrahedron.* **1983**, *24*, 143.

(64) Alder, A., Buehler, N., and Bellus, D. *Helv. Chim. Acta.* **1982**, *65*, 2405.

(65) Kanaoka, Y. and Hatanaka, Y. *J. Org. Chem.* **1976**, *41*, 400.

(66) Oda, K., Machida, M., and Kanaoka, Y. *Synthesis.* **1986**, 768.

(67) Machida, M., Oda, K., and Kanaoka, Y. *Chem. Pharm. Bull.* **1985**, *33*, 3552.

(68) Oda, K., Machida, M., Aoe, K., Nishikata, Y., Sato, Y., and Kanaoka, Y. *Chem. Pharm. Bull.* **1986**, *34*, 1411.

(69)(a) Machida, M., Oda, K., Yoshida, E., Wakao, S., Ohno, K., and Kanaoka, Y. *Heterocycles.* **1985**, *23*, 1615. (b) Machida, M., Oda, K., Yoshida, E., and Kanaoka, Y. *Tetrahedron.* **1986**, *42*, 4619.

(70) Okada, K., Okamoto, K., and Oda, M. *J. Am. Chem. Soc.* **1988**, *110*, 8736.

(71) Hasebe, M. and Tsuchiya, T. *Tetrahedron Lett.* **1987,** *28,* 6207.

(72) Hasebe, M. and Tsuchiya, T. *Tetrahedron Lett.* **1986,** *27,* 3239.

(73) Bauld, N. L., Harirchian, B., Reynold, D. W., and White, J. C. *J. Am. Chem. Soc.* **1988,** *110,* 8111.

(74) For reviews, see (a) Lenz, G. R. *Synthesis.* **1978,** 489. (b) Campbell, A. L. and Lenz, G. R. *Synthesis.* **1987,** 421. (c) Ninomiya, I. and Naito, T. *Heterocycles.* **1981,** *15,* 1433. (d) Ninomiya, I. and Naito, T. In *The Alkaloids.* Brossi, A., ed. Academic Press, New York, 1983, Vol. 22, p. 189. (e) For an asymmetric induction in the vinylogous amide photo-cycloaddition reaction, see Winkler, J. D., Scott, R. D., and Williard, P. G. *J. Am. Chem. Soc.* **1990,** *112,* 8971.

(75) Ninomiya, I., Yamamauchi, S., Kiguchi, T., Shinohara, A., and Naito, T. *J. Chem. Soc., Perkin Trans. 1.* **1974,** 1747.

(76) Cleveland, P. G. and Chapman, O. L. *J. Chem. Soc., Chem. Commun.* **1967,** 1064.

(77)(a) Ninomiya, I. and Naito, T. *J. Chem. Soc., Chem. Commun.* **1973,** 137. (b) Iida, H., Aoyagi, S., and Kibayashi, C. *J. Chem. Soc., Chem. Commun.* **1974,** 499.

(78) Ninomiya, I., Naito, T., and Kiguchi, T. *J. Chem. Soc., Perkin Trans. 1.* **1973,** 2257.

(79) Ninomiya, I., Kiguchi, T., Yamauchi, S., and Naito, T. *J. Chem. Soc., Perkin Trans. 1.* **1980,** 197, and references therein.

(80) Naito, T., Tada, Y., Nishiguchi, Y., and Ninomiya, I. *J. Chem. Soc., Perkin Trans. 1.* **1985,** 487.

(81) Ninomiya, I. and Naito, T. *Photochemical Synthesis.* Academic Press: New York, 1989, p. 176.

(82) Naito, T., Hirata, Y., Miyata, O., and Ninomiya, I. *J. Chem. Soc., Perkin Trans. 1.* **1988,** 2219, and references therein.

(83) Ninomiya, I., Hashimoto, C., Kiguchi, T., and Naito, T., *J. Chem. Soc., Perkin Trans. 1.* **1985,** 941, and references therein.

(84) For the amide-imidole tautomerism in benzanilide, see Tang, G.-Q., McInnis, J., and Kasha, M. *J. Am. Chem. Soc.* **1987,** *109,* 2531.

(85) For the oxidative photocyclization of N^6-benzoyl adenosine derivatives, see Shimada, K., Sako, M., Hirota, K., and Maki, Y. *Tetrahedron Lett.* **1987,** *28,* 207.

(86) Kessar, S. V., Singh, G., and Balakrishnan, P. *Tetrahedron Lett.* **1974,** 2269.

(87) Koch, T. H., Anderson, D. R., Burns, J. M., Crockett, G. C., Howard, K. A., Keute, J. S., Rodehorst, R. M., and Sluski, R. J. *Topics Curr. Chem.* **1978,** *75,* 1978.

(88) For an example of the preparative procedure, see Crockett, G. C. and Koch, T. H. *Org. Synth.* **1988,** Col. Vol. 6, 226.

Chapter 11

PHOTOCHEMISTRY OF NITROGEN-CONTAINING COMPOUNDS

PHOTOCHEMISTRY OF NITROGEN-CONTAINING COMPOUNDS

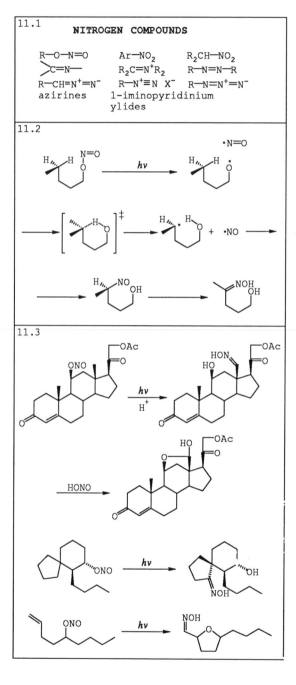

11.1

NITROGEN COMPOUNDS

R—O—N=O   Ar—NO₂   R₂CH—NO₂

>C=N—   R₂C=N⁺R₂   R—N=N—R

R—CH=N⁺=N⁻   R—N⁺≡N X⁻   R—N=N⁺=N⁻

azirines   1-iminopyridinium

ylides

11.2

11.3

11.1. The photochemistry of several nitrogen-containing chromophores will be discussed in this chapter. These include nitrites, nitro compounds, imines, iminium salts, azocompounds, diazocompounds, diazonium salts, azides, and some heterocyclic compounds.

11.2. In the photolysis of nitrites (*Barton reaction*)[1,2] a hydrogen atom in the γ-position is replaced by a nitroso group and a hydroxy group is formed. Thus, the Barton reaction is an example of the remote functionalization of an unactivated saturated carbon atom. The reaction starts from the $S_1(n,\pi^*)$ state ($\lambda = 310$–390 nm) with the cleavage of the RO—NO bond ($E_{diss} = 150$–190 kJ/mol, 39–45 kcal/mol) to produce nitric oxide and an activated alkoxy radical.[3] This is followed by abstraction of hydrogen via a six-membered cyclic transition state and the formation of the hydroxy group and a C-radical that combines with nitric oxide. The nitroso compound dimerizes or preferentially isomerizes to the oxime.

11.3. The Barton reaction is potentially useful in the synthesis of natural products, such as hormones and alkaloids. The photolysis of the 11-nitrite of corticosterone acetate has been utilized on an industrial scale in the partial synthesis of aldosterone 21-acetate.[3-5] The key step in the synthesis of the alkaloid perhydrohistorionicotoxin is the photolysis of the nitrite of a spirocyclic alcohol having three different γ-hydrogen atoms. The transition state of optimal geometry is formed between both rings of the spirocyclic molecule.[6] When the double bond of an alkenyl nitrite is in close proximity to the alkoxy radical formed by irradiation, the sole product is a result of its addition to the C=C bond with the formation of a cyclic ether.[7]

11.4

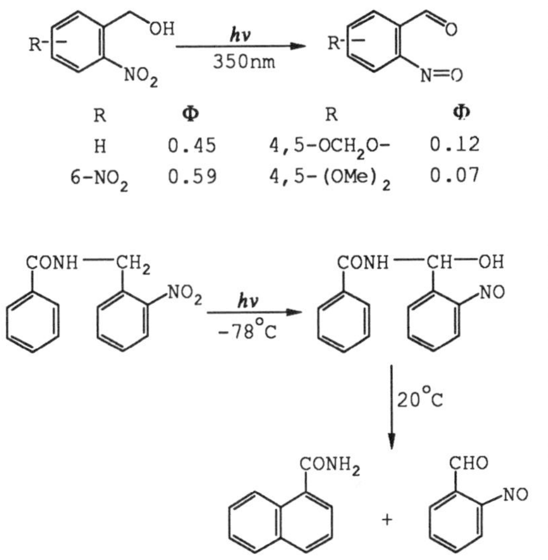

R	Φ	R	Φ
H	0.45	4,5-OCH$_2$O-	0.12
6-NO$_2$	0.59	4,5-(OMe)$_2$	0.07

11.5

14%

81%

11.6

Ar=Ar'=Ph[17a];
Ar=β-naphthyl;
Ar'=α-naphthyl

ref.20

11.4. Aromatic nitrocompounds in their $T_1(n,\pi^*)$ state ($\lambda = 350$ nm) abstract a hydrogen atom and possibly transfer one of the oxygen atoms of the nitro group of the C-radical center to form a hydroxy group.[8] This reaction is highly efficient and involves an intramolecular arrangement, as demonstrated by o-nitrobenzyl compounds.[9,10] It is used to protect the carboxyl group in peptides,[11] particularly in the form of a 2'-nitrobenzhydryl polystyrene resin, which permits the synthesis of peptides with BOC-amino acids. The final cleavage of the attached peptide proceeds under neutral conditions.[12] The principle of photolytic deprotection of the carbonyl group by an internal photoredox reaction of o-nitroaromatic compounds[13] is shown.

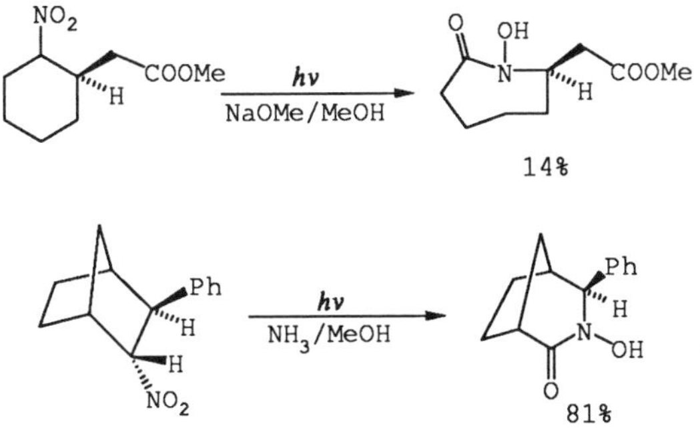

11.5. Irradiation of nitronate anions with low-pressure Hg lamps proceeds to the corresponding hydroxamic acids with high regioselectivity (only one of the possible diastereomers is formed).[14] The rearrangement is controlled by the number of substituents on the β-carbon atom, in which the more subsituted one migrates to the nitrogen atom. If the number of substituents is the same, the one with an electron-withdrawing group migrates preferentially.[15] This reaction is used in the stereospecific cleavage of cyclic compounds.

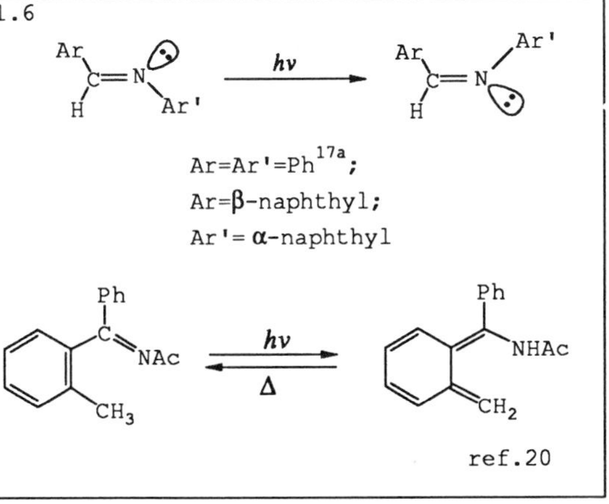

11.6. The photochemistry of the $>C=N-$ chromophore in a diverse set of compounds involves many different photochemical reactions,[16] depending on their electron configuration and the multiplicity of the lowest excited state involved. These compounds undergo photochemical reactions common to either the $C=C$ bond (*syn-anti* photoisomerization,[17] photocyclization reaction[18]) or the carbonyl group, e.g., cycloaddition reactions[19] and intramolecular hydrogen abstractions.[20] A practical example is the photochromism of salicylaldehyde anil (see 16.14).

11.7

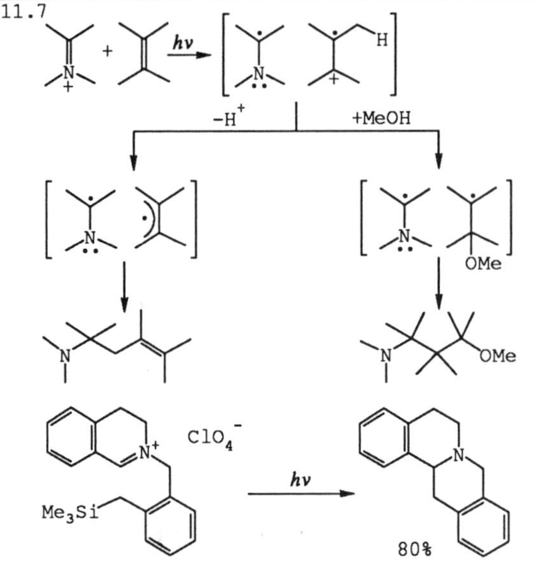

11.8

11.9

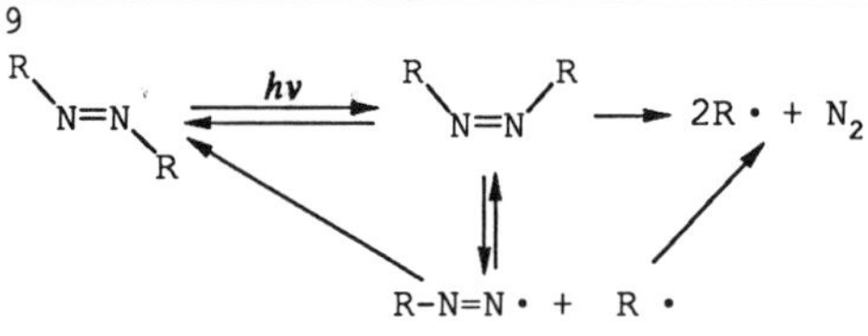

R	ϕ_{N2} (direct)	ϕ_{N2} (sens)
CH_3	0.15	0.009
n-C_3H_7	0.007	
i-C_3H_7	0.025	0.001
t-C_4H_9	0.46	0.016
$PhCMe_2$	0.36	0.31

11.7. In recent years photochemical transformations (*cis-trans* isomerization and [2+2] cycloaddition) of iminium salts (alkylideneammonium salts, $R_2C={}^+NR_2$) have received considerable attention.[21] It has been suggested that, since these salts are powerful electron acceptors, the initiating step is a single-electron transfer (SET) from the donor (alkene) to the excited iminium salt with the formation of an α-aminoradical-radical-cation pair. These recombine either directly through their radical positions or after trapping the radical-cation by a nucleophile, e.g., MeOH. The alkene/iminium salt photoaddition reaction has been applied to the synthesis of N-heterocyclic systems.[22]

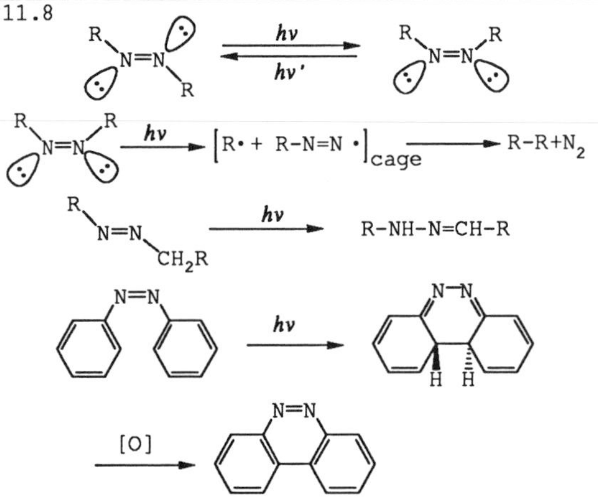

11.8. Photochemical reactions of acyclic azocompounds (1,2-*diazenes*) are either unimolecular [*cis-trans* isomerization, N_2 extrusion, H-atom shift, photocyclization (diaryldiazenes)] or bimolecular (photoreduction, photocycloaddition).[23] For most of the photochemical reactions involving direct excitation, the only important state is the low-lying singlet n,π^* state, since the π,π^* state is very high in energy.[24] The n,π^* state, unlike the π,π^* state, retains substantial double bond character and, to some extent, resists rotation about the N—N bond. It was suggested, therefore, that *cis-trans* isomerization involves a rapid $^1(n,\pi^*) \rightarrow {}^3(\pi,\pi^*)$ intersystem crossing.[25] The lack of fluorescence of dialkyldiazenes supports this mechanism.[26]

11.9. Acyclic *cis*-azoalkanes differ significantly in stability. Many undergo a facile thermal extrusion of N_2, depending on the size of the alkyl groups (R) and the stability of the radicals (R·) formed.[23(a),27] The "*deazatization*" is a stepwise process in which the *diazenyl radical* is a bound intermediate. The racemization observed with azoalkanes having chiral alkyl groups supports this stepwise mechanism.[28]

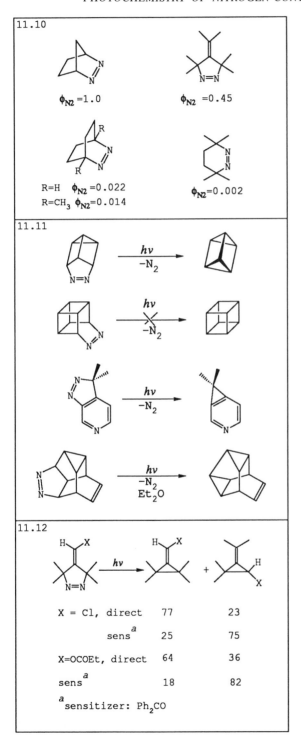

11.10

$\phi_{N2} = 1.0$

$\phi_{N2} = 0.45$

R=H $\phi_{N2} = 0.022$
R=CH$_3$ $\phi_{N2} = 0.014$

$\phi_{N2} = 0.002$

11.11

$\xrightarrow[-N_2]{h\nu}$

$\xrightarrow[-\cancel{N_2}]{h\nu}$

$\xrightarrow[-N_2]{h\nu}$

$\xrightarrow[\underset{Et_2O}{-N_2}]{h\nu}$

11.12

$\xrightarrow{h\nu}$ +

X = Cl,	direct	77	23
	sens[a]	25	75
X=OCOEt,	direct	64	36
	sens[a]	18	82

[a] sensitizer: Ph$_2$CO

11.10. For cyclic azoalkanes[29] the stepwise mechanism via a π-type diazenyl biradical has been experimentally proven.[30] While acyclic azoalkanes do not lose nitrogen efficiently upon triplet sensitization, the photoextrusion of nitrogen from cyclic azoalkanes occurs by both direct irradiation and triplet sensitization, often with different quantum yields and different products (*spin-correlation effect*, see 11.12). The quantum yield is strongly dependent on the structure of the azocompound and varies from 1 (N=N group is a part of a five-membered ring) to close to 0 (six-membered ring, for so-called "*reluctant azoalkanes*," e.g., 2,3-diazabicyclo[2.2.2]oct-2-ene).[23(a),31]

11.11. For this reason, nitrogen extrusion occurs upon irradiation of pyrazolines, which are readily available from the 1,3-dipolar cycloaddition of a diazoalkane to an activated double bond. It is a synthetically useful, general method for the synthesis of cyclopropane derivatives[32,33] and has been used not only for the synthesis of a variety of cage compounds,[29,34–36] such as the polycyclic hydrocarbons *snoutene* and *prismane* shown, but also for natural products, particularly sesquiterpenes,[37] which have a cyclopropane ring in the molecule. The loss of nitrogen occurs stereospecifically from the S$_1$ state in a concerted process. However, there is a significant or total loss of stereospecificity in both the thermal and triplet-sensitized reactions.

11.12. The spin-correlation effect is frequently observed in photoextrusions of N$_2$ from cyclic azoalkanes as a result of a very inefficient intersystem crossing (in contrast to acyclic azoalkanes).[38,39] The longer lifetime of an allylic triplet biradical apparently accounts for the increased formation of the rearranged product.

11.13

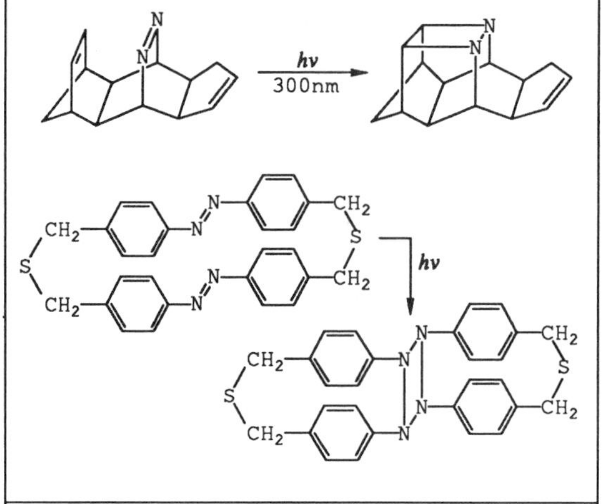

EtOOC-N=N-COOEt + CH₂=CH-OEt

$$EtOOC-N=N-COOEt + CH_2=CH-OEt$$

11.13. A particularly interesting bimolecular reaction is a photochemically allowed $[_\pi 2_s + _\pi 2_s]$ cycloaddition between an azo group and an alkene with the formation of a 1,2-diazetidine or between two azo groups to yield a tetrazetidine. The azo/alkene photocycloaddition results in a cage compound via an intramolecular pathway by irradiation of a rigid unsaturated polycyclic azo compound.[40] In contrast, all attempts to obtain a tetrazetidine (where the four-membered ring is composed only of nitrogen atoms) by azo/azo photocycloaddition have been unsuccessful so far. However, the tetrazetidine ring is assumed to be a short-lived intermediate formed by the interaction of two *trans*-diaryldiazene structures during the *cis-trans* photoisomerization of an azabenzenophane.[41]

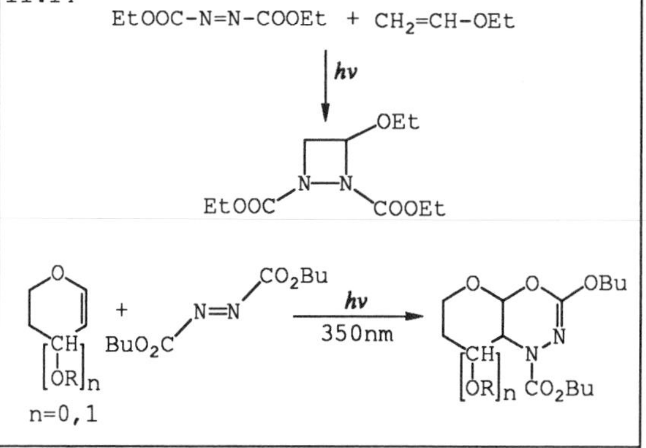

11.14. The azo/alkene photocycloaddition is especially successful when the azo compound contains electron-withdrawing groups and the alkene has an electron-donor substituent. [4+2] and/or [2+2] photocycloadditions of azodicarboxylate with acyclic vinyl ethers have been reported.[42] The [4+2] photocycloaddition of azodicarboxylate to glycals is an efficient method for the preparation of 2-aminoglycosides.[43]

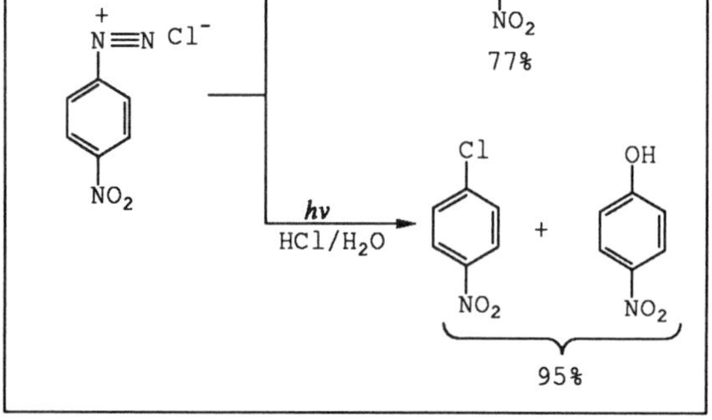

11.15. The photolysis of solutions of diazonium salts causes the facile loss of molecular nitrogen by two independent primary processes: heterolytic and homolytic.[44] The sum of the quantum yields of these two processes is close to 1 and their participation in the reaction depends on the structure of the diazonium salt, and particularly on the polarity of the solvent. Heterolytic cleavage to a carbocation proceeds in an aqueous solution, while homolytic cleavage to a phenyl radical plays an important role in an alcohol solution.[45] The photochemical decomposition of diazonium salts is the principal of the diazo type process (see 16.13).

11.16

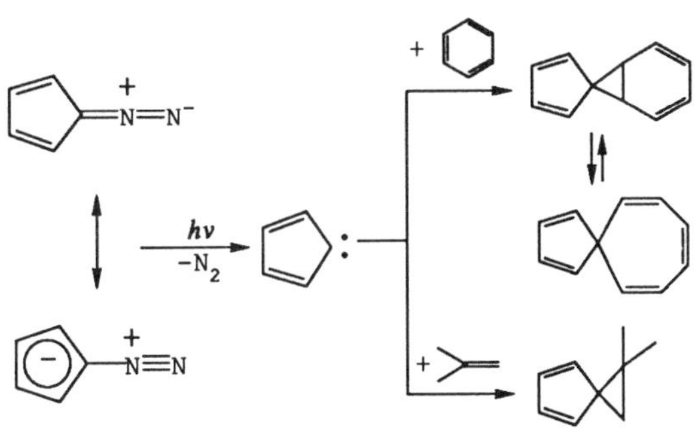

11.16. The mesomeric form of diazocyclopentadiene (diazofluorene, etc.) represents an internal diazonium salt. Irradiation of diazocyclopentadiene in either a benzene[46] or an alkene[47] solution results in the formation of spirocyclic compounds. These diazo compounds can also serve as photodeoxygenation reagents, e.g., irradiation of 9-diazafluorene with epoxides yields fluorenone and an alkene is formed from deoxygenation of the epoxide.[49] A carbonyl ylide that can be trapped by a dipolarophile is formed from diazomethane and acetone.[50]

11.17

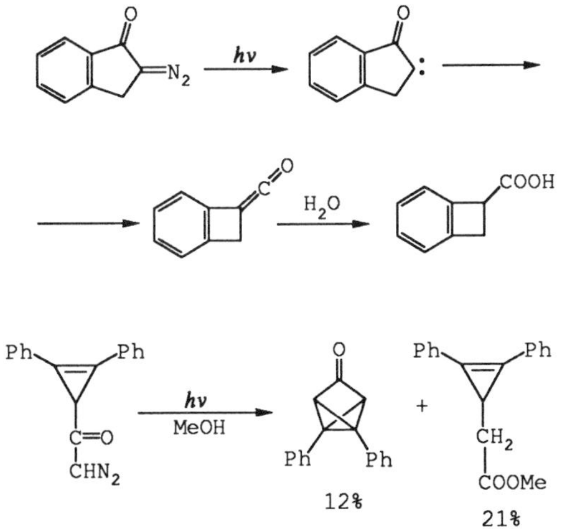

11.17. Other examples of internal diazonium salts are quinone diazides and diazoketones. Their irradiation results in the preparatively significant *photo-Wolff rearrangement*.[51] The singlet carbene formed by the loss of nitrogen (apparently via an oxirene[52]) is stabilized by rearrangment to a ketene. Cyclic diazoketones undergo a concomitant one carbon atom ring contraction; a molecule of solvent is added to the resulting ketene.[53,54] When a C=C bond is present in the molecule of a diazoketone, addition of the carbene to the double bond competes with the Wolff rearrangement.[55]

11.18

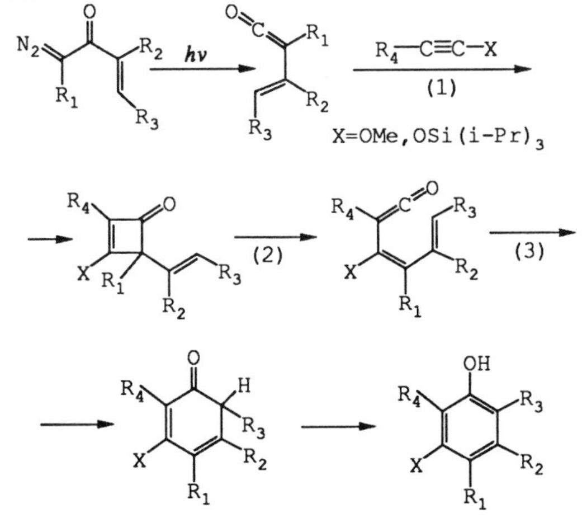

11.18. (Hetero)aromatic compounds can be synthesized by the photo-Wolff rearrangement when a vinyl or aryl α-diazoketone is irradiated in the presence of an acetylene derivative. After an initial rearrangement to an aryl or vinylketene, three consecutive pericyclic reactions occur: (1) thermal regiospecific [2+2] cycloaddition of the acetylene to yield a 4-substituted cyclobutenone; (2) photochemical (or thermal, *vide infra*) 4π-electron ring opening to a dienylketene; and (3) 6π-electron ring closure to 2,4-cyclohexadienone, which tautomerizes to the aromatic product. A variety of phenols, naphthalenes, and benzo derivatives of five-membered heterocycles can be prepared in good yield by this method.[56]

11.19

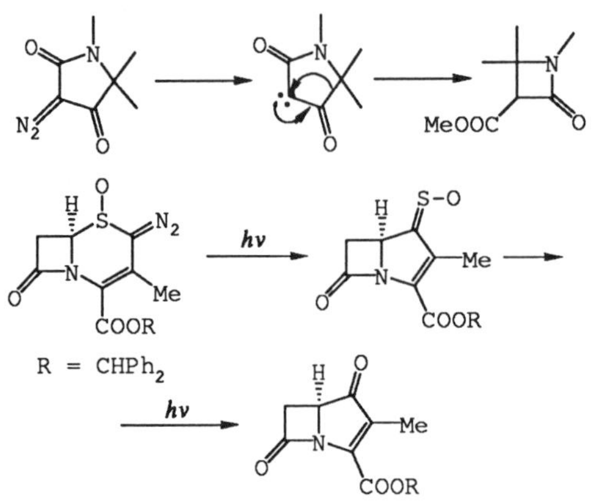

R = CHPh$_2$

11.20

11.21

11.19. The photo-Wolff rearrangement has been used in the synthesis of β-lactams,[57] which occur in penicillins and cephalosporins, and in the transformation of cephalosporins to carbapenems. In the latter, the readily available 2-diazoceph-3-em-1-oxide rearranges to the thiocarbonyl S-oxide, which loses sulfur upon further irradiation to give the desired product.[58] Another versatile photochemical method for the preparation of β-lactams is produced in 9.2.

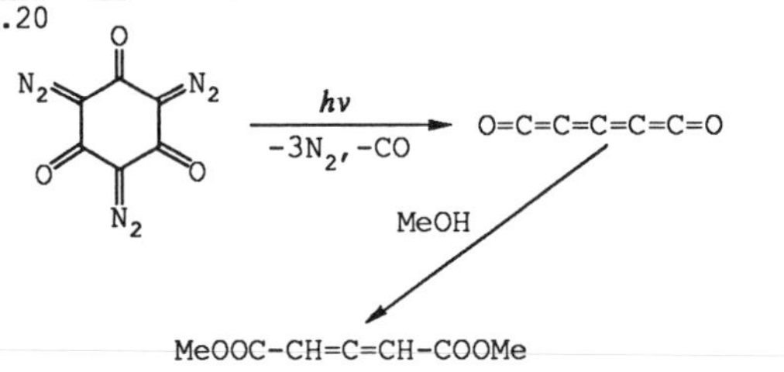

11.20. The longest carbon oxide, C_5O_2, was prepared by the photolysis of a cyclic *tris*-diazoketone. It is stable in solution at ambient temperature and from its IR spectrum has been identified as a (quasi)linear molecule. It undergoes reaction with methanol to form dimethyl allenedicarboxylate.[59]

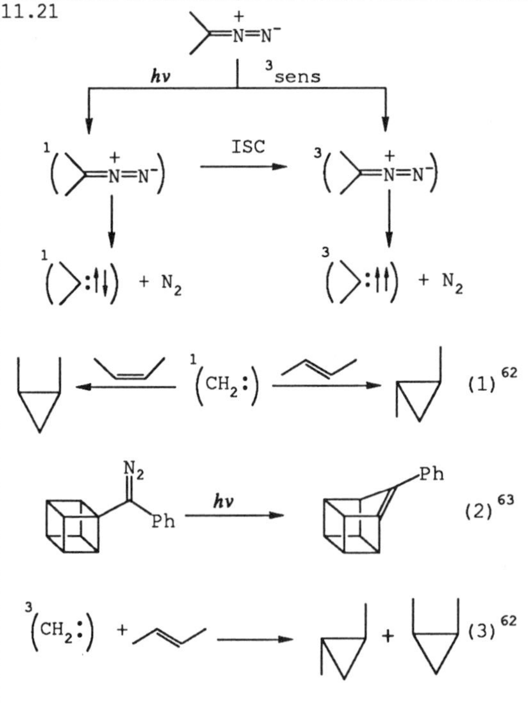

11.21. Diazo compounds exhibit a weak absorption band in the visible region (400–500 nm) attributed to the $n \rightarrow \pi^*$ transition of the diazo group.[60] Their photolysis represents a valuable synthetic route to carbenes. It is assumed that direct irradiation produces an electronically excited singlet state that rapidly extrudes N_2 with conservation of spin to produce singlet carbene. In contrast, triplet sensitization produces a triplet state of the diazo compound that leads directly to triplet carbenes. The most important reactions of singlet carbenes are stereospecific cycloaddition to a π bond (1), insertion into a σ-bond (2), [1,2]sigmatropic shift, and addition to an unshared electron pair with the formation of an ylide (see 11.16). A triplet carbene, in contrast, undergoes a nonstereospecific cycloaddition to a π bond (3) and H-abstraction with the formation of a radical. Despite the relatively fast intersystem crossing to its triplet ground state, the initially formed singlet state can be trapped in an intermolecular reaction by a second component, e.g., an alkene present in excess.[61]

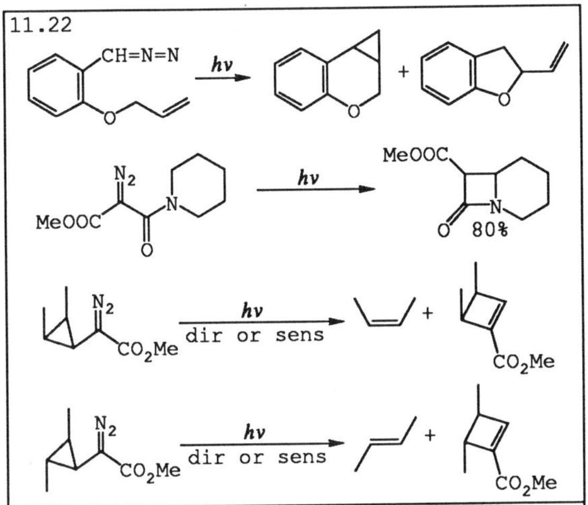

11.22. Intramolecular insertion of a carbene into an allylic C—H bond competes with its addition to the double bond. If no double bond is present, carbene insertion into the C—H bond occurs exclusively.[64] Lack of a significant difference in the stereospecificity of the direct or fluorenone-sensitized photolysis of a cyclopropyldiazoacetate is caused by the higher reactivity of the singlet state, assumed to be in equilibrium with the triplet state.[65,66]

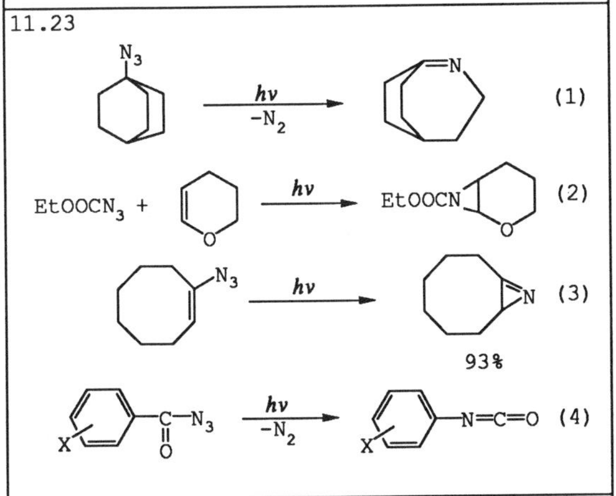

11.23. Similar to diazo compounds, azides lose molecular nitrogen in their excited singlet or triplet states with the formation of singlet or triplet nitrenes in a spin-specific way. The nitrenes exhibit reactions similar to carbenes.[67] Irradiation of azides can lead to a [1,2] hydrogen or alkyl shift (1)[68]; with a C=C or N=N bond the corresponding aziridines (2)[69] or triaziridines[70] are formed. Direct irradiation of vinylazides produces derivatives of 2H-azirine by the intramolecular addition of a nitrene to the C=C bond (3).[71] Acyl azides can be subjected to a photochemical Curtius reaction, which is not limited to carbocyclic compounds (4).[72]

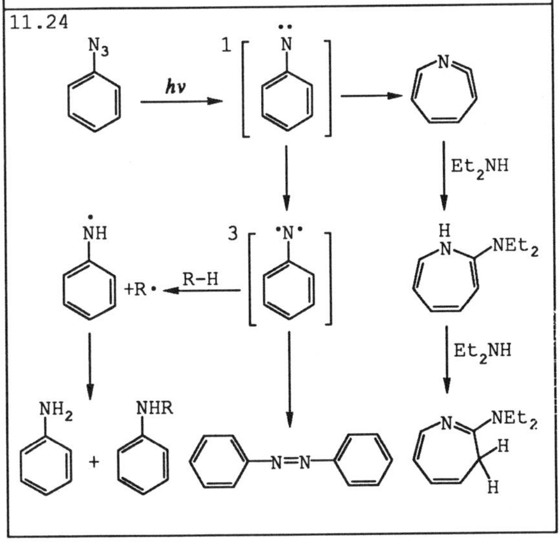

11.24. Irradiation of phenylazide in the presence of an aliphatic amine produces a derivative of 3H-azepine by ring expansion.[73] The following mechanism is generally accepted: the molecule of phenylazide in its S_1 (n,π^*) state eliminates N_2 to form a singlet phenylnitrene that rearranges to 1,2-didehydroazepine. The addition of the amine yields 1H-azepine, and a [1,3] hydrogen shift results in the formation of the final product, 3H-azepine.[74,75] The occurrence of triplet phenyl nitrene and its consequent products have also been observed. The photolysis of naphthyl azides does not lead to ring expansion, but to the exclusive formation of azirines.[76] The photochemistry of aryl azides plays an important role in photolabeling of biological macromolecules,[77] as well as in photolithographic applications.[78]

11.25

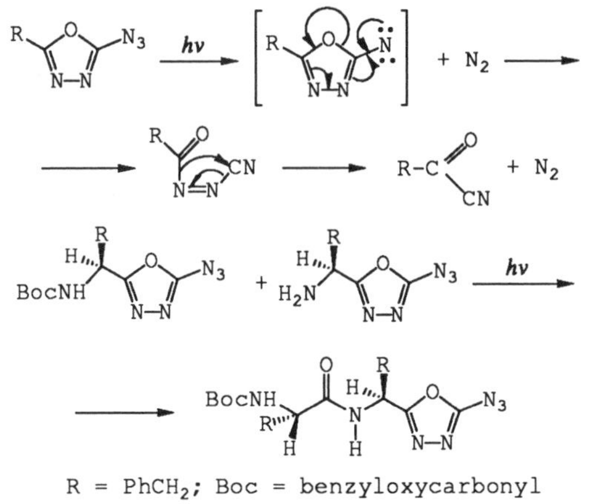

R = PhCH₂; Boc = benzyloxycarbonyl

11.26

11.27

11.28

11.25. Irradiation of 2-substituted 5-azido-1,3,4-oxadiazoles produces nitrenes which, upon rearrangement to unstable azo compounds, lose a second molecule of nitrogen to yield acylcyanides. This serves as a mild acylating method, e.g., in peptide synthesis.

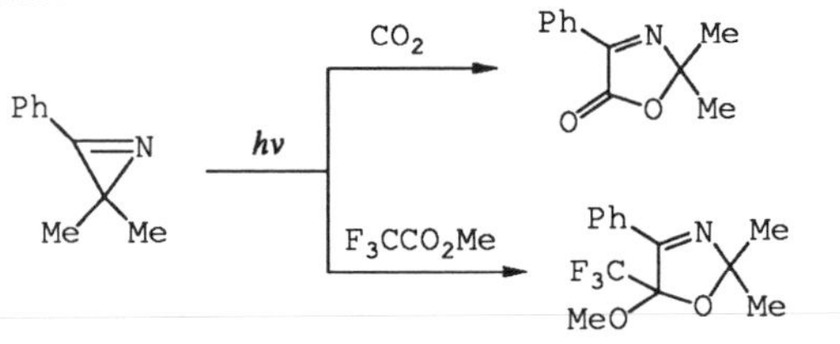

11.26. Irradiation of 2H-azirines causes cleavage of a C—C bond to form a reactive intermediate, a nitrile ylide that has been isolated in a matrix at $-185°C$.[79] It readily undergoes 1,3-dipolar cycloadditions both inter- and intramolecularly with a variety of electron-deficient dipolarophiles to form five-membered ring heterocycles.[80,81] In the presence of ketones the irradiation of 2H-azirines leads to oxazolines;[82] with carbon dioxide an oxazolinone derivative is formed.[83]

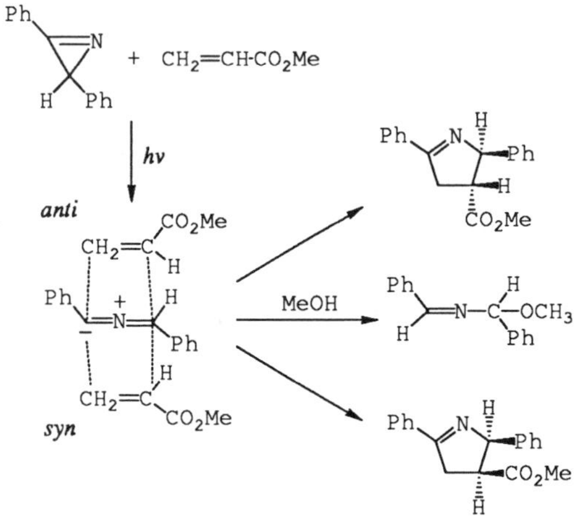

11.27. The intermolecular photocycloaddition of electron-deficient alkenes to arylazirines proceeds in a concerted fashion (stereo- and regiospecifically).[84] The regioselectivity of the process can be easily rationalized with the frontier-molecular orbital approach in which the HOMO of the nitrile ylide and the LUMO of the dipolarophile interact to stabilize the transition state. A bond is formed between the unsubstituted carbon atom of the alkene and the less-substituted carbon atom of the nitrile ylide, i.e., atoms with the highest frontier-electron densities.[85] In the intramolecular arrangement, as demonstrated by 2-styryl-2H-azirine, irradiation leads to ring expansion with the formation of a benzazepine.[86]

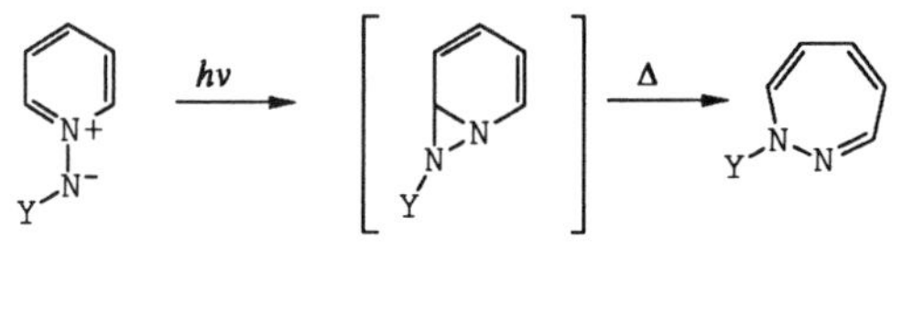

11.28. Irradiation of 1-iminopyridinium ylides leads to thermally unstable 1,2-diazanorcaradienes, which immediately form 1,2-diazepines by disrotatory ring opening.[87] 1,2-diazepines are suitable intermediates for the synthesis of polyheterocyclic compounds, such as homodiazepines, bicyclic β-lactams, etc.[88]

REFERENCES

(1) Barton, D. H. R., Beaton, J. M., Geller, L. E., and Pechet, M. M. *J. Am. Chem. Soc.* **1960**, *82*, 2640.

(2) For reviews on the Barton reaction, see (a) Brun, P. and Waegell, B. In *Reactive Intermediates*. Plenum: New York, 1983, Vol. 3, p. 367, and references therein. (b) Akhtar, A. *Adv. Photochem.* **1964**, *2*, 263. (c) Boar, R. B. In *Methoden der Organischen Chemie Houben-Weyl*. G. Thieme: Stuttgart, 1975, Vol. 4/5a, p. 717. (d) Barton, D. H. R. *Pure Appl. Chem.* **1968**, *16*, 1.

(3)(a) Barton, D. H. R., Beaton, J. M., Geller, L. E., and Pechet, M. M. *J. Am. Chem. Soc.* **1961**, *83*, 4076. (b) For a theoretical study of the Barton reaction, see Dorigo, A. E., McCarrick, M. A., Loncharich, R. J., and Houk, K. N. *J. Am. Chem. Soc.* **1990**, *112*, 7508.

(4) Barton, D. H. R. and Beaton, J. M. *J. Am. Chem. Soc.* **1960**, *82*, 2641.

(5) For another example of the Barton reaction in steroid chemistry, see Nussbaum, A. L., Carlon, F. E., Olivevo, E. P., Townley, E., Kabasakalian, P., and Barton, D. H. R. *Tetrahedron*. **1962**, *18*, 373.

(6) Corey, E. J., Arnett, J. F., and Widiger, G. N. *J. Am. Chem. Soc.* **1975**, *97*, 430.

(7)(a) Nougier, R. and Surzur, J.-M. *Bull. Soc. Chim. France*. **1973**, 2399. (b) Rieke, R. D. and Moore, N. A. *J. Org. Chem.* **1972**, *37*, 413.

(8) For an excellent review, see Morrison, A. H. In *The Photochemistry of Nitro and Nitroso Groups*. Fever, H., ed. Wiley-Interscience: New York, 1969, Vol. 1, p. 165.

(9) Cummings, R. T. and Kraft, G. A. *Tetrahedron Lett.* **1988**, *29*, 65.

(10) For recent reports on the photochemistry of o-nitrobenzyl compounds, see (a) Gravel, D., Hebert, J., and Thoraval, D. *Can. J. Chem.* **1982**, *61*, 400. (b) Schupp, H., Wong, W. K., and Schnabel, W. *J. Photochem.* **1987**, *36*, 85.

(11) Peyser, J. R. and Flechtner, T. W. *J. Org. Chem.* **1987**, *52*, 4645.

(12) Ajayaghosh, A. and Pillai, V. N. R. *J. Org. Chem.* **1987**, *52*, 5714.

(13) Pillai, V. N. R. *Synthesis*. **1980**, 1.

(14) Yamada, K., Tanaka, S., Naruchi, K., and Yamamoto, M. *J. Org. Chem.* **1982**, *47*, 5283.

(15) Yamada, K., Kishikawa, K., and Yamamoto, M. *J. Org. Chem.* **1987**, *52*, 2327.

(16) For an exhaustive review, see Padwa, A. *Chem. Rev.* **1977**, *77*, 37.

(7)(a) Anderson, D. G. and Wettermark, G. *J. Am. Chem. Soc.* **1965**, *87*, 1433. (b) Fischer, E. and Frey, Y. *J. Chem. Phys.* **1957**, *27*, 808.

(18) For reviews, see (a) Blackburn, E. V. and Timmons, C. J. *Quart. Rev.* **1969**, *23*, 482. (b) Stermitz, F. R. *Org. Photochem.* **1967**, *1*, 248.

(19)(a) Koch, T. H. and Howard, K. H. *J. Am. Chem. Soc.* **1975**, *97*, 7288. (b) Koch, T. H. and Rodehorst, R. M. *J. Am. Chem. Soc.* **1975**, *97*, 7298.

(20) Huffmann, K. R., Loy, M., and Ullman, E. F. *J. Am. Chem. Soc.* **1965**, *87*, 5417.

(21) For reviews, see (a) Mariano, P. S. *Org. Photochem.* **1987**, *9*, 1. (b) Mariano, P. S. *Acc. Chem. Res.* **1983**, *16*, 130. (c) Mariano, P. S. *Tetrahedron*. **1983**, *39*, 3845. (d) Mattay, J. *Angew. Chem., Int. Ed. Engl.* **1987**, *26*, 825.

(22)(a) Cho, I. S., Tu, C. L., and Mariano, P. S. *J. Am. Chem. Soc.* **1990**, *112*, 3594. (b) Ohga, K., Yoon, U. C., and Mariano, P. S. *J. Org. Chem.* **1984**, *49*, 213.

(23) For reviews of acyclic diazenes, see (a) Engel, P. S. *Chem. Rev.* **1980**, *80*, 99. (b) Engel, P. S. and Nalepa, C. J. *Pure Appl. Chem.* **1980**, *52*, 2621. (c) Engel, P. S. and Steel, C. *Acc. Chem. Res.* **1973**, *6*, 275. (d) For a theoretical study of the stabilities of several acyclic diazenes, see McKee, M. L. *J. Am. Chem. Soc.* **1990**, *112*, 7957. (e) For the *cis-trans* isomerization of [2.2](4,4')azobenzenophane, see Tamaoki, N., Koseki, K., and Yamaoka, T. *Angew. Chem., Int. Ed. Engl.* **1990**, *29*, 105. See also Tamaoki, N., Ogata, K., Koseki, K., and Yamaoka, T. *Tetrahedron*. **1990**, *46*, 5931.

(24)(a) Engel, P. S. *J. Am. Chem. Soc.* **1961**, *91*, 6903. (b) Clark, W. D. K. and Steel, C. *J. Am. Chem. Soc.* **1971**, *93*, 6347.

(25)(a) Abram, I. I., Milne, G. S., Solomon, B. S., and Steel, C. *J. Am. Chem. Soc.* **1969**, *91*, 1220. (b) El-Sayed, M. A. *Acc. Chem. Res.* **1968**, *1*, 8.

(26) Collier, S. S., Slater, D. H., and Calvert, J. G. *Photochem. Photobiol.* **1968**, *7*, 737.

(27)(a) Chae, W. K., Baugham, S. A., Engel, P. S. Bruch, M., Özmeral, C., Szilagyi, S., and Timberlake, J. W. *J. Am. Chem. Soc.* **1981**. *103*, 4824. (b) Azocyclopropane does not lose N_2 thermally or photochemically. See Engels, P. S. and Bodager, G. A. *J. Org. Chem.* **1988**, *53*, 4748.

(28) Porter, N. A., Landis, M. E., and Marnett, L. J. *J. Am. Chem. Soc.* **1971**, *93*, 795.

(29) For a review of the photochemistry of cyclic 1,2-diazenes, see Adam, W. and DeLucchi, O. *Angew. Chem., Int. Ed. Engl.* **1980**, *19*, 762.

(30)(a) Adam, W. and Dörr, M. *J. Am. Chem. Soc.* **1987**, *109*, 1240. (b) Reedich, D. E. and Sheridan, R. S. *J. Am. Chem. Soc.* **1988**, *110*, 3697. (c) For a review of localized triplet biradicals, see Adam, W., Grabonski, S., and Wilson, R. M. *Acc. Chem. Res.* **1990**, *23*, 165.

(31) For recent reports of photoextrusion of N_2 from cyclic azolkanes, see (a) Adam, W., Hannemann, K., Peters, E. M., Peters, K., von Schnering, H. G., and Wilson, R. M. *J. Am. Chem. Soc.* **1987**, *109*, 5250. (b) Jain, R., McElwee-White, L., and Dougherty, D. A. *J. Am. Chem. Soc.* **1988**, *110*, 152. (c) For a recent example of photolysis of polycyclic azoalkanes, see Christl, M., Freund, S., Hennenberger, H., Kraft, A., Hauck, J., and Irngartinger, H. *J. Am. Chem. Soc.* **1988**, *110*, 3263. (d) For a photodissociation study of 2,3-diazabicyclo[2.2.1]hept-2-ene in the vaporphase, see Adams, J. S., Weisman, R. B., and Engel, P. S. *J. Am. Chem. Soc.* **1990**, *112*, 9115.

(32) van Auken, T. V. and Rinehart, K. L. *J. Am. Chem. Soc.* **1962**, *84*, 3736.

(33) The method has also been used for the preparation of cycloproparenes: See (a) Billups, W. E., Rodin, W. A., and Haley, M. M. *Tetrahedron.* **1988**, *44*, 1305 (a review). (b) Bambal, R., Fritz, H., Rihs, G., Tschamber, T., and Streith, J. *Angew. Chem., Int. Ed. Engl.* **1987**, *26*, 668 (cyclopropapyridine).

(34) For exhaustive reviews, see (a) Meier, H. and Zeller, K.-P. ''Thermische und Photochemische Stickstoff-Cycloeliminierungen,'' *Angew. Chem.* **1977**, *89*, 876. (b) Wentrup, C. ''Rearrangements and Interconversions of Carbenes and Nitrenes,'' *Topics Curr. Chem.* **1976**, *62*, 173.

(35)(a) For the synthesis of [3.5.4]fenestrane by photolysis of diazatricyclodecene, see Brinker, U. H., Schrievers, T., and Xu, L. *J. Am. Chem. Soc.* **1990**, *112*, 8609. (b) For the synthesis of triaxane by photolysis of polycyclic azoalkanes, see Garat, P. J. and White, J. F. *J. Org. Chem.* **1977**, *42*, 1733. For a recent example, see Paquette, L. A., Kesselmayer, M. A., and Rogers, R. D. *J. Am. Chem. Soc.* **1990**, *112*, 284. (c) However, photolysis of 1,3,4-oxadiazolines leads to diazoalkanes. See Majchzak, M. W., Bekhazi, M., Tse-Sheepy, I., and Warkentin, J. *J. Org. Chem.* **1989**, *54*, 1842.

(36)(a) Katz, T. J. and Acton, N. *J. Am. Chem. Soc.* **1973**, *95*, 2738. (b) Turro, N. J., Renner, C. A.,

Waddell, W. H., and Katz, T. J. *J. Am. Chem. Soc.* **1976**, *98*, 4320.

(37) Piers, E., Gerghty, M. B., Smilie, R. D., and Soucy, M. *Can. J. Chem.* **1975**, *53*, 2849.

(38)(a) For X = Cl, see Andrews, S. D. and Day, A. C. *J. Chem. Soc. (B).* **1968**, 1271. (b) For X = OCOEt, see Day, A. C. and Whiting, M. C. *J. Chem. Soc. (C).* **1966**, 464.

(39) For a spin correlation effect on product distribution in the photolysis of a polycyclic azo compound, see Zimmerman, H. E., Boetcher, R. J., Buehler, N. E., and Keck, G. E. *J. Am. Chem. Soc.* **1975**, *97*, 5635.

(40) Berning, W., Hünig, S., and Prokschy, F. *Chem. Ber.* **1984**, *119*, 1455.

(41) Ritter, G., Häfelinger, G., Lüddecke, E., and Rau, H. *J. Am. Chem. Soc.* **1989**, *111*, 4627.

(42)(a) Firl, J. and Sommer, S. *Tetrahedron Lett.* **1972**, 4713. (b) von Gustorf, E. K., White, D. V., Kim, B., Hess, D., and Leitich, J. *J. Org. Chem.* **1970**, *35*, 1155.

(43) LeBlanc, Y., Fitzsimmons, B. J., Springer, J. P., and Rokach, J. *J. Am. Chem. Soc.* **1989**, *111*, 2995.

(44) Becker, H. G. O., Hoffmann, G., and Israel, G. *J. Prakt. Chem.* **1977**, *319*, 1021.

(45) Lee, W. E., Calvert, J. G., and Malmberg, E. W. *J. Am. Chem. Soc.* **1961**, *83*, 1928.

(46)(a) Dürr, H., and Kober, H. *Tetrahedron Lett.*, **1972**, 1259. (b) For the formation of the cycloheptatriene/norcaradiene systems in the decomposition of diaryldiazomethanes in benzene, see Hannemann, K. *Angew. Chem., Int. Ed. Engl.* **1988**, *27*, 284.

(47)(a) Zimmerman, H. E., Juers, D. F., McCall, J. M., and Schreder, B. *J. Am. Chem. Soc.* **1971**, *93*, 3662. (b) For a mechanistic study, see Wilson, R. M. and Schnapp, K. A. *J. Am. Chem. Soc.* **1988**, *110*, 982.

(48) For a preparative example, see LaBar, R. A. *Org. Photochem. Synth.* **1976**, *2*, 90.

(49) Shields, C. J. and Schuster, G. B. *Tetrahedron Lett.* **1987**, *28*, 853.

(50) Turro, N. J. and Cha, Y. *Tetrahedron Lett.* **1987**, *28*, 1723.

(51) For reviews of the photo-Wolff rearrangement, see (a) Meier, H. and Zeller, K. P. *Angew. Chem., Int. Ed. Engl.* **1975**, *14*, 32. (b) Ando, W. In *The Chemistry of the Diazonium and Diazo Groups*, Patai, S., ed. Wiley: New York, 1978, Part 1, p. 458.

(c) Regitz, M. and Maas, G. *Diazo Compounds: Properties and Synthesis*. Academic Press: Orlando, Florida, 1986, p. 185.

(52)(a) Maier, G., Reisenauer, H. P., and Sayrac, T. *Chem. Ber.* **1982**, *115*, 2192. (b) Torres, M., Bourdelande, J. L., Clement, A., and Strausz, O. P. *J. Am. Chem. Soc.* **1983**, *105*, 1698.

(53)(a) Cava, M. P. and Moroz, E. *J. Am. Chem. Soc.* **1962**, *84*, 115. (b) Hodakowski, L. and Cupas, C. A. *Tetrahedron Lett.* **1973**, 1009. (c) Grychtol, K., Musso, H., and Oth, J. F. M. *Chem. Ber.* **1972**, *105*, 1798. (d) Wiberg, K. B. and Ubersax, R. W. *J. Org. Chem.* **1972**, *37*, 3827. (e) Wagner, P. J. and Chiu, C. *J. Am. Chem. Soc.* **1979**, *101*, 7134. (f) For the photolysis of matrix-isolated α-diazoketones, see Bachmann, C., N'Guessan, T. Y., Debü, F., Monnier, M., Pourcin, J., Aycard, J.-P., and Bodot, H. *J. Am. Chem. Soc.* **1990**, *112*, 7488.

(54)(a) For the synthesis of [7] paracyclophane, see Allinger, N. L. and Walter, T. J. *J. Am. Chem. Soc.* **1972**, *94*, 9267. (b) For the synthesis of [6](1,4)naphthalenophane and [6](1,4)anthracenophane, see Tobe, Y., Takashashi, T., Ishikawa, T., Yoshimura, M., Suwa, M., Kobiro, K., Kakiuchi, K., and Gleiter, R. *J. Am. Chem. Soc.* **1990**, *112*, 8889. (c) For the synthesis of perannulane, see Marshall, J. A., Peterson, J. C., and Lebioda, L. J. *J. Am. Chem. Soc.* **1983**, *105*, 6515. (d) For the synthesis of diasterane, see Otterbach, A. and Musso, H. *Angew. Chem., Int. Ed. Engl.* **1987**, *26*, 554. (e) For the photolysis of 1,3-*bis*(diazo)indan-2-one in Ar matrix at 10 K, see Murata, S., Yamamoto, T., Tomioka, H., Lee, H. K., Kim, H. R., and Yabe, A. *J. Chem. Soc., Chem. Commun.*, **1990**, 1258.

(55)(a) Trotter, J., Gibbons, C. S., Nakatsuka, N., and Masamune, S. *J. Am. Chem. Soc.* **1967**, *89*, 2793. (b) Closs, G. L. and Larrabee, R. B. *Tetrahedron Lett.*, **1965**, 267. (c) For the preparative procedure, see Nakatsuka, N. and Masamune, S. *Org. Photochem. Synth.* **1976**, *2*, 57.

(56) Danheiser, R. L., Brisbois, R. G., Kowalczyk, J. J., and Miller, R. F. *J. Am. Chem. Soc.* **1990**, *112*, 3093.

(57)(a) Hlubucek, J. R. and Lowe, G. *J. Chem. Soc., Chem. Commun.* **1974**, 419. (b) Lowe, G. and Ridley, D. D. *J. Chem. Soc., Chem. Commun.* **1973**, 328. (c) For a tandem photochemical synthesis of N-amino-β-lactams for pyrazolidin-3-ones, see Perri, S. T., Slater, S. C., Toske, S. G., and White, J. D. *J. Org. Chem.* **1990**, *55*, 6035.

(58) Rosati, R. L., Kapili, L. V., Morrissey, P., Bordner, J., and Subramanian, E. *J. Am. Chem. Soc.* **1982**, *104*, 4262.

(59)(a) Maier, G., Reisenauer, H. P., Schäfer, U., and Balli, H. *Angew. Chem., Int. Ed. Engl.* **1988**, *27*, 566. (b) For the first dioxide of carbon with an even number of carbon atoms (C_4O_2), see Maier, G., Reisenauer, H. P., Balli, H., Brandt, W., and Janoschek, R. *Angew. Chem., Int. Ed. Engl.* **1990**, *29*, 905.

(60) Li, Y. Z. and Schuster, G. B. *J. Org. Chem.* **1987**, *52*, 4460.

(61) For reviews of the photochemistry of diazo compounds, see (a) Dürr, H. In *Methoden der Organischen Chemie/(Houben-Weyl)*. G. Thieme: Stuttgart, 1975, Vol. 4/5b, p. 1158. (b) Dürr, H. *Topics Curr. Chem.* **1976**, *55*, 87. (c) Regitz, M. and Maas, G. *Diazo Compounds: Properties and Synthesis*. Academic: New York, 1986.

(62) Kopecky, K. R., Hammond, G. S., and Leersmakers, P. A. *J. Am. Chem. Soc.* **1962**, *84*, 1015.

(63) Eaton, P. E. and Hoffmann, K.-L. *J. Am. Chem. Soc.* **1987**, *109*, 5285.

(64)(a) Lowe, G. and Parker, J. *Chem. Commun.* **1971**, 577. (b) Brunwin, D. M., and Lowe, G. *Chem. Commun.* **1972**, 589. (c) For utilization in the synthesis of [3.4.5]fenestrane, see Brinker, U. H., Schrievers, T., and Xu, L. *J. Am. Chem. Soc.* **1990**, *112*, 8609.

(65)(a) Shevlin, P. B. and McKee, M. L. *J. Am. Chem. Soc.* **1989**, *111*, 519. (b) Gallucci, R. R. and Jones, M., Jr. **1976**, *98*, 7704.

(66) For the photocycloaddition reactions of simple carbonyl compounds with diazomalonate, see L'Esperance, R. P., Ford, T. M., and Jones, M. *J. Am. Chem. Soc.* **1988**, *110*, 209.

(67) For reviews of the photochemistry of azides, see (a) Scriven, E. F. V. and Turnbull, K. *Chem. Ber.* **1988**, *88*, 297. (b) Scriven, E. F. V. ed. *Azides and Nitrenes, Reactivity and Utility*. Academic Press: New York, 1984. (c) Dürr, H. and Kober, H. *Topics Curr. Chem.* **1976**, *66*, 89.

(68)(a) Michl, J., Radziszewski, J. G., Downing, J. W., Kopecký, J., Kaszynski, P., and Miller, R. B. *Pure Appl. Chem.* **1987**, *59*, 1613. (b) For the photolysis of 1-azido-4-methylcubane, see Eaton, P. E. and Hormann, R. E. *J. Am. Chem. Soc.* **1987**, *109*, 1268. (c) For the photolysis of the matrix-isolated 3-azidonoradamantane leading to a highly twisted bridgehead imine, see Radziszewski, G., Downing, J. W., Wentrup, C., Kaszynski, P., Jawdosiuk, M., Kovacic, P., and Michl, J. *J. Am. Chem. Soc.* **1984**, *106*, 7996.

(69)(a) Carbonyl nitrenes, sulfonyl nitrenes, alkoxy nitrenes, phosphoryl nitrenes, etc., produce aziridines. See Ref. 68(b). (b) For functionalization of olefins by alkoxyimidoyl nitrenes, see Subbaraj, A., Rao, S. O., and Lwowski, W. *J. Org. Chem.* **1989**, *54*, 3945. (c) For functionalization of the benzene ring by imidoyl nitrenes, see Dabbagh, H. A. and Lwowski, W. *J. Org. Chem.* **1989**, *54*, 3952.

(70)(a) Marterer, W., Fritz, H., and Prinzbach, H. *Tetrahedron Lett.* **1987**, 5497.   (b) Prinzbach, H. and Klinger, O. *Angew. Chem., Int. Ed. Engl.* **1987**, *26*, 566.

(71)(a) Ford, R. G. *J. Am. Chem. Soc.* **1977**, *99*, 2389. (b) Hassner, A. and Fowler, F. W. *Tetrahedron Lett.* **1967**, 1545.

(72) Eibler, E., Käsbauer, J., Pohl, H., and Saver, J. *Tetrahedron Lett.*, **1987**, *28*, 1097.

(73)(a) Leyva, E., Platz, M. S., Persy, G., and Wirz, J. *J. Am. Chem. Soc.* **1986**, *108*, 3783, and references therein.   (b) Albini, A., Bettinetti, G., and Minoli, G. *J. Org. Chem.* **1987**, *52*, 1245.

(74)(a) Li, Y. Z., Kirby, J. P., George, M. W., Poliakoff, M., and Schuster, G. B. *J. Am. Chem. Soc.*, **1988**, *110*, 8092.   (b) Drzaic, P. S. and Brauman, J. I. *J. Am. Chem. Soc.* **1984**, *196*, 3443.   (c) Drzaic, P. S. and Brauman, J. I. *J. Phys. Chem.* **1984**, *88*, 5285.

(75) For quantum mechanical calculations, see (a) Frenking, G. and Schmidt, J. *Tetrahedron.* **1984**, *40*, 2123.   (b) Mauridis, A. and Harrison, J. F. *J. Am. Chem. Soc.* **1980**, *102*, 7651.

(76)(a) Leyva, E. and Platz, M. S. *Tetrahedron Lett.* **1987**, *28*, 11.   (b) Autrey, T. and Schuster, G. B. *J. Am. Chem. Soc.* **1987**, *109*, 5814.   (c) Hayes, J. C. and Sheridan, R. S. *J. Am. Chem. Soc.* **1990**, *112*, 5879.

(77)(a) Kanakarajan, K., Goodrich, R., Young, M. J. T., Soundararajan, S., and Platz, M. S. *J. Am. Chem. Soc.* **1988**, *110*, 6536.   (b) Sigman, M. E., Autrey, T., and Schuster, G. B. *J. Am. Chem. Soc.* **1988**, *110*, 4297.

(78) Meijer, E. W., Nijhuis, S., and van Vroonhoven, F. C. B. M. *J. Am. Chem. Soc.* **1988**, *110*, 7209.

(79)(a) Sieber, W., Gilgen, P., Chaloupka, S., Hansen, H. J., and Schmid, H. *Helv. Chim. Acta.* **1973**, *56*, 1679.   (b) Orahovats, A., Heimgartner, H., Schmid, H., and Heinzelmann, W. *Helv. Chim. Acta.* **1975**, *58*, 2662.

(80)(a) For a review, see Padwa, A. *Acc. Chem. Res.* **1976**, *9*, 371.   (b) Anderson, D. F. and Hassner, A. *Synthesis.* **1975**, 483.   (c) Padwa, A. *Angew. Chem., Int. Ed. Engl.* **1976**, *15*, 123.

(81) Gilgen, P., Heimgartner, H., and Schmid, H. *Heterocycles.* **1977**, 143.

(82) For the preparative procedure, see Vebelhart, P., Gilgen, P., and Schmid, H. *Org. Photochem. Synth.*, **1976**, *2*, 72.

(83) For the preparative procedure, see Padwa, A. and Wetmore, S. I., Jr. *Org. Photochem. Synth.* **1976**, *2*, 87.

(84) Padwa, A., Dharan, M., Smolanoff, J., and Wetmore, S. I., Jr. *J. Am. Chem. Soc.* **1973**, *95*, 1945, 1954.

(85) Padwa, A. and Smolanoff, J. *Tetrahedron Lett.* **1973**, 342.

(86)(a) Fleming, I. *Frontier Orbitals and Organic Chemical Reactions.* Wiley: New York, 1976. (b) Padwa, A. and Smolanoff, J. *Tetrahedron Lett.* **1974**, 33.

(87)(a) Streith, J. *Heterocycles.* **1977**, *6*, 2021. (b) For the experimental procedure for 1-ethoxycarbonyl-1H-1,2-diazepine, see Nastasi, M., Schilling, E. S., and Streith, J. *Org. Photochem. Synth.* **1976**, *2*, 60.

(88) Snieckus, V. and Streith, J. *Acc. Chem. Res.* **1981**, *14*, 348.

Chapter 12

PHOTOCHEMISTRY OF HALOGEN-CONTAINING COMPOUNDS

PHOTOCHEMISTRY OF HALOGEN-CONTAINING COMPOUNDS

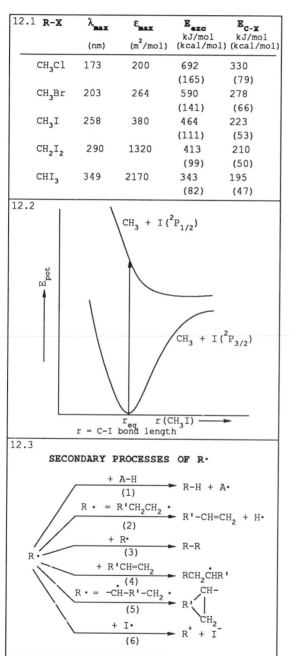

12.1 R-X	λ_{max} (nm)	ε_{max} (m²/mol)	E_{exc} kJ/mol (kcal/mol)	E_{C-X} kJ/mol (kcal/mol)
CH_3Cl	173	200	692 (165)	330 (79)
CH_3Br	203	264	590 (141)	278 (66)
CH_3I	258	380	464 (111)	223 (53)
CH_2I_2	290	1320	413 (99)	210 (50)
CHI_3	349	2170	343 (82)	195 (47)

12.2

$CH_3 + I(^2P_{1/2})$

E_{pot}

$CH_3 + I(^2P_{3/2})$

r_{eq} $r(CH_3I) \longrightarrow$
r = C-I bond length

12.3

SECONDARY PROCESSES OF R·

+ A-H
(1) → R-H + A·

R· = R'CH₂CH₂·
(2) → R'-CH=CH₂ + H·

+ R·
(3) → R-R

+ R'CH=CH₂
(4) → RCH₂ĊHR'

R· = ·ĊH-R'-CH₂·
(5) → R'⟨CH-CH₂⟩

+ I·
(6) → R⁺ + I⁻

12.1. The longest wavelength bands in the UV spectra of haloalkanes correspond to $n \rightarrow \sigma^*$ transitions, i.e., the excitation of an n electron of the halogen (X) atom to the σ^* MO of the C—X bond. As the atomic number of X increases, the absorption band shifts toward the red (bathochromic shift) and its intensity increases with the number of halogen atoms. The photolysis of alkyl halides leads to effective homolytic cleavage of the C—X bond, in agreement with the observation that the excitation energy (E_{exc}) is substantially higher than the bond dissociation energy (E_{C-X}).[1]

12.2. Homolytic cleavage of the C—X bond in haloalkanes in their n,σ^* state produces a halogen atom and an alkyl radical in the primary photochemical step. The halogen atom can be formed in either an electronically excited state or the ground state. In the photolysis of methyl iodide the formation of the two states occurs in a ratio of 3.5:1. The alkyl radical possesses the excess energy (E_{exc}) as increased translational energy, reflected in its collisions with, and hydrogen abstraction from, surrounding molecules, especially those of the solvent.[2]

12.3. Most photochemical reactions of alkyl halides proceed by a photoinitiated free-radical reaction mechanism.[3,4] The alkyl radical from the homolytic dissociation of the C—X bond in the primary process undergoes the following secondary reactions: (1) reduction to the corresponding alkane, usually by abstraction of hydrogen atoms from the solvent; (2) elimination of a hydrogen atom to form an alkene and a hydrogen halide; (3) recombination with another alkyl radical to form an alkane; (4) addition to an alkene; (5) intramolecular cyclization; and (6) electron transfer to produce a carbocation.

12.4

$$F_2BrC-Br + CH_2=CH-SiMe_3 \xrightarrow{h\nu}$$

$$\longrightarrow \underset{\underset{Br}{|}}{F_2BrCCH_2CHSiMe_3} \quad 79\% \text{ ref.5}$$

$$Br_3C-Br + Ph-CH=CH_2 \xrightarrow{h\nu}$$

$$\longrightarrow \underset{\underset{Br}{|}}{Ph-CHCH_2CBr_3} \quad 96\% \text{ ref.6}$$

$$Cl_3C-Br + CH_2=CH-OCOCH_3 \xrightarrow{h\nu}$$

$$\longrightarrow \underset{\underset{Br}{|}}{Cl_3CCH_2CH-OCOCH_3} \quad 90\% \text{ ref.7}$$

12.5

12.6

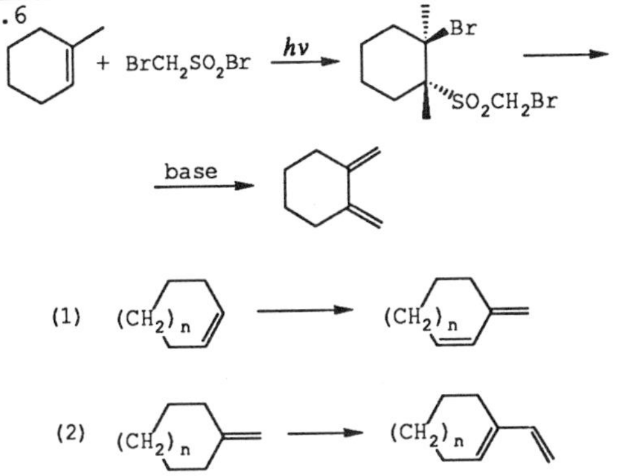

	1	*2*
n=1	41%	53%
2	31%	74%
3	49%	75%

12.4. Of the above photoreactions of haloalkanes, (4), (5), and (6) are particularly significant. Photoinitiated addition to double bonds is the most important reaction of *polyhalogenated alkanes*. Because they absorb at longer wavelengths and have larger extinction coefficients, polyhaloalkanes in which the weakest C—X bond breaks preferentially photolyze more readily than monohaloalkanes. They typically undergo photoaddition to terminal alkenes in reactions similar to peroxide-initiated processes.[4(a)] This addition usually occurs with a high yield and exhibits anti-Markovnikov orientation.

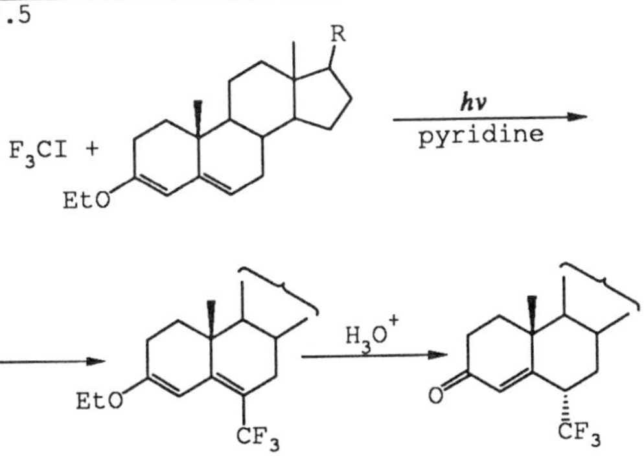

12.5. The order of reactivity of some tetrahalomethanes is: CBr_4 > Cl_3CBr > F_3CI > CCl_4.[5] Telomers are common byproducts in these reactions with CCl_4 unless a large excess is used. However, only a slight excess of CBr_4 is necessary to avoid telomer formation. Because of the higher reactivity of CBr_4, its photoaddition can be effected in CCl_4 using VIS radiation.[4(d)] There is also a difference in the bond scission in $CHBr_3$ and $CHCl_3$. The weakest bond in bromoform is the C—Br bond, while in chloroform it is the C—H bond. When the photoaddition of a haloalkane to an alkene is carried out in the presence of a base, the corresponding substituted alkene is obtained directly.

12.6. The recently reported photoaddition of bromomethanesulfonyl bromide (BMSB) to an alkene and the subsequent use of the Ramberg-Backlund reaction[8] (treatment of the adduct with a strong base) provides a simple general method for the synthesis of conjugated dienes with one carbon atom more than the original alkene.[9] The carbon atoms of the double bond of the alkene are atoms 2 and 3 of the diene system. Potassium *t*-butoxide or preferably, triethylamine followed by t-BuOK, are the best bases for this reaction.

12.7

$$X = (CH_2)_n; \quad Y = Br, I$$

n	Y	ROH	1	2
1	Br	CH_3OH	55%	30%
	I		11%	89%
2	Br		31%	69%
	I		2%	98%
1	Br	$(CH_2OH)_2$	6%	92%
	I		trace	99%

12.8

12.9

12.7. Kropp[10] has demonstrated the complexity of the photolysis of polycyclic bromides and iodides in polar solvents. Although the reaction starts with homolytic cleavage of the C—X bond, irradiation of methanolic solutions of bicyclic bromides or iodides with the halogen atom on the bridgehead carbon gives products from the polar cleavage of the C—C bond, in addition to radical reaction products. The ratio of polar to radical products is directly proportional to the solvent viscosity (the rate of diffusion of radicals out of the solvent cage is less than the rate of electron transfer).[4(g)]

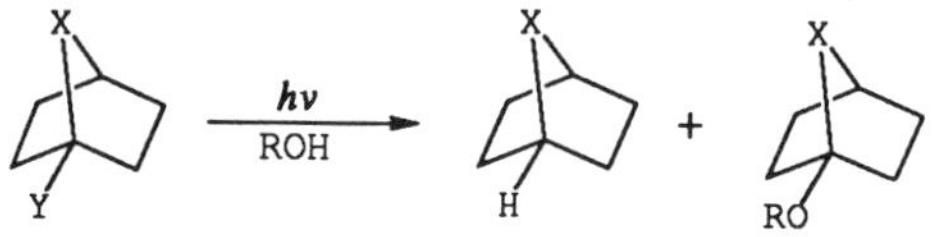

12.8. The homolytic cleavage of the C—X bond, producing a radical pair in the solvent cage, is a primary step.[11] Two competing events can occur: (1) escape of the radical pair from the solvent cage to yield radical-reaction products; (2) electron transfer between the radical pair, leading to an ion pair that forms polar bond association products. The ratio of polar to radical products may be related to the relative energies of the bridgehead cation and radical.[4(g), 12] However, the high ratio (>95:5) of products from 1-iodocubane photomethanolysis is not in agreement with this assumption.[13] The mechanism of the electron transfer is not fully understood, e.g., why it proceeds more readily with iodo than with the more electronegative bromo derivatives.

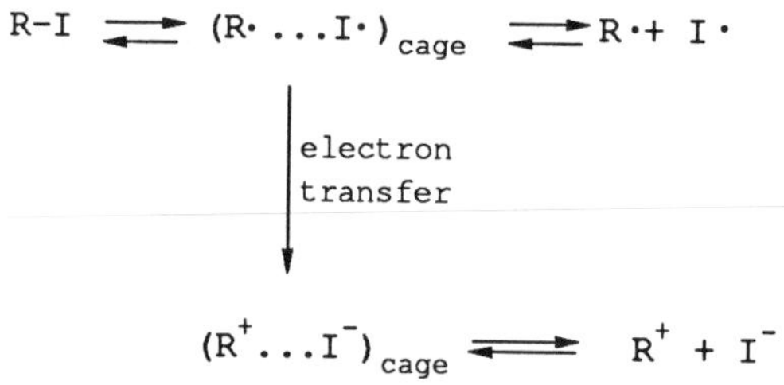

12.9. The photostimulated reaction of 1-haloadamantane[14] or bicyclo-[2.2.2]octane[15] with carbanionic nucleophiles (enolate ions of acetone, acetophenone, or propiophenone)[14] presumably occurs by an $S_{RN}1$ mechanism (inhibition by p-dinitrobenzene) involving an electron transfer. With nitromethane anion, the substitution takes place only in the presence of acetone or acetophenone (entrainment reaction). Small amounts of reduction product are formed and the yield of substitution products increases in the presence of a crown ether.

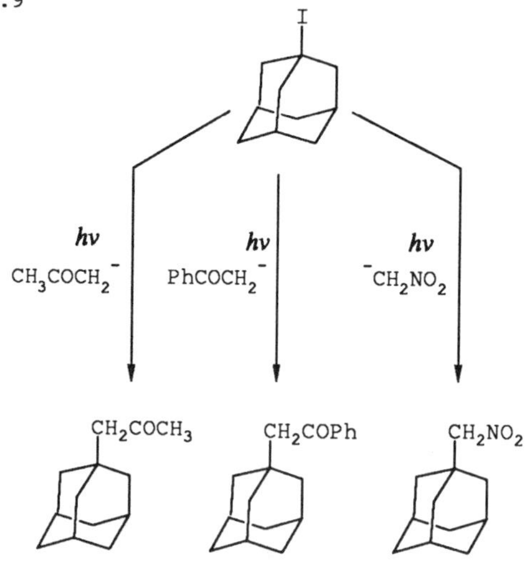

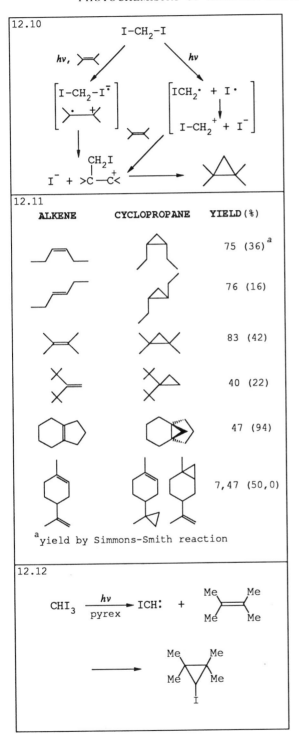

12.10

12.11

ALKENE	CYCLOPROPANE	YIELD(%)
		75 (36)[a]
		76 (16)
		83 (42)
		40 (22)
		47 (94)
		7,47 (50,0)

[a]yield by Simmons-Smith reaction

12.12

$$CHI_3 \xrightarrow[\text{pyrex}]{hv} ICH: + \text{(tetramethylethylene)}$$

12.10. Irradiation of diiodomethane in the presence of an alkene produces the corresponding cyclopropane. The fact that the CH_2 group is not inserted into the $C-H$ bond, along with the high degree of selectivity in the addition to sterically hindered alkenes, excludes the possibility of carbene involvement. The results are compatible with two different mechanisms. First, after the homolytic cleavage of one $C-I$ bond, an electron transfer between the carbon radical and the iodine atom produces the carbocation $I-CH_2^+$, which attacks the double bond.[4(g),16] Second, a photostimulated electron transfer occurs with the formation of a radical-ion pair, with CH_2I_2 as the acceptor. The $CH_2I_2^-$ radical-anion decomposes to yield an iodide ion and an iodomethyl radical (ICH_2), which combines with the alkene radical-cation. Both intermediates have all of the properties of a reagent that transfers a methylene group, i.e., both are electrophilic, do not insert into a $C-H$ bond, and are not sterically demanding.

12.11. This photocyclopropanation method occurs in a wide variety of alkenes and is particularly significant for cyclopropanation of sterically hindered alkenes, where the Simmons–Smith reaction fails or produces low yields. The reaction is stereospecific; *cis(trans)*-alkenes produce *cis(trans)*-cyclopropanes. There is a significant difference between the products of cyclopropanation and the Simmons–Smith method in the cyclopropanation of limonene. Thus, the reaction is preparatively significant and offers a suitable alternative for cyclopropanation in organic synthesis.[16]

12.12. The photolysis of iodoform in the presence of an alkene gives a high yield of an iodocyclopropane.[17] Whether the cyclopropanation proceeds via a carbene, as shown, or by one of the mechanisms discussed above, remains unresolved.

12.13

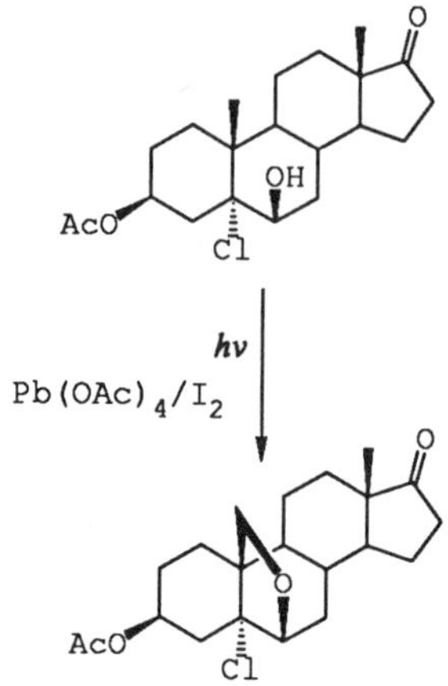

12.14

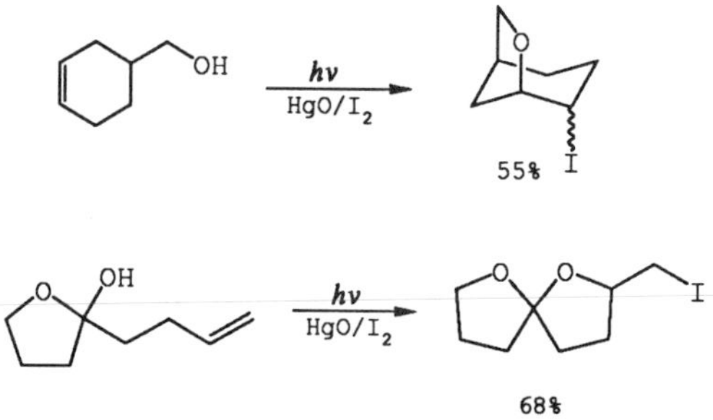

55% I

68%

12.15

35% ref.22

11% ref.23

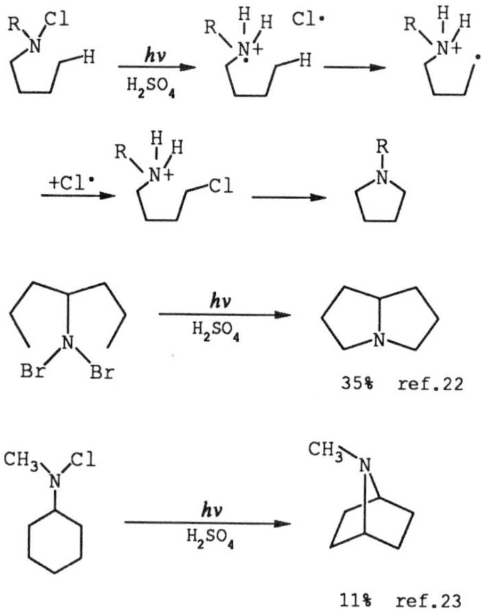

12.13. The photolysis of hypoha-lites is analogous to the Barton reaction (11.2) and yields halogenated alcohols (or cyclic ethers after treatment with a base).[18] The reaction is limited to hy-pohalites in which homolytic cleavage of the O—X bond gives an alkoxy rad-ical that is structurally similar to that obtained by irradiation of nitrites, i.e., one in which a 1,5-hydrogen atom shift from a carbon to an oxygen atom is possible. The reaction is particularly significant for hypoiodites prepared *in situ* by the irradiation of the corre-sponding alcohol with lead tetraacetate or mercuric oxide and iodine in cyclo-hexane, usually in the presence of a base. This reaction has been used to functionalize angular methyl groups in steroids.[19] A disadvantage is its gener-ally low yield; the example shown is an exception.

12.14. The hypoiodite formed *in situ* by the irradiation of a benzene so-lution of an unsaturated alcohol with an excess of HgO + I_2 at 0°C (higher temperatures lower the yield) produces an alkoxy radical whose cyclization re-sults in a five-membered cyclic ether. The reaction is general, leads to tetra-hydrofuran derivatives in satisfactory to excellent yields, and has been uti-lized in the synthesis of a spiroketal with a carbohydrate molecular frag-ment.[20]

12.15. The thermal cyclization of N-haloamines to pyrrolidine deriva-tives, known as the *Hoffmann–Löffler–Freytag* reaction, can be also effected photochemically.[21] The irradiation of a solution of an N-chloroamine in sul-furic or trifluoroacetic acid first pro-duces an aminium radical and a chlo-rine atom by homolytic cleavage. This is followed by a 1,5-hydrogen shift via a cyclic transition state that results in the formation of a C radical, which combines with the chlorine atom to produce a δ-chloroaminium salt. Fi-nally, treatment with a base results in ring closure to an N-alkylpyrrolidine.

12.16

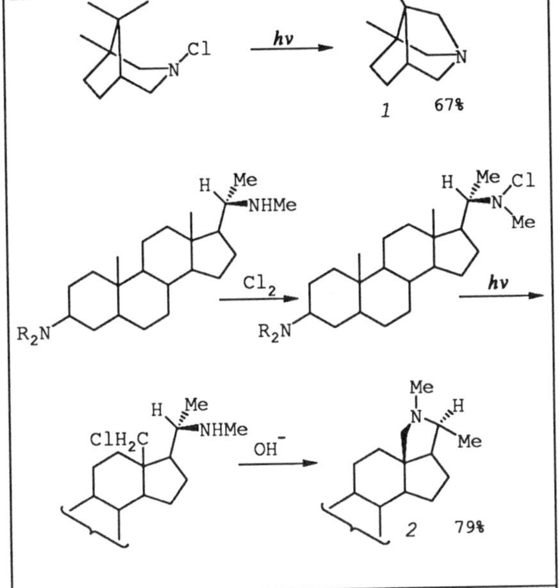

1 67%

2 79%

12.17

12.18

X=Br,I; Y=O,S; R=H,Me,OMe

12.16. The photocyclization reactions of hypohalites and haloamines provide additional examples of remote functionalization of nonactivated saturated carbon atoms. A variety of heterocyclic compounds have been synthesized by this method. In spite of the low yields from the photocyclization, the preparation of certain alkaloids, such as cyclocamphidine[24] and dihydroconessine,[25] makes it a useful synthetic method.[26,27]

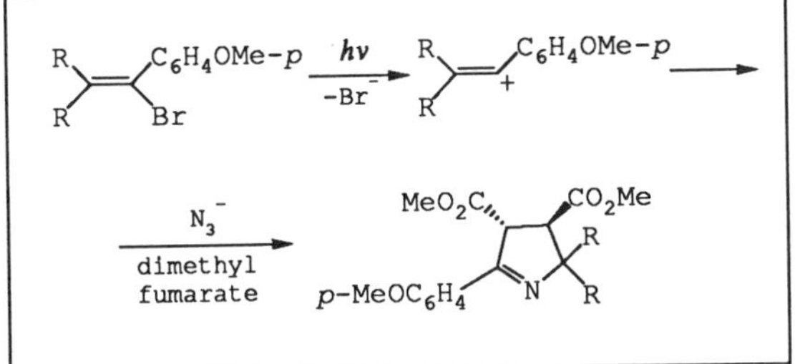

12.17. The mechanism of the reaction of a vinyl bromide with an azide anion, in the presence of dimethyl fumarate, is unresolved. It is questionable whether the reaction begins with the formation of a vinyl carbocation, which is then trapped by the azide anion, or whether it proceeds by an $S_{RN}1$ mechanism (photoinduced electron transfer) with the formation of a vinyl bromide radical-anion (an electron acceptor), which could decompose into a bromide ion and a vinyl radical. Recombination of the vinyl radical with the azide radical would yield the vinyl azide. The irradiation of vinyl azides is a general method for the preparation of 2H-azirines, which subsequently undergo a photochemical reaction with an alkene to give 1,3-dipolar addition products. The overall process requires the participation of three photons.

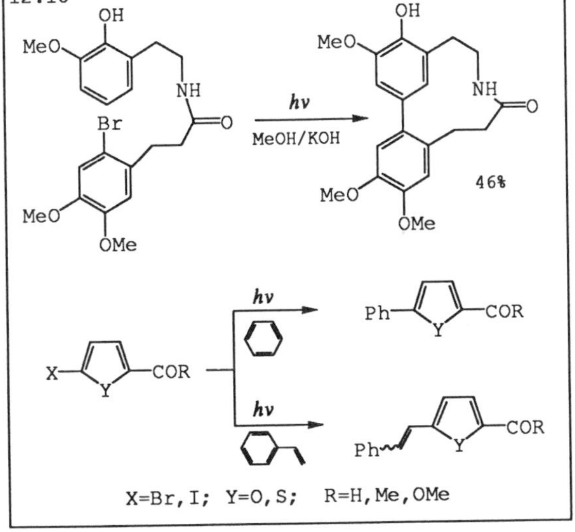

12.18. Photochemical *intra-* and *inter*molecular coupling reactions via *de*hydrohalogenation have been observed between aromatic (benzenoid and heterocyclic) halocompounds and arenes or arylalkenes. The *intra*molecular path has been widely utilized as a key step in the synthesis of various alkaloids in spite of low yields (dehydrohalogenation without coupling is almost always observed).[28,29] With halofurans or thiophenes *inter*molecular photosubstitution products are obtained in high yields, apparently through a photostimulated electron transfer from the benzenoid to the heterocyclic compound with the intermediate formation of radical-ion pairs.[30]

REFERENCES

(1)(a) Porret, D. and Goodeve, C. G. *Proc. Roy. Soc. A.* **1958**, *165*, 31.   (b) Table adapted from 2(b).

(2)(a) Figure adapted from *Faraday Disc. Chem. Soc. No. 53*, **1972**, 132.   (b) Wagniere, G. H. In *The Chemistry of the Carbon-Halogen Bond*, Patai, S., ed. Wiley: New York, 1973, Part 1, p. 1.

(3)(a) Dannenberg, J. J. and Dill, K. *Tetrahedron Lett.* **1972**, 1571.   (b) Bakale, D. K. and Gillis, H. A. *J. Phys. Chem.* **1970**, *74*, 2074.

(4) For reviews on the photochemistry of organic halogen compounds, see   (a) Walling, C. and Huyser, E. S. *Org. Reactions.* **1963**, *13*, 91.   (b) Majer, J. R. and Simmons, J. P. *Adv. Photochem.* **1964**, *2*, 137.   (c) Sosnovsky, G. *Free Radical Reactions in Preparative Organic Chemistry.* Macmillan: New York, 1964.   (d) Elad, D. *Org. Photochem.* **1969**, *2*, 168.   (e) Sammes, P. G. In *The Chemistry of the Carbon-Halogen Bond*. Patai, S., ed. Wiley: New York, 1974, Vol. 2, p. 747.   (f) Dürr, H. In *Methoden der Organischen Photochemie Houben-Weyl*, G. Thieme: Stuttgart, 1975; Vol. 4/5a: Photochemie, p. 628.   (g) Kropp, P. J. *Acc. Chem. Res.* **1984**, *17*, 131.

(5)(a) Haszeldine, R. N. and Steele, B. R. *J. Chem. Soc.* **1955**, 3005.   (b) For the krypton monofluoride laser induced telomerization of CF_3Br with alkenes, see Zhang, L, Fuss, W., and Kompa, K. L. *Ber. Bunsen-Ges. Phys. Chem.* **1990**, *94*, 867, 874.

(6) Geyer, A. M. et. al. *J. Chem. Soc.* **1957**, 4472.

(7) Kharasch, M. S., Jenson, E. V., and Urry, W. H. *J. Am. Chem. Soc.* **1947**, *69*, 1100.

(8)(a) Block, E. and Aslam, M. *J. Am. Chem. Soc.* **1983**, *105*, 6164.   (b) Block, E., Aslam, M., Eswarakrishnan, V., and Wall, A. *J. Am. Chem. Soc.* **1983**, *105*, 6165.   (c) Block, E., Aslam, M., Iyer, R., and Hutchinson, J. *J. Org. Chem.* **1984**, *49*, 3664.   (d) Block, E., Eswarakrishnan, V., and Gegreyes, K. *Tetrahedron Lett.* **1984**, *25*, 5469.

(9) Hartman, G. D. and Hartman, R. D. *Synthesis.* **1982**, 504.

(10) Kropp, P. J., Poindexter, G. S., Pienta, N. J., and Hamilton, D. C. *J. Am. Chem. Soc.* **1976**, *98*, 8135.

(11)(a) Kropp, P. J., Jones, T. H., and Poindexter, G. S. *J. Am. Chem. Soc.* **1973**, *95*, 5420.   (b) A similar mechanism was proposed for the photosolvolysis of 9-fluorenol. See Gaillard, E., Fox, M. A., and Wan, P. *J. Am. Chem. Soc.* **1989**, *111*, 2180.

(12) Perkins, R. R. and Pincock, R. E. *Tetrahedron Lett.* **1975**, 943.

(13) Reddy, D. S., Sollott, G. P., and Eaton, P. E. *J. Org. Chem.* **1989**, *54*, 722.

(14) Borosky, G. L., Pierini, A. B., and Rossi, R. A. *J. Org. Chem.* **1990**, *55*, 3705.

(15) Santiago, A. N., Iyer, V. S., Adcock, W., and Rossi, R. A. *J. Org. Chem.* **1985**, *50*, 3016.

(16) Kropp, P. J., Pienta, N. J., Sawyer, J. A., and Polniaszek, R. P. *Tetrahedron.* **1981**, *37*, 3229.

(17) For the preparative procedure, see Marolewski, T. A. and Yang, N. C. *Org. Synth.* **1972**, *52*, 132.

(18) Heusler, K. and Kalvoda, J. *Ang. Chem., Int. Ed. Eng.* **1964**, *3*, 525.

(19) Uberwasser, H., Heusler, K., Kalvoda, J., Meystre, C., Wieland, P., Anner, G., and Wettstein, A. *Helv. Chim. Acta.* **1963**, *46*, 344.

(20)(a) Kraus, G. A. and Thurston, J. *Tetrahedron Lett.* **1987**, *28*, 4011. For an example of utilization of this reaction in the   (b) formation of cyclic anhydrides from 2-hydroxycycloalkanones, see Suginome, H., Satoh, G., Wang, J. B., Yamada, S., and Kobayashi, K. *J. Chem. Soc., Perkins Trans. 1.* **1990**, 1239.   (c) synthesis of estrone and 19-nortestosterone, see Suginome, H., Senboku, H., and Yamada, S. *J. Chem. Soc., Perkins Trans. 1.* **1990**, 2199.

(21) Wawzopec, S. and Norstrom, J. D. *J. Org. Chem.* **1962**, *27*, 3726.

(22) Schmitz, E. and Murawski, D. *Chem. Ber.* **1960**, *93*, 754.

(23) Corey, E. J. and Hertler, W. R. *J. Am. Chem. Soc.* **1960**, *82*, 1657.

(24) Hertler, W. R. and Corey, E. J. *J. Org. Chem.* **1959**, *24*, 572.

(25) Corey, E. J. and Hertler, W. R. *J. Am. Chem. Soc.* **1959**, *81*, 5209.

(26) For a preparative procedure for this reaction, see Adam, G. and Schreiber, K. *Org. Photochem. Synth.* **1976**, *2*, 25.

(27) For the photochemical conversion of N-chloroazasteroids into electrophilic N-acylimines and their potential use in photolabeling, see Back, T. G. and Brummer, K. *J. Org. Chem.* **1989**, *54*, 1904.

(28) For the example shown, see Hoshino, O., Ogasawara, H., Takahashi, A., and Umezawa, B. *Heterocycles.* **1987**, *25*, 155; **1985**, *23*, 1943.

(29) For other examples, see Gilbert, A. In *Photochemistry in Organic Synthesis*, Coyle, J. D., ed. The Royal Society of Chemistry: Burlington House, London, 1986.

(30) D'Auria, M., Piancatelli, G., and Ferri, T. *J. Org. Chem.* **1990**, *55*, 4019.

Chapter 13

PHOTOCHEMICAL OXYGENATIONS—SINGLET OXYGEN

PHOTOCHEMICAL OXYGENATIONS—SINGLET OXYGEN

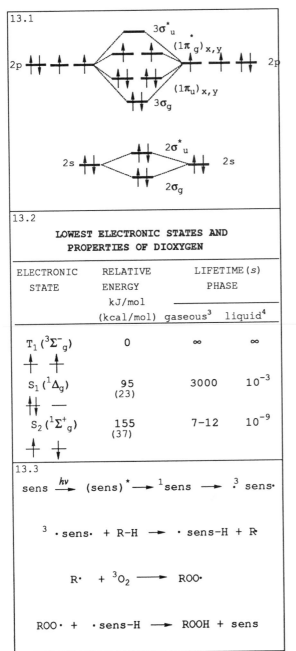

13.1

13.2

LOWEST ELECTRONIC STATES AND PROPERTIES OF DIOXYGEN

ELECTRONIC STATE	RELATIVE ENERGY kJ/mol (kcal/mol)	LIFETIME (s) PHASE gaseous[3]	liquid[4]
$T_1 (^3\Sigma^-_g)$	0	∞	∞
$S_1 (^1\Delta_g)$	95 (23)	3000	10^{-3}
$S_2 (^1\Sigma^+_g)$	155 (37)	7-12	10^{-9}

13.3

$$\text{sens} \xrightarrow{h\nu} (\text{sens})^* \longrightarrow {}^1\text{sens} \longrightarrow {}^3 \text{sens}\cdot$$

$$^3\cdot\text{sens}\cdot + \text{R-H} \longrightarrow \cdot\text{sens-H} + \text{R}\cdot$$

$$\text{R}\cdot + {}^3\text{O}_2 \longrightarrow \text{ROO}\cdot$$

$$\text{ROO}\cdot + \cdot\text{sens-H} \longrightarrow \text{ROOH} + \text{sens}$$

13.1. The oxygen molecule (dioxygen) possesses a high degree of symmetry ($D_{\infty h}$) and degenerate frontier orbitals (see 2.18). Since symmetry does not permit interaction between the polar structures, the frontier orbitals of the ground state are singly occupied by electrons with parallel spins (unpaired electrons), in accordance with Hund's rule. Therefore, the ground state of the molecule is a triplet and dioxygen is a biradical. Since there are more low-lying orbitals available to the electrons in an open-shell system with degenerate orbitals, the description, either in terms of MO or VB structures, of the electronic states of dioxygen is somewhat complex and will not be discussed here.[1]

13.2. The ground state of dioxygen is a triplet T_1 ($^3\Sigma^-_g$) state with the symbol $^3O_2$. The lowest energy excited states are two singlet states with antiparallel electron spins, i.e., S_1 ($^1\Delta_g$) and S_2 ($^1\Sigma^+_g$). The $^1\Sigma^+_g \rightarrow {}^3\Sigma^-_g$ and $^1\Delta_g \rightarrow {}^3\Sigma^-_g$ electronic transitions are forbidden. Consequently, both singlet states of dioxygen have relatively long lifetimes. Dioxygen in the triplet ground state is a biradical and undergoes *one-electron* reactions. In its $^1\Delta_g$ state it is called "*singlet oxygen*" with the symbol $^1O_2$, and in valence-bond theory it is usually described by zwitterionic structures. Since one of the antibonding π^* MOs is empty, the $^1O_2$ is electrophilic in nature and undergoes *two-electron* reactions.[2]

13.3. Depending on the electronic state of dioxygen, there are two different types of photooxygenation reactions: with $^3O_2$ and with $^1O_2$. In the former, a sensitizer is employed for the formation of a radical center that combines with $^3O_2$.[5]

13.4

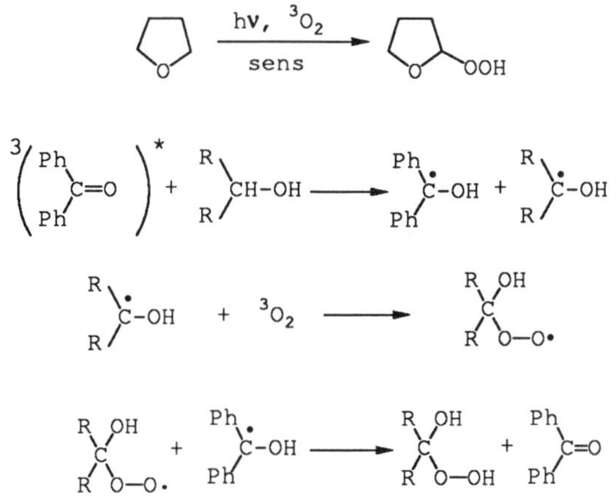

$$\xrightarrow[\text{sens}]{h\nu,\ {}^3O_2}$$

$$\left({}^3\!\!\begin{array}{c}Ph\\Ph\end{array}\!\!C{=}O\right)^{*} + \begin{array}{c}R\\R\end{array}\!\!CH{-}OH \longrightarrow \begin{array}{c}Ph\\Ph\end{array}\!\!\overset{\bullet}{C}{-}OH + \begin{array}{c}R\\R\end{array}\!\!\overset{\bullet}{C}{-}OH$$

$$\begin{array}{c}R\\R\end{array}\!\!\overset{\bullet}{C}{-}OH + {}^3O_2 \longrightarrow \begin{array}{c}R\\R\end{array}\!\!C\!\!\begin{array}{c}OH\\O{-}O\bullet\end{array}$$

$$\begin{array}{c}R\\R\end{array}\!\!C\!\!\begin{array}{c}OH\\O{-}O\bullet\end{array} + \begin{array}{c}Ph\\Ph\end{array}\!\!\overset{\bullet}{C}{-}OH \longrightarrow \begin{array}{c}R\\R\end{array}\!\!C\!\!\begin{array}{c}OH\\O{-}OH\end{array} + \begin{array}{c}Ph\\Ph\end{array}\!\!C{=}O$$

13.5

$$\xrightarrow{h\nu} \left[\ \right]^{*} \xrightarrow{{}^3O_2}$$

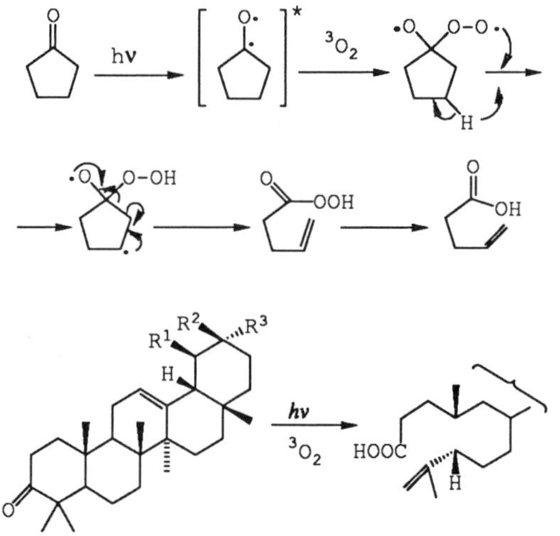

(1) R^1, R^3 = CH_3; R^2 = H

(2) R^1 = H; R^2, R^3 = CH_3

13.6

$$(DYE)_0 \longrightarrow {}^1(DYE)^{*} \qquad (1)$$

$$ {}^1(DYE)^{*} \xrightarrow{ISC} {}^3(DYE)^{*} \qquad (2)$$

$$ {}^3(DYE) + {}^3O_2 \longrightarrow (DYE)_0 + {}^1O_2 \qquad (3)$$

$$REACTANT + {}^1O_2 \longrightarrow REACTANT.O_2 \qquad (4)$$

13.4. The primary step in photooxygenation by ${}^3O_2$ is the formation of a C-radical by hydrogen-atom abstraction from a molecule of the reactant by a triplet sensitizer, e.g., benzophenone. The reaction of the radical with triplet dioxygen produces a peroxy radical that abstracts a hydrogen atom from either the radical of the reduced sensitizer or another molecule of the substrate and forms a hydroperoxide. As shown in the figure, the ground state sensitizer is regenerated and not consumed in the course of the reaction.[6]

13.5. The photooxygenation of cyclic ketones in the presence of air proceeds by the addition of ${}^3O_2$ to the C atom of the carbonyl group in the $T(n,\pi^*)$ state. The resulting biradical is stabilized by a hydrogen shift from the β–C atom followed by a cleavage to a peroxy acid, which is further deoxygenated to an unsaturated carboxylic acid.[7] The synthesis of roburic acid by the photolysis of α-amyrone (1) and nictanthic acid by the photooxygenation of β-amyrone (2) are examples of this *aerobic Norrish Type I cleavage*.[8]

13.6. The second type of photoxygenation involves singlet oxygen.[9] Several photochemical thermal methods have been developed for its preparation.[10] The generation of ${}^1O_2$ by energy transfer from a photosensitizer, which can be either a natural (chlorophyll) or synthetic dye (Rose Bengal, eosin, porphyrins), is the most common method.[11] For ease of separation from the reaction mixture, it is advantageous to anchor the dye to a polymeric support[12] or to silica gel.[13] Polynuclear aromatic hydrocarbons can also be used as sensitizers. *Triplet-triplet annihilation* [Eq. (3)] is the most important step in the entire process. ${}^1O_2$ sensitizers must have a low tendency for self-oxygenation (dye-bleaching) and a triplet energy not far above the energy of the ${}^1\Delta_g$ or ${}^1\Sigma_g^+$ states of O_2 for an efficient energy transfer.[14]

13.7

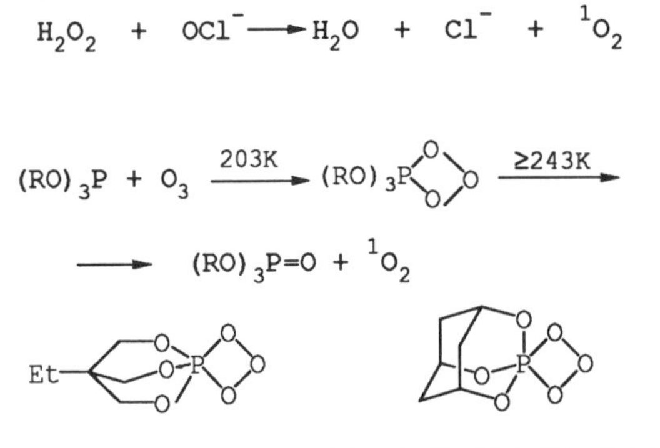

$$H_2O_2 + OCl^- \longrightarrow H_2O + Cl^- + {}^1O_2$$

$$(RO)_3P + O_3 \xrightarrow{203K} (RO)_3P\overset{O}{\underset{O}{\bigcirc}}O \xrightarrow{\geq 243K}$$

$$\longrightarrow (RO)_3P{=}O + {}^1O_2$$

13.8

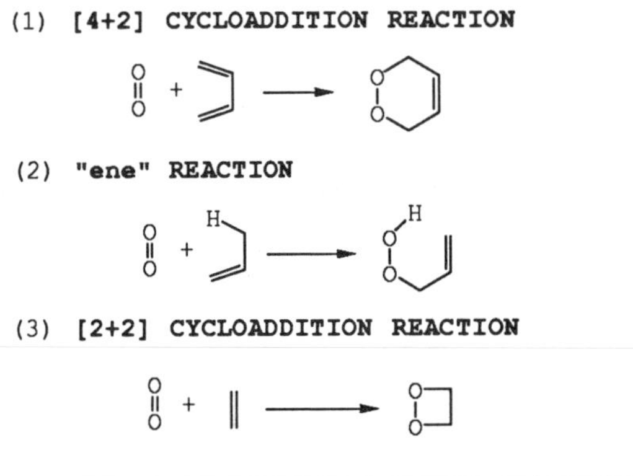

(1) **[4+2] CYCLOADDITION REACTION**

(2) **"ene" REACTION**

(3) **[2+2] CYCLOADDITION REACTION**

13.9

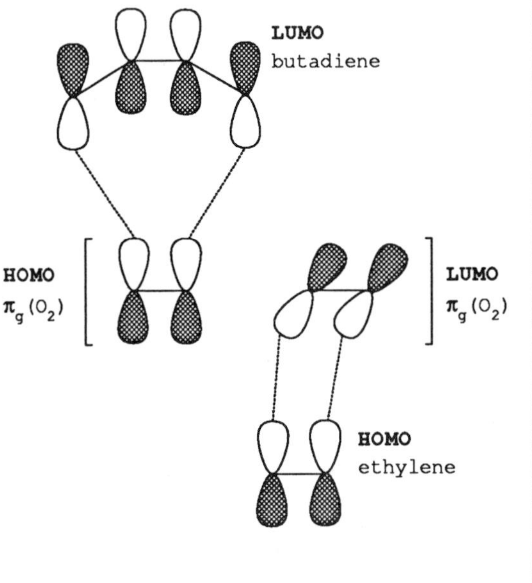

LUMO
butadiene

HOMO
$\pi_g(O_2)$

LUMO
$\pi_g(O_2)$

HOMO
ethylene

13.7. ${}^1O_2$ is generated by a thermal chemiluminescent reaction by: (1) a hydrogen peroxide/hypochlorite system;[15] (2) a hydrogen peroxide/molybdate system;[16] (3) the thermal decomposition of arene endoperoxides (see 13.17);[17] (4) the thermal decomposition of ozonides of various phosphites,[18] which leads to the quantitative decomposition of the complex with the formation of the corresponding phosphate and ${}^1O_2$; or (5) triethylsilyl trioxide.[19] When the decomposition is performed in the presence of a suitable singlet oxygen acceptor, the oxygenation occurs in high yield. However, photosensitized generation of ${}^1O_2$ offers easier product isolation, especially with polymer-anchored sensitizers.

13.8. The reactions of singlet oxygen $({}^1O_2)$[1,20] with unsaturated organic compounds, e.g., alkenes, dienes, polynuclear aromatic hydrocarbons, etc., can be divided into three classes: (1) [4+2] cycloaddition reactions with 1,3-dienes and polynuclear aromatics to produce endoperoxides; (2) the *ene* reaction with alkenes having an allylic hydrogen atom to form hydroperoxides; and (3) [2+2] cycloaddition reactions with electron-rich alkenes with no allylic hydrogen atoms to produce 1,2-dioxetanes. Singlet oxygen in these reactions behaves either as a dienophile or as an electrophile, depending on the structure of the acceptor.

13.9. When dioxygen in its ${}^1\Delta_g$ state interacts with its environment, the degenerate levels are split and the resulting HOMO and LUMO are symmetrical with respect to a plane that bisects the O—O bond. These orbitals are capable of providing bonding (stabilizing) interactions in reactions with (4n) and (4n+2) π electron systems. Thus, both reaction types are symmetry allowed. Its large electron affinity favors [2+2] cycloaddition in which ${}^1O_2$ is an acceptor of electrons. However, reactions in which six electrons participate, e.g., [4+2] cycloadditions, are also important.

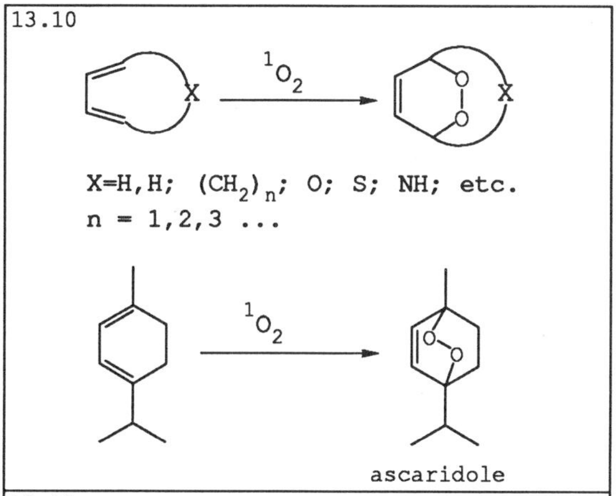

13.10

X=H,H; (CH$_2$)$_n$; O; S; NH; etc.
n = 1,2,3 ...

ascaridole

13.11

Me$\overset{H}{\diagup}$ ... Me$\overset{H}{\diagdown}$

1 +

100%   83%   17%
1   *2*

13.12

E$_A$ = 10kJ/mol

E$_A$ = 4.6kJ/mol

OH

1

OH

60%

CS(NH$_2$)$_2$ (*1*)

13.10. The [4+2] cycloaddition of $^1O_2$ to aliphatic, isocyclic, and hetero-cyclic 1,3-dienes and aromatic or het-erocyclic substrates yields cyclic or bi-cyclic endoperoxides[21] in a concerted fashion via a six-membered transition state.[22] However, polar (zwitterionic) or radical intermediates that rearrange to endoperoxides have been proposed for some structurally restricted dienes.[23] Endoperoxides are key intermediates in many syntheses (see 13.13). In a few cases the competing *ene* reaction, which produces diallylic hydroperox-ides, has been observed.[24] The natu-rally occurring endoperoxide ascari-dole was prepared from α-terpinene.[25]

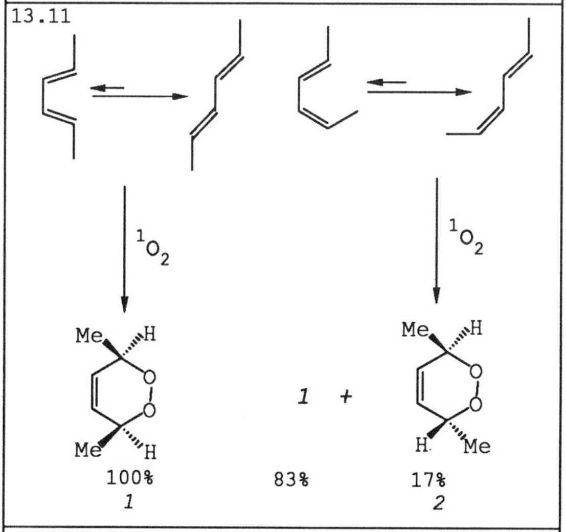

13.11. Acyclic dienes react with $^1O_2$ to produce 1,2-dioxenes in good yield,[20(b)] especially if the diene is ac-tivated by alkyl groups. However, if a 1,3-diene contains an abstractable al-lylic H-atom or if it cannot assume an *s-cis* configuration, the *ene* reaction competes. Although the *s-cis* ⇌ *s-trans* equilibrium in solution is strongly shifted in the direction of *s-trans*, the oxygenation of *trans, trans*-2,4-hexadiene is stereospecific and forms only *1*, while the oxygenation of the *cis-trans* isomer gives dioxenes *1* and *2* in a ratio of 83:17. It appears that $^1O_2$ induces transformation of the *s-trans* to the *s-cis* isomer.[26]

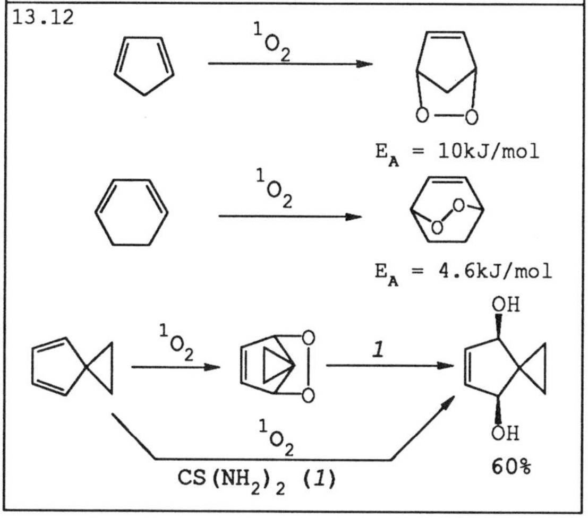

13.12. These results indicate that the reaction of $^1O_2$ with 1,3-dienes is concerted. In addition, the observation that this reaction, especially with cyclic dienes that are restricted to the *s-cis* configuration, are rapid, and have a low activation enthalpy (~ 25 kJ/mol, ~ 6 kcal/mol) suggests that the formation of a 1,4-endoperoxide is an allowed concerted process. Cyclic 1,3-dienes yield bicyclic 1,4-endoperoxides, which are easily reduced by thiourea (1) to the corresponding 1,4-diols.[27] Oxy-genation in the presence of (1) can be effected at ambient temperature, since the often unstable endoperoxides are immediately reduced to diols.[28]

13.13

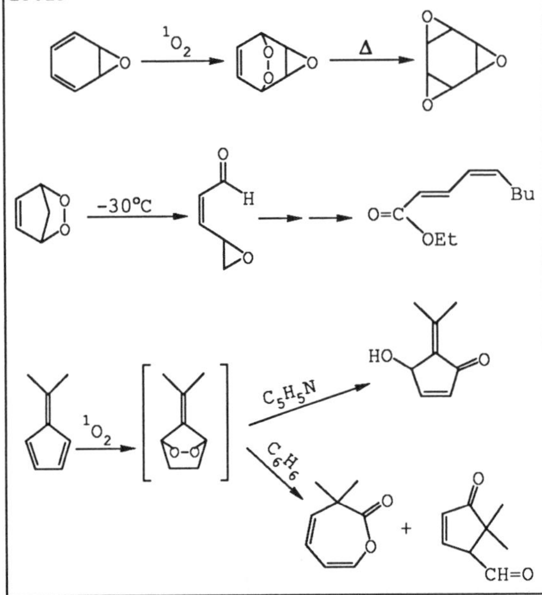

13.13. Bicyclic endoperoxides, i.e., endoperoxides from cyclic dienes, rearrange either thermally or by catalysis with cobalt complexes to form *cis*-dioxiranes via alkenedioxyl biradicals.[29] This method has been utilized in the synthesis of benzene trioxide produced from the thermal rearrangement of the endoperoxide of benzene oxide.[30] Similar rearrangements have been observed for other endoperoxides, e.g., the thermally unstable endoperoxide of cyclopentadiene. It readily undergoes rearrangement at $-30°C$ to an epoxy-aldehyde, which serves as the starting material for the synthesis of the flavor in the liqueur "le bon pair Williams."[31] The rearrangement to oxirane derivatives is not the only possible reaction path, as shown by the endoperoxide of dimethylfulvene.[32]

13.14

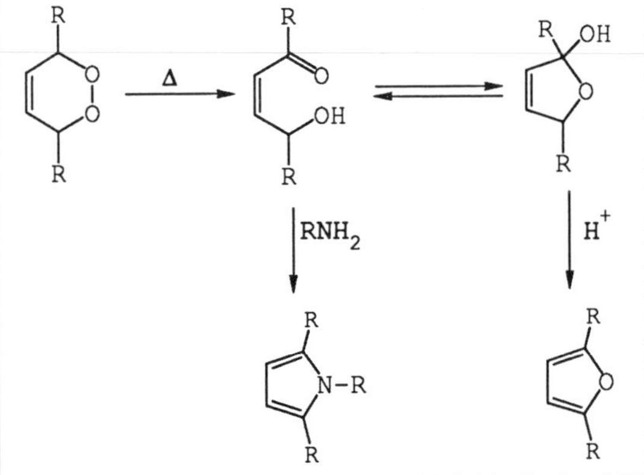

13.14. An important transformation of monocyclic endoperoxides is the base-catalyzed (KOH in ethanol) conversion to furan[33] or, in the presence of ammonia or primary amines, to pyrrole derivatives.[34] The reduction of the peroxide bond by LiAlH$_4$ is significant since conclusions can be drawn about the configuration of the endoperoxide from the stereochemistry of the resulting unsaturated diol. Using this reaction, the stereospecific pathway of the endoperoxide formation has been elucidated.[35]

13.15

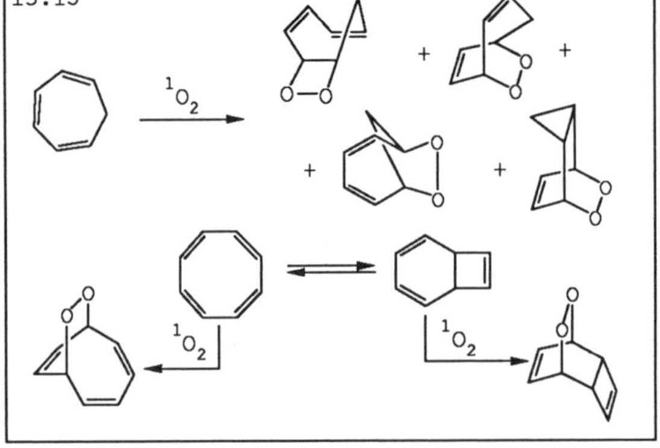

13.15. Cyclic polyenes, such as cycloheptatriene and cyclooctatetraene, also undergo oxygenation by singlet oxgyen. The oxygenation of cycloheptatriene results in the formation of four products: the [2+2], [4+2], and [6+2] cycloadducts singlet oxygen to tropylidene and the [4+2] cycloadduct of norcaradiene, the valence isomer of tropylidene.[36] Cyclooctatetraene and its bicyclic valence isomer undergo singlet oxygen oxygenation reluctantly to produce 1,4-endoperoxides; the reaction requires 14 days.[37]

13.16

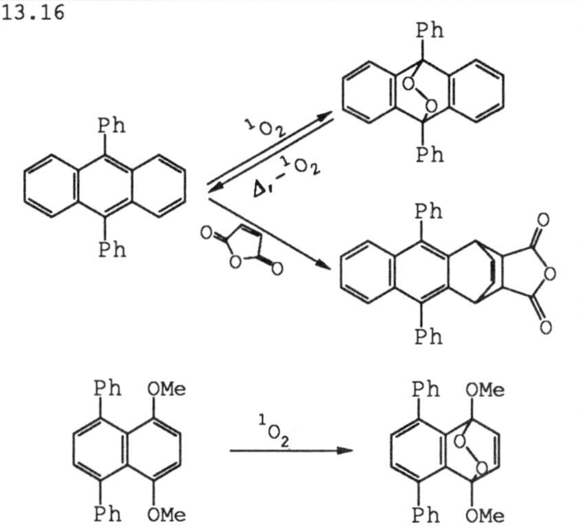

13.17

13.18

SOLVENT	τ_{rel} ($^1O_2$)	k_{rel}	1:2
ethanol	1	1	1.5
benzene	2	1.7	2.7
2-butanone	2.1	3.3	2.3
chloroform	5	15	2.0
tetrachloromethane	58	24	3.2

τ_{rel} = relative lifetime
k_{rel} = relative rate

13.16. [4+2] cycloaddition reactions of singlet oxygen with polynuclear aromatics are analogous to the diene reaction. However, the analogy is only a formal one since the reaction centers of aromatic hydrocarbons in reaction with singlet oxygen are not the same as those in the diene addition. 9,10-disubstituted anthracenes and larger polycyclic aromatic systems react with singlet oxygen almost quantitatively. The reaction does not require a sensitizer since autosensitization occurs (see 13.7).[38] Sufficiently activated naphthalenes[16] and some derivatives of benzene[39] undergo a [4+2] cycloaddition reaction to $^1O_2$. However, photooxygenation of these relatively unreactive substrates requires prolonged irradiation times and a high oxygen flux. Better results have been obtained with the H_2O_2/MoO_4^{2-} system, which can be used for the perepoxidation of a water-soluble singlet oxygen carrier, such as sodium 2-(4-methylnaphthalene)propionate.[16]

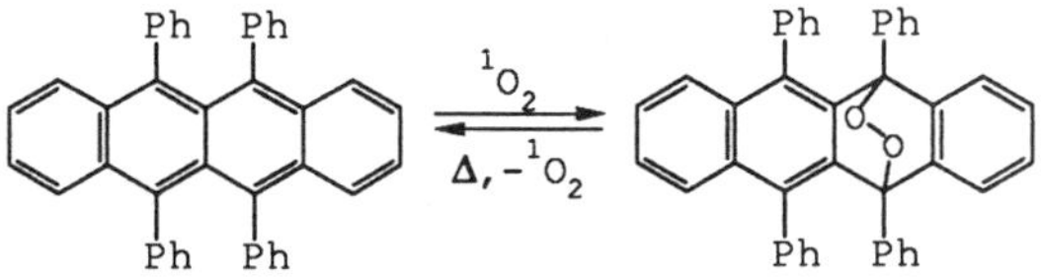

13.17. This reaction is reversible and the aromatic 1,4-endoperoxides, upon warming to moderate temperatures, decompose quantitatively into their components and dioxygen in its singlet excited state. The E_{act} of the decomposition is comparable to E_{exc} for singlet oxygen. Accordingly, the process can be regarded as a spin-controlled, IR chemiluminescent reaction (emission recorded at 1.25 μm) with a quantum yield of approximately 1. The rate and yield of the thermolysis depend on the resonance energy of the aromatic system; the higher the resonance energy, the better the yield of thermal decomposition products.[40]

13.18. The oxygenation of vinyl arenes with $^1O_2$ yields both [4+2] and *ene* reaction products. The observed dependence of reaction rate and product ratios on the solvent suggests that the reaction rate, especially that of the [4+2] reaction, increases significantly with the lifetime of singlet oxygen in the solvent.[41]

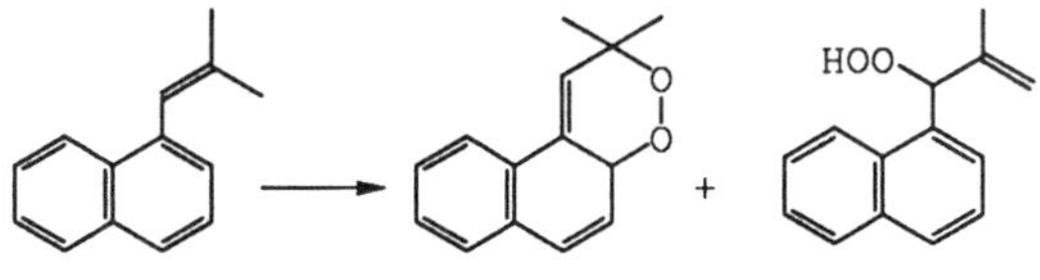

13.19

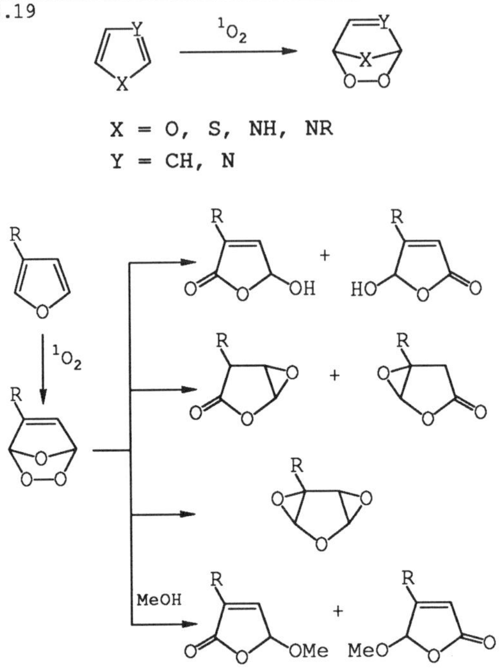

X = O, S, NH, NR
Y = CH, N

MeOH

13.20

R = —⟨benzene ring⟩—(CH$_2$)$_9$COOH

13.21

(1)
(2)
(3)
(4)

* =radical or charge

13.19. Five-membered ring heterocycles, such as furan, thiophene, pyrrole, and azoles, are also oxygenated by $^1O_2$ to form 1,4-endoperoxides, most of which are unstable and rearrange or solvolyze to other products.[42] The endoperoxides of oxazoles have been isolated and characterized at low temperature.[43] A number of products are formed from furan, including hydroxybutenolides, epoxylactones, diepoxides, and solvent (MeOH) addition products.[44,45]

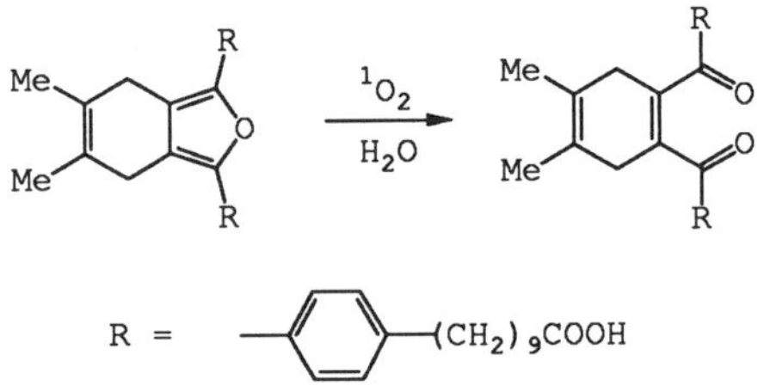

13.20. Singlet oxygen is usually detected by identification of stereospecific oxygenation products of some carbocyclic and heterocyclic compounds. Of considerable importance is the detection of $^1O_2$ in biological systems in aqueous media in which the water-soluble potassium salt of 1,3-*bis* [4-(9-carboxylnonylphenyl)]-4,7-dihydro - 5,6 - dimethylisobenzofuran, rather than water-insoluble 1,3-diphenylisobenzofuran, is used.[46]

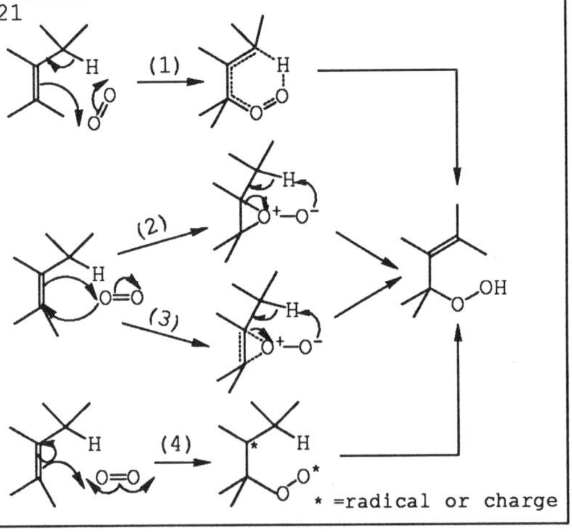

13.21. The most important synthetic reaction of $^1O_2$ is the *ene* reaction that occurs with alkenes with an allylic H atom.[47] $^1O_2$ bonds to one of the C atoms of the C=C, the double bond shifts, and the allylic H atom migrates to the peroxide group to produce an allyl hydroperoxide. This represents a general method for functionalization in the allylic position, e.g., for the preparation of allyl alcohols by the reduction of hydroperoxides. The mechanism of the *ene* reaction is not yet fully elucidated. In fact, experimental results, as well as theoretical calculations, support four possible mechanisms; (1) concerted, with a cyclic transition state; (2) formation of a polar (zwitterionic) or biradical intermediate with the preferential formation of a C—O bond; (3) formation of a perepoxide intermediate; (4) formation of an exciplex with geometry similar to that of (3). Most studies support the concerted mechanism; however, the fact that the reaction appears to be substrate dependent must be considered.[48]

13.22

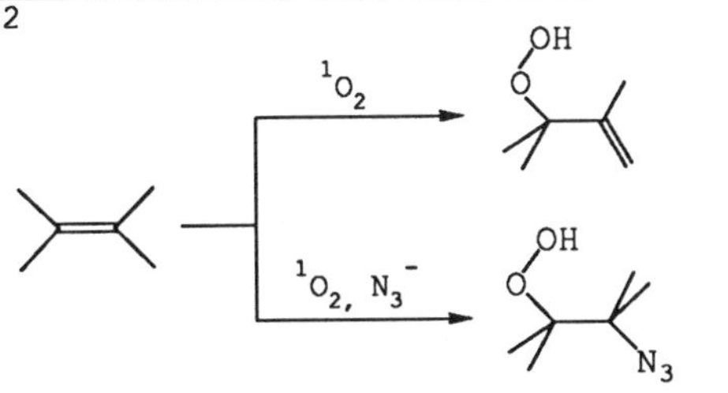

13.23

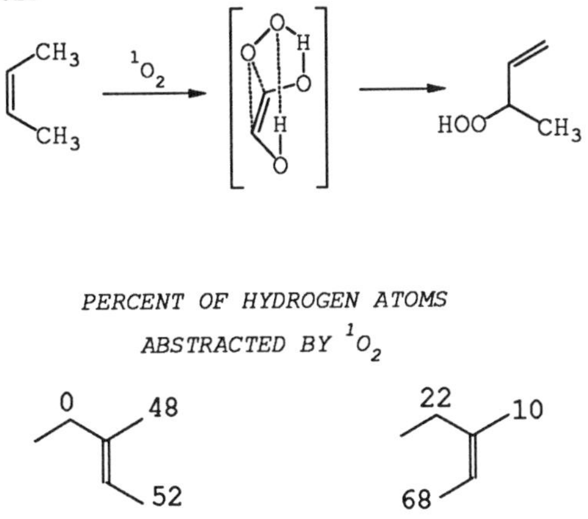

PERCENT OF HYDROGEN ATOMS
ABSTRACTED BY $^1O_2$

13.24

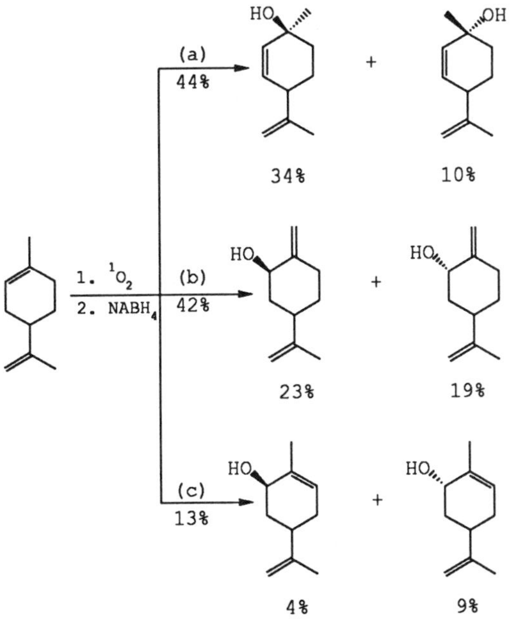

13.22. The concerted mechanism is supported by the observation that the *ene* reaction is a stereospecific *syn* addition (no racemization is observed with pure chiral alkenes) and is fairly insensitive to substituent, isotope, and solvent effects. The fact that strong nucleophiles or aldehydes open the intermediate peroxirane ring to produce azidohydroperoxides or 1,2,4-trioxanes is evidence for the peroxirane mechanism.[49] *Ab initio* calculations justify a biradical mechanism.[50]

13.23. Mechanistic and theoretical studies favor the two-step peroxirane mechanism. In the first step, suprafacial CT interaction occurs between the HOMO of the electron donor (the alkene) and LUMO of the acceptor ($^1O_2$). The transition state has C_s symmetry which could account for the *syn* course of the reaction. The O—H—C interaction plays an important role in the transition state, as shown for 2-butene. In the second step, the C—O bond is formed with the simultaneous disappearance of the C—H bond. Oxygenation occurs on the more sterically hindered side of trisubstituted alkenes, which explains the regiospecificity of the *ene* reaction.[51]

13.24. The characteristics of the *ene* reaction are demonstrated by the oxygenation of limonene.[52] (1) When nonconjugated double bonds are present, the most electron-rich (more highly substituted) double bond is attacked (there is a 10–100-fold increase in reactivity per alkyl group at the double bond). (2) Primary and secondary allylic hydrogen atoms migrate at a similar rate (compare paths *a* and *b*), while in $^3O_2$ oxygenations the abstraction of a secondary H atom proceeds faster than that of a prmary one. (3) The reaction has certain steric requirements, illustrated by the comparison of paths *a* and *b* relative to *c*. The oxidation of a chiral alkene[52] does not lead to racemization or *cis-trans* isomerization.[53] However, asymmetric induction in these oxidations has been observed.[53(b)]

13.25

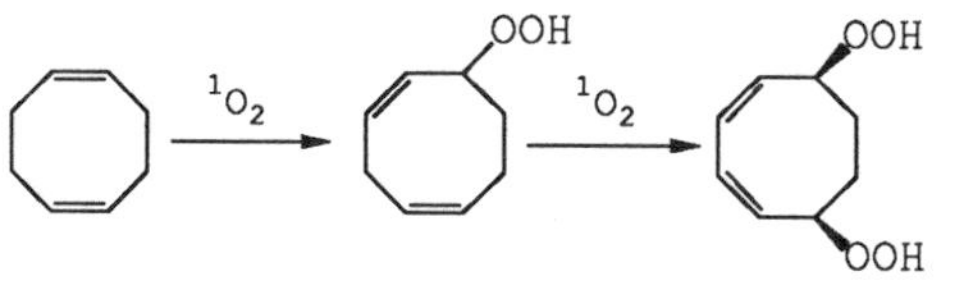

13.26

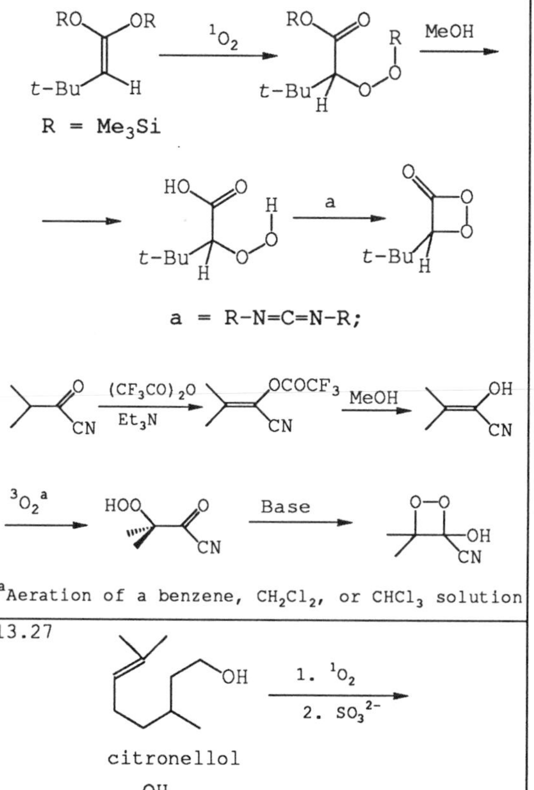

R = Me₃Si

a = R-N=C=N-R;

ᵃAeration of a benzene, CH₂Cl₂, or CHCl₃ solution

13.27

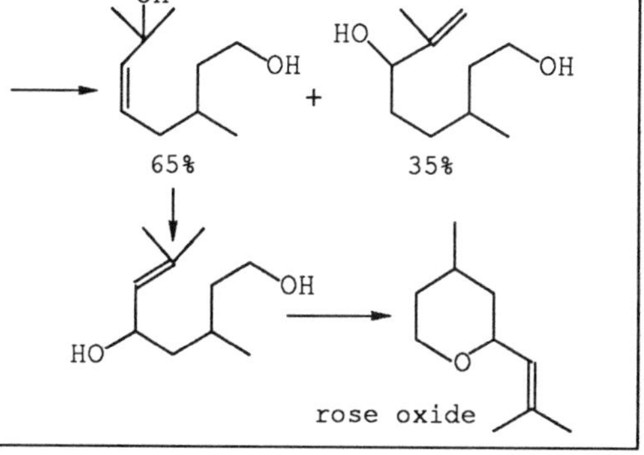

citronellol

65%   35%

rose oxide

13.25. 1,5-cyclooctadiene undergoes a double *ene* oxygenation reaction with singlet oxygen to form a *bis*-hydroperoxide. Although the product is a 1,3-diene, it resists further reaction with singlet oxygen by a [4+2] cycloaddition reaction to produce a 1,4-endoperoxide.[54] In a few instances, competition exists between endoperoxide and diallylic hydroperoxide formation in the oxygenation of cyclic conjugated dienes, e.g., phellandrene. The product pattern can be explained by the presumption of a common intermediate.[24]

13.26. The migration is observed not only with an allylic hydrogen atom, but also with an allylic trimethylsilyl group, e.g., in trimethylsilylvinyl ethers. This reaction is utilized in the synthesis of the unstable α-peroxylactone, a derivative of 1,2-dioxetane that decomposes thermally in a chemiluminescent process.[55] α-peroxylactones (1,2-dioxetan-3-ones), which are much less stable than the corresponding 1,2-dioxetanes, are interesting because of their similarity to intermediates in bioluminescence reactions.[56] These compounds are also formed by the reaction of ³O₂ with an isolable enol and produce excitation via chemically initiated electron exchange luminescence (CIEEL) (see 16.35).

13.27. An important industrial application of the *ene* reaction is the production of artificial flavors and fragrances. The synthesis of rose oxide and nerol oxide, important constituents of Bulgarian rose oil, by the Rose Bengal sensitized oxygenation of citronellol or nerol, is an example.[57] After reduction of the mixture of hydroperoxides from the β-citronellol oxygenation with sulfite and an allylic rearrangement of the principal oxygenation product, followed by dehydration, a mixture of stereoisomeric rose oxides is obtained. The reaction has also been employed in the stereospecific oxidation of tetracycline.[58]

13.28

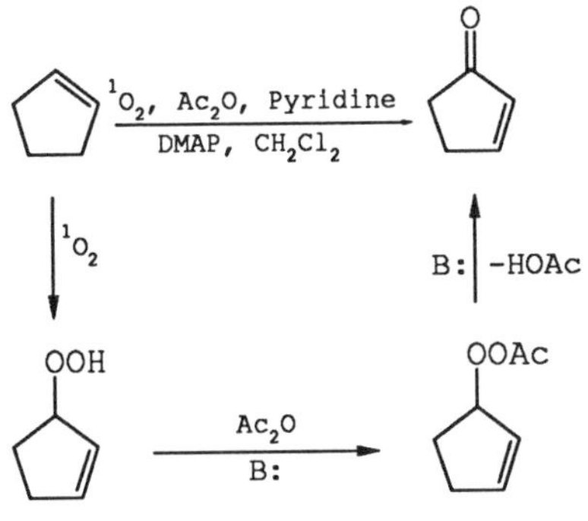

cyclohexene → cyclohexenone 78%

cycloheptene → cycloheptenone 85%

cyclooctene → cyclooctenone 85%

methoxymethylenecycloheptane → methyl-1-
(1-cycloheptenyl)carboxylate 77%

13.29

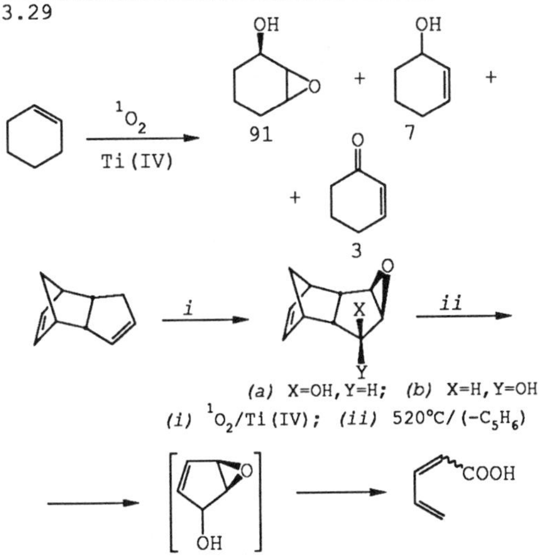

(a) X=OH,Y=H; (b) X=H,Y=OH
(i) $^1O_2$/Ti(IV); (ii) 520°C/($-C_5H_6$)

13.30

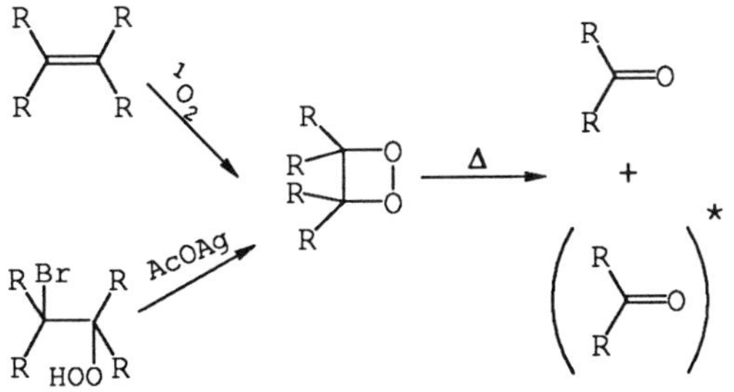

13.28. Allylic hydroperoxides with an α H atom are converted into conjugated enones by a weak base, e.g., pyridine in the presence of acetic anhydride (Kornblum–DeLaMare dehydration).[59] As a one-pot reaction, the dye-sensitized photooxygenation of a methylene chloride solution of an alkene in the presence of acetic anhydride/pyridine is a facile method for the transformation of alkenes into conjugated enones. It has been applied to both cyclic and acyclic alkenes, even those normally too unreactive toward $^1O_2$ to be synthetically useful, e.g., cyclohexene.[60]

13.29. The photooxygenation of alkenes in the presence of titanium or vanadium catalysts is an effective one-pot method for the synthesis of epoxy alcohols. After photooxygenation to an allylic hydroperoxide, intermolecular oxygen transfer occurs. This results exclusively in an epoxy alcohol that is easily separated from the catalyst. The oxygen donor (epoxy hydroperoxide) and the oxygen acceptor (allylic hydroperoxide) are located in different molecules, but are coordinated at the same metal center. This accounts for the diastereo- and enantioselectivity of this reaction.[61]

13.30. The [2+2] cycloaddition of $^1O_2$ to form 1,2-dioxetanes[62] occurs with electron-rich alkenes (vinyl ethers, enamines).[63] If the alkene has an activated double bond and an allylic H atom, the *ene* reaction competes with dioxetane formation. This competition is strongly solvent dependent, i.e., the ratio of *ene* to dioxetane from dihydropyran is 10.1:1 in benzene and 0.059:1 in acetonitrile.[64] 1,2-dioxetanes are also prepared by dehydrobromination of 2-bromohydroperoxides.[65] They have the unique ability to generate electronically excited carbonyl compounds, usually in the triplet state,[66] by thermal activation, apparently via an exciplex[67] or encounter complex.[68]

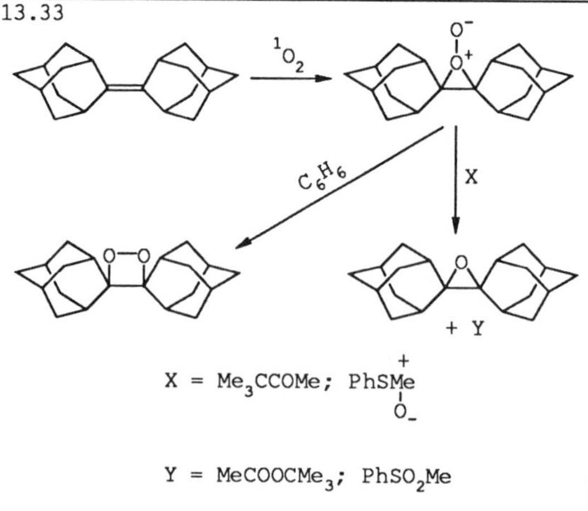

13.31. Many 1,2-dioxetanes with heteroatoms (electron donors) on the four-membered ring have been isolated. Their thermal stability and the efficiency of their triplet carbonyl production is altered by electronic and steric factors. While O-substituted dioxetanes are moderately stable at ambient temperature, the corresponding N- and S-substituted compounds are unstable and decompose immediately with chemiluminescence into two carbonyl compounds.[62(a),66(b),69]

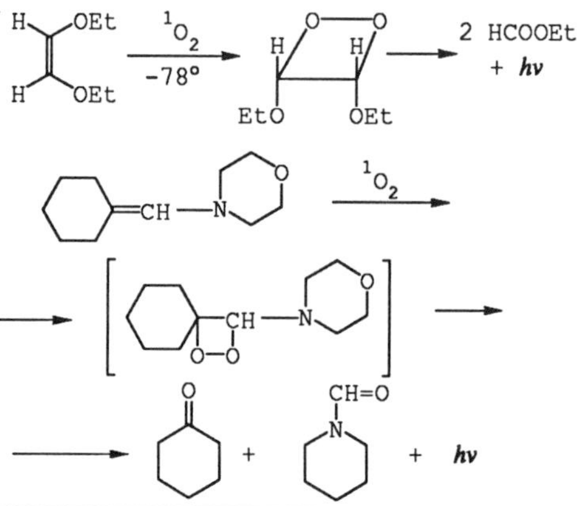

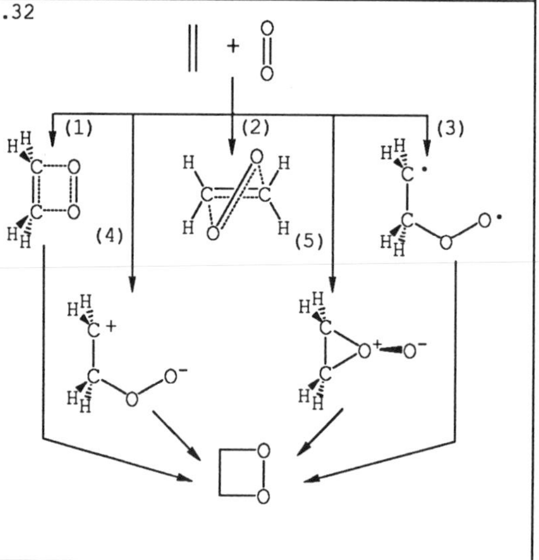

13.32. The proposed mechanisms for 1,2-dioxetane formation include both concerted and nonconcerted approaches, such as: (1) a supra-supra attack (with a possible charge-transfer from the alkene to a π^* orbital of dioxygen); (2) a supra-antara attack; (3) the involvement of a 1,4-biradical intermediate; (4) the involvement of a 1,4-zwitterionic intermediate; and (5) the formation of an intermediate peroxirane. Since the biradical may have some ionic character, mechanisms (3) and (4) are similar. Theoretical and experimental studies support different mechanisms.[70]

13.33. Although a concerted mechanism is supported by the stereospecificity of the oxygenation of *cis*- and *trans*-alkenes, the peroxirane mechanism was suggested by a study of the oxygenation of bisadamantylidene with $^1O_2$.[71] Here, the *ene* reaction cannot take place because the formation of the product would violate Bredt's rule. Pinacolone and phenyl methyl sulfoxide were used as traps for the peroxirane intermediate. With the former, the nucleophilic attack by the carbonyl group of pinacolone produced a *t*-butyl acetate (Bayer–Villiger reaction); in the latter, phenyl methyl sulphone was formed by a nucleophilic transfer of an oxygen atom.[71(b)] In both cases the subsequent oxygenation product was the corresponding oxirane. In the absence of a trap, oxygenation in benzene yielded the expected 1,2-dioxetane.

13.34

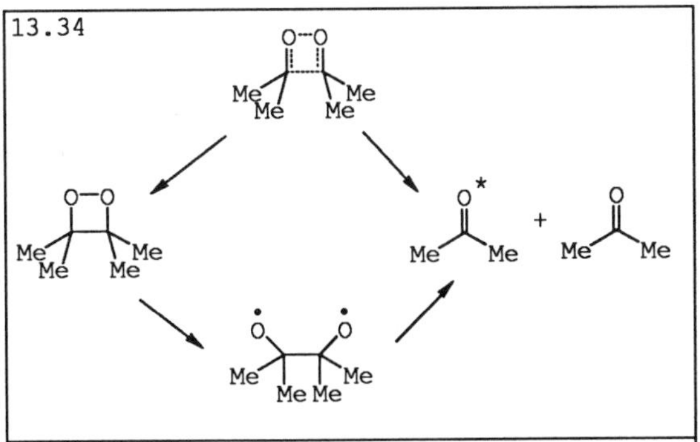

13.35

ratio $T_1/S_1 \sim 300$

acenaphthylene

[acenaphthylene]* ⟶ dimer

syn:anti = 1:3

yield 7%

13.36

hv/O_2 Rose Bengal
−78°C

silica gel
o-xylene

An =

13.34. A striking property of 1,2-dioxetanes is their thermal decomposition to two molecules of carbonyl compounds, one of which is formed in an electronically excited n,π^* state, almost exclusively T_1 (*chemienergized electronic excitation*). The quantum yield of excited acetone is 50% with $\Phi(S_1) = 0.5\%$ and $\Phi(T_1) = 50\%$ from the thermolysis of tetramethyl-1,2-dioxetane.[72] Two mechanisms have been proposed for the thermal decomposition of 1,2-dioxetanes: a nonconcerted biradical mechanism and a concerted mechanism. The former mechanism is supported by molecular mechanical (MM2) calculations, while the latter is suggested to proceed via an exciplex or an encounter complex. In both cases spin flipping occurs.[73] For the reaction coordinate of dioxetane thermolysis, see 4.14.

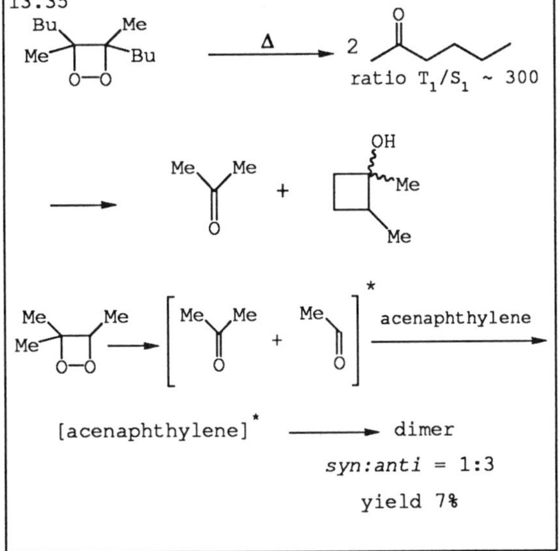

13.35. The formation of triplet states of carbonyl compounds from the thermal decomposition of 1,2-dioxetanes has been utilized in both intramolecular- and intermolecular-chemisensitized reactions (photochemistry without light). The thermolysis of 3,4-dibutyl-3,4-dimethyl-1,2-dioxetane yields both Norrish Type II cleavage products and a cyclobutanol. The thermal decomposition of trimethyldioxetane, in the presence of acenaphthylene, leads to heptacyclenes (see 5.22).[74]

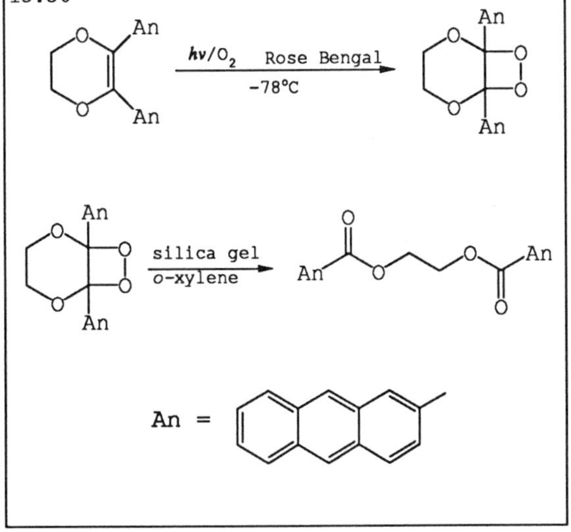

13.36. The dye-sensitized oxygenation of 2,3-di(2-anthryl)-1,4-dioxene produces a dioxetane that chemiluminesces at 84.1°C with an efficiency of 0.2%. The addition of silica gel to an *o*-xylene solution of the dioxetane dramatically increases the intensity of the emitted radiation (by a factor of 10^4). The fact that the catalyst can be easily removed by filtration makes this heterogeneous system a practical chemical source of light that can be turned on and off as desired.[75]

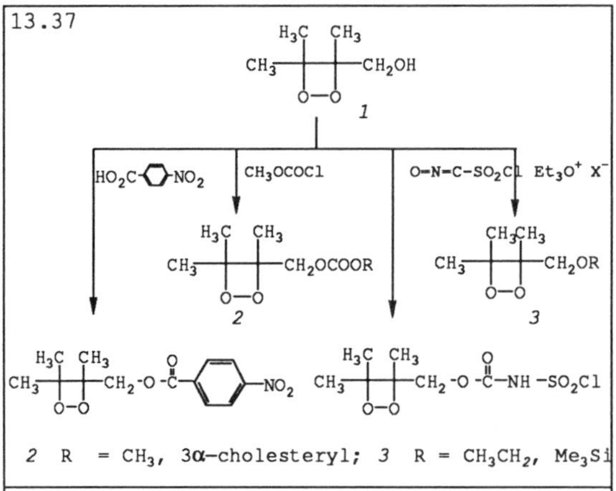

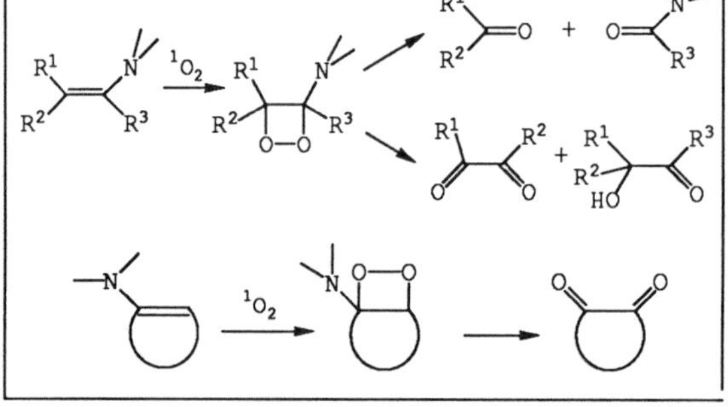

13.37

2   R = CH₃, 3α-cholesteryl; 3   R = CH₃CH₂, Me₃Si

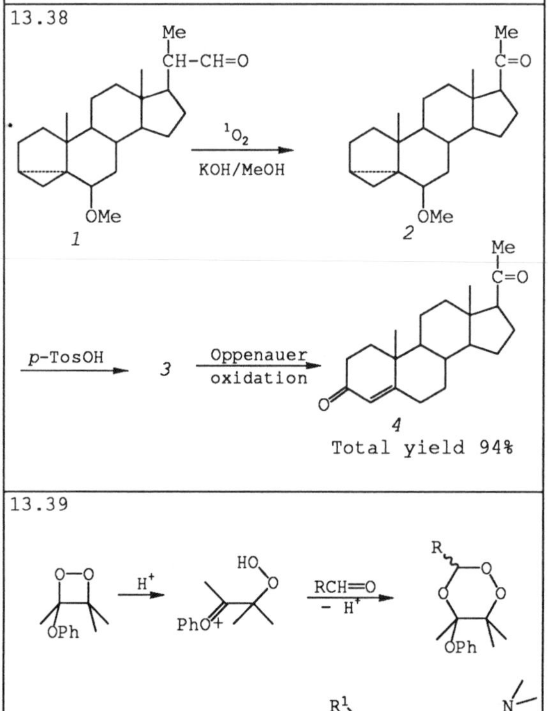

13.38

Total yield 94%

13.39

13.37. Functionalized derivatives of dioxetane can be formed from 3-(hydroxymethyl)-3,4,4-trimethyl-1, 2-dioxetane (1) by the use of appropriate electrophiles, such as carboxylic acids, chlorocarbonates, isocyanates, trialkylsilyl chlorides, and trialkyloxonium salts. These dioxetane derivatives represent chemical sources of triplet carbonyl compounds having applications in photobiology and photomedicine, e.g., as therapeutic agents.[76] It has been shown that 1,2-dioxetanes undergo a facile quantitative reduction by thiols, e.g., glutathione, cysteine, etc., to form vicinal diols. This represents an effective method for the detoxification of these genotoxic agents in the cell.[77]

13.38. In methanolic KOH solution, the oxygenation of aldehydes, in which the aldehydic group is bonded to a secondary carbon atom, produces high yields of ketones with one less carbon atom. The reaction is a key step in the synthesis of progesterone from stigmasterol. When stigmasterol with a protected C(5)-double bond and a 3β—OH group is used, the aldehyde (1) is prepared by degradation. The sensitized oxygenation of 1 in methanolic KOH produces ketone 2 in high yield. The most probable intermediate is a dioxetane from the [2+2] cycloaddition of singlet oxygen to the enol form of the aldehyde. The remaining steps include hydrolysis to pregnenolone (3) and oxidation to progesterone (4).

13.39. Several transformations of 1,2-dioxetanes are known. However, their use in synthesis is generally limited. One important reaction involving 1,2-dioxetanes is the photooxygenation of enamines to 1,2-dicarbonyl or 2-hydroxycarbonyl compounds. Another example is restricted to 1,2-dioxetanes with a phenoxy group in position 3. These undergo a reaction with aldehydes in weakly acidic media to produce 1,2,4-trioxanes. The intermediate is a 2-hydroperoxyphenoxonium cation.[78]

13.40

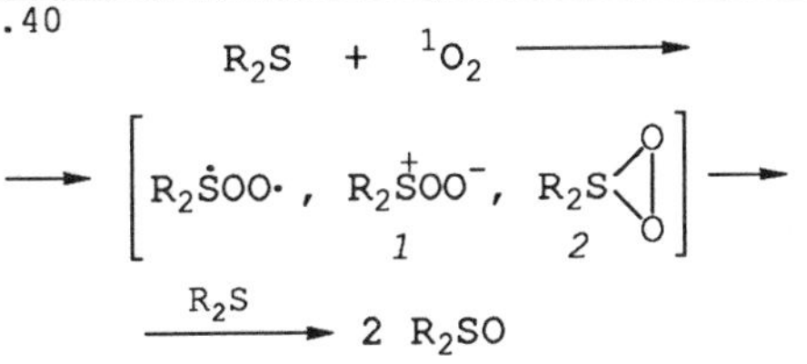

13.41

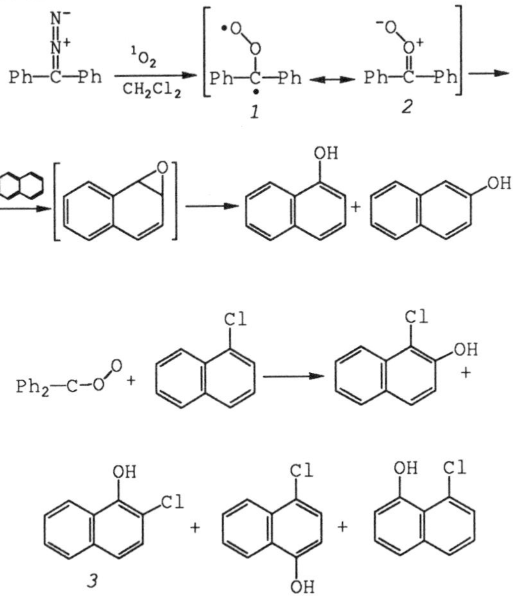

3

13.42

**SOME PHOTOTOXIC AND PHOTOCARCINOGENIC
COMPOUNDS**

	PHOTO-TOXIC	PHOTOCAR-CINOGENIC	IN DARK
Acridine orange	+	−	−
Eosine	+	−	−
$CHCl_3$	+	+	+
CCl_4	+	+	−

13.40. Sulfides undergo photosensitized oxygenation to produce 2 moles of a sulphoxide for each mole of oxygen used, although sulfones are obtained under certain conditions.[79,80] It was suggested that, depending on the polarity of the solvent, the reaction proceeds either via *1* or *2*.[81] However, experimental (IR spectroscopy in a dioxygen matrix at 13 K)[82] and theoretical studies[83] have shown the persulfoxide to be the major intermediate, although participation of a sulfurane intermediate cannot be excluded.[84,85] Thiadioxirane does not appear to be a valid intermediate.

13.41. The oxygenation of diphenyldiazomethane by singlet oxygen yields the carbonyl oxide of benzophenone, a strong oxidizing agent, which simulates biological oxygenations catalyzed by flavine monoxygenase.[86] Carbonyl oxides[87] oxidize saturated hydrocarbons to alcohols, alkenes to oxiranes, and aromatic hydrocarbons to phenols via an arene oxide (sometimes with an NIH shift, i.e., a 1,2 shift of the atom or group bonded at the position where the oxidation occurs, e.g., in *3*). The electrophilic nature of the carbonyl oxide in the oxygen transfer favors the polar biradical structure *1* over the zwitterionic state *2*.[88]

13.42. Although both solar radiation and oxygen are absolute necessities for life, their combination can provoke severe health problems and even death. These phenomena are referred to as the *photodynamic effect*[89] and are caused by the production of singlet oxygen. Because of its high reactivity, $^1O_2$ is highly toxic. Biological systems are protected from this toxicity by carotenes, strong quenchers of singlet oxygen. Many compounds that are not toxic in the dark are highly toxic under the influence of light (they sensitize the transformation of $^3O_2$ to $^1O_2$).[90] Also, the increased incidence of skin cancer in mariners and mountain climbers is attributed to the effects of singlet oxygen.

13.43

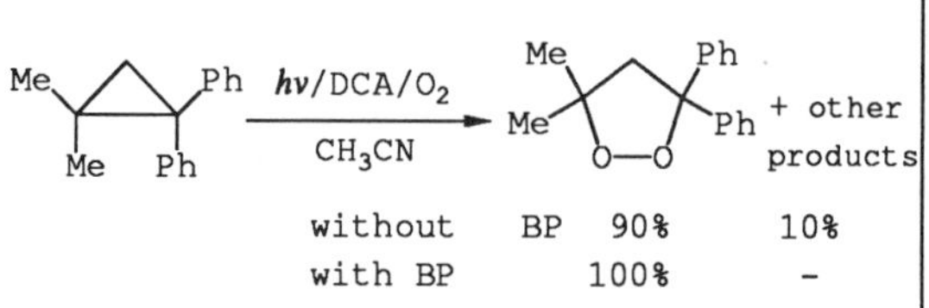

	BP	
without	BP	90% 10%
with	BP	100% —

$$DCA \xrightarrow{} {}^1DCA^* \tag{1}$$

$${}^1DCA^* + CP \longrightarrow DCA^{\bullet-} + CP^{\bullet} \tag{2}$$

$$CP^{\bullet+} + {}^3O_2 \longrightarrow (CPO_2)^{\bullet+} \tag{3}$$

$$(CPO_2)^{\bullet+} + CP \longrightarrow 1,2\text{-dioxolane} + CP^{\bullet} \tag{4}$$

$$(CPO_2)^{\bullet+} + DCA^{\bullet-} \longrightarrow 1,2\text{-dioxolane} + DCA \tag{5}$$

13.44

13.45

13.43. Aryl-substituted cyclopropanes (CP), upon irradiation in an oxygen-saturated polar solvent such as acetonitrile, and in the presence of a sensitizer, e.g., 9,10-dicyanoanthracene (DCA) or quinones, undergo an electron-transfer-induced oxygenation with the formation of 1,2-dioxolane derivatives, allylhydroperoxides, and other products. The formation of the byproducts are suppressed in the presence of a co-sensitizer, such as biphenyl (BP). Of several mechanisms that have been proposed, the experimental results are most consistent with a chain mechanism involving ground state dioxygen and the 1,3-radical cation, CP$^-$ (reactions 1–5).[91]

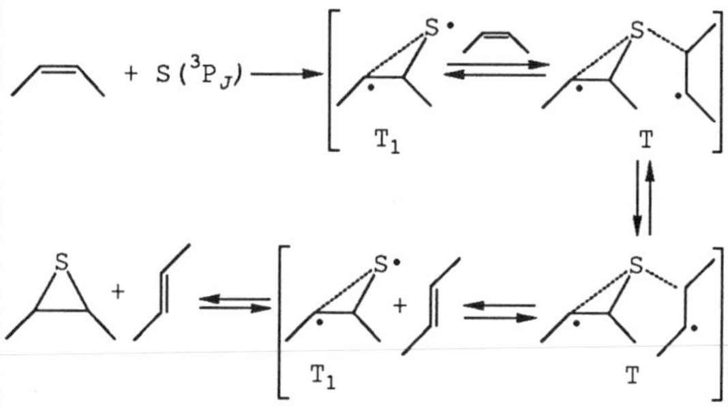

13.44. The sulfur atom in its lowest excited triplet state, S(^3P_J), formed by photolysis of carbonyl sulfide (COS), undergoes reaction with alkenes to form thiiranes in their lowest triplet state. The gas-phase addition of S(^3P_J) to *cis-*or *trans-*2-butene proceeds totally stereospecifically with triplet thiirane to cause geometrical isomerization of the butene with which it forms a collisional complex. A theoretical study promotes the proposed mechanism.[92]

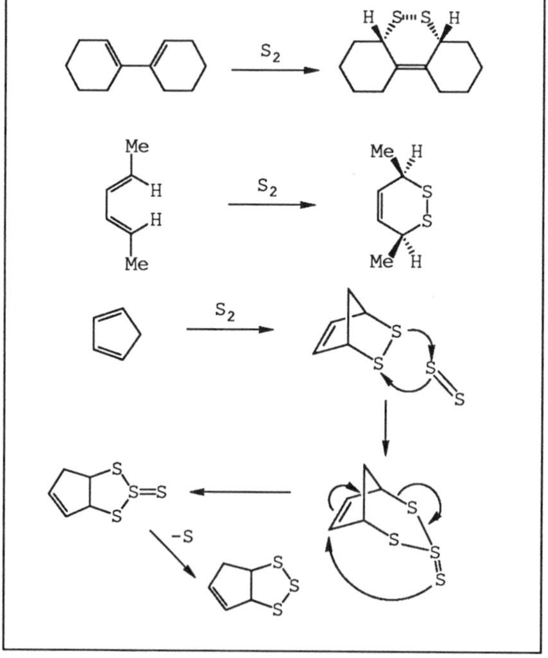

13.45. Like $^1O_2$, disulfur, S_2, undergoes a Diels–Alder type of cycloaddition with acyclic 1,3-dienes to form cyclic disulfides with stereochemistry consistent with the Woodward–Hoffmann rules. However, with bridged dienes, as well as C_5 and C_7 cyclic dienes (but not 1,3-cyclohexadiene), epitrisulfides are formed. This is in striking contrast to the products obtained by the analogous singlet dioxygen chemistry. The following mechanism has been proposed. The epitrisulfide product is produced as a consequence of sulfur deposition from the insertion of a second molecule of S_2 into the highly strained S—S bond of the corresponding thietane precursor intermediate.[93]

REFERENCES

(1)(a) Salem, L. *Electrons in Chemical Reactions: First Principle*, Wiley: New York, 1982, p. 67. (b) For MO description, see Kasha, M. and Brabham, D. E. In *Singlet Oxygen*. Wasserman, H. H., and Murray, R. W., eds. Academic Press: New York, 1979, Chapter 1. (c) For VB description, see Moss, B. J., Bobrowicz, E. S., and Goddard, W. A. *J. Chem. Phys.*, 1975, *63*, 4632.

(2)(a) Gorman, A. A. and Rodgers, M. A. J. In *CRC Handbook of Organic Photochemistry*, Scaiano, J. C., ed. CRC: Boca Raton, Florida, 1989, Vol. 2, p. 229. (b) Gorman, A. A. and Rodgers, M. A. J. *Chem. Soc. Rev.* 1981, *10*, 205. (c) Frimmer, A. A. *Singlet O₂*. CRC: Boca Raton, Florida, 1984. (d) Kearns, D. R. *Chem. Rev.* 1971, *71*, 395. (e) Herzberg, G. *Molecular Spectra and Molecular Structure*. Van Nostrand Reinhold: Princeton, 1950. (f) For the dependence on the lifetime of $^1O_2$ of pressure and temperature in different solvents, see Schmidt, R., Seikel, K., and Brauer, H. D. *Ber. Bunsen-Ges. Phys. Chem.* 1990, *94*, 1100.

(3) Badger, R. M., Wright, A. C., and Whitlock, R. F. *J. Chem. Phys.* 1965, *43*, 4345.

(4) Merker, P. B. and Kearns, D. R. *J. Am. Chem. Soc.* 1972, *94*, 1030, 7244.

(5) Gollnick, K. and Schenck, G. O. *Pure Appl. Chem.* 1964, *9*, 507, and references therein.

(6)(a) For the sensitized oxidation of secondary alcohols, see Schenck, G. O., Becker, H. D., Schulte-Elte, K. H., and Krauch, C. H. *Chem. Ber.* 1963, *96*, 509. (b) For the photochemical oxidation of cycloalkanes promoted by ceric ammonium nitrate, see Baciocchi, E., Del Giacco, T., and Sebastiani, G. V. *Tetrahedron Lett.* 1987, *28*, 1941.

(7) For a review, see Quinkert, G. *Angew. Chem., Int. Ed. Engl.* 1965, *4*, 211.

(8) Arigoni, D., Barton, D. H. R., Bernasconi, R., Djerassi, C., Mills, J. S., and Wolff, R. E. *J. Chem. Soc.* 1960, 1900.

(9) Considerable attention has been given recently to oxygenations by photoinduced electron transfer. For a review, see Lopez, L. *Topics Curr. Chem.* 1990, *156*, 117.

(10) Same as 18(b).

(11)(a) Foote, C. S. and Wexler, S. *J. Am. Chem. Soc.* 1964, *86*, 3880. (b) Wintgens, V., Scaiano, J. C., Linden, S. M., and Neckers, D. C. *J. Org. Chem.*

1989, *54*, 5242. (c) Murasecco-Suardi, P., Gassmann, E., Braun, A. M., and Oliveros, E. *Helv. Chim. Acta.* 1987, *70*, 1760. (d) For phase transfer catalysts promoting photooxidations with $^1O_2$, see Guarini, A. and Tundo, P. *J. Org. Chem.* 1987, *52*, 3501, and references therein. (e) For a review of Rose Bengal, See Neckers, D. C. *J. Chem. Ed.* 1987, *64*, 649.

(12) Schaap, A. P., Thayer, A. L., Blossey, E. C., and Neckers, D. C. *J. Am. Chem. Soc.* 1975, *97*, 3741.

(13) Tamagaki, S., Liesner, C. E., and Neckers, D. C. *J. Org. Chem.* 1980, *45*, 1573.

(14) For a recent bleaching study, see Linden, S. M. and Neckers, D. C. *J. Am. Chem. Soc.* 1988, *110*, 1257.

(15) Foote, C. S. and Wexler, S. *J. Am. Chem. Soc.* 1964, *86*, 3879.

(16) Aubry, J. M., Cazin, B., and Duprat, F. *J. Org. Chem.* 1989, *54*, 726, and references therein.

(17) DiMascio, P. and Sies, H *J. Am. Chem. Soc.* 1989, *111*, 2909, and references therein (decomposition of a water-soluble endoperoxide).

(18)(a) Murray, R. W. and Kaplan, M. L. *J. Am. Chem. Soc.* 1969, *91*, 5358. (b) Murray, R. W. In *Singlet Oxygen*. Wassermann, H. H. and Murray, R. W., eds. Academic Press: New York, 1979, Chapter 3.

(19)(a) Corey, E. J., Mehrotra, M. M., and Khan, A. U. *J. Am. Chem. Soc.* 1986, *108*, 2472, and references therein. (b) Posner, G. H., Weitzberg, M., Nelson, W. M., Murr, B. L., and Seliger, H. H. *J. Am. Chem. Soc.* 1987, *109*, 278.

(20) For reviews of $^1O_2$ chemistry, see (a) Adam, W. In *Methoden der Organischen Chemie Houben-Weyl*. G. Thieme: Stuttgart, 1975, Vol. 4/5b, p. 1465. (b) Denny, R. W. and Nickon, A. *Org. Reactions.* 1973, *20*, 133. (c) Ohloff, G. *Pure Appl. Chem.* 1975, *43*, 481. (d) Pfoertner, K. H. In *Photochemistry in Organic Synthesis*, Coyle, J. D., ed. The Royal Society of Chemistry: London, 1986, p. 189. (e) Rånby, Rabek, J. E. F., eds. *Singlet Oxygen*, Wiley: Chichester, 1978. (f) Schaap, A. P., ed. In *Singlet Molecular Oxygen, Benchmark Papers in Organic Photochemistry*. Dowden, Hutchinson and Ross: Stroudsburg, 1976, Vol. 5. (g) Jiang, Z. *Res. Chem. Intermed.* 1990, *14*, 185.

(21) For reviews, see (a) Gollnick, K. and Kuhn, H. J. See ref. 1(b), p. 287. (b) Monroe, B. M. *J. Am. Chem. Soc.* 1981, *103*, 7253, and references therein.

(22)(a) Gollnick, K., Haisch, D., and Schade, G. *J. Am. Chem. Soc.* 1972, *94*, 1747. (b) Machin, P. J.,

Porter, A. E., and Sammes, P. G. *J. Chem. Soc., Perkins Trans. 1.* **1973**, *404*, 407.

(23)(a) O'Shea, K. E. and Foote, C. S. *J. Am. Chem. Soc.* **1988**, *110*, 7167.   (b) Jensen, F. and Foote, C. S. *J. Am. Chem. Soc.* **1987**, *109*, 6376.   (c) Manring, L. E. and Foote, C. S. *J. Am. Chem. Soc.* **1983**, *105*, 4710.

(24)(a) Atkins, R. and Carless, H. A. J. *Tetrahedron Lett.* **1987**, *28*, 6093, and references therein. (b) Matusch, R. and Schmidt, G. *Angew. Chem., Int. Ed. Engl.* **1988**, *27*, 717.

(25)(a) Schenck, G. O. and Ziegler, K. *Naturwiss.* **1944**, *32*, 157.   (b) For preparation of a bicyclic peroxide related to ascariodole, see Schenck, G. O. *Angew Chem.*, **1949**, *61*, 434.

(26)(a) Gollnick, K. and Griesbeck, A. *Tetrahedron Lett.* **1988**, *29*, 3303.   (b) See also Ref. 22(a). (c) For the photooxygenation of an acetoxy diene as the key step in the synthesis of cyclic peroxy ketals, see Snider, B. B. and Shi, Z. *J. Org. Chem.* **1990**, *55*, 5669.

(27) For photooxidation in the absence of thiourea, see Takeshita, H. Kanamori, H., and Hatsui, T. *Tetrahedron Lett.* **1973**, 3139.

(28) The 2-cyclopentene-1,4-diol formed is a useful intermediate in the synthesis of prostaglandins and jasmones. See Kaneko, C., Sugimoto, A., and Tanaka, S. *Synthesis.* **1974**, 876.

(29) CoTPP-catalyzed rearrangement is suggested to be superior to thermolysis, see: (a) Boyd, J. D., Foote, C. S., and Imagana, D. K. *J. Am. Chem. Soc.* **1980**, *102*, 3641.   (b) Akbulut, N., Menzer, A., and Balci, M. *Tetrahedron Lett.* **1987**, *28*, 1689.   (c) For the Ru(II) catalyzed rearrangement of 1,4-epiperoxides, see Suzuki, M., Ohtake, H., Kameya, Y., Hamanaka, N., and Noyori, R. *J. Org. Chem.* **1989**, *54*, 5292.

(30) Foster, C. H. and Berchtold, G. A. *J. Am. Chem. Soc.* **1972**, *94*, 7939.

(31) Foster, C. H. and Berchtold, G. A. *J. Am. Chem. Soc.* **1972**, *94*, 7939.

(32)(a) Skorianetz, W., Schulte-Elte, K. H., and Ohloff, G. *Helv. Chim. Acta.* **1971**, *54*, 1913. (b) Harada, N., Suzuki, S., Uda, H., and Ueno, H. *J. Am. Chem. Soc.* **1972**, *94*, 1777.   (c) Burger, U. and Jefford, C. *Chimia.* **1971**, *25*, 304.

(33) Pfoertner, K. H. In *Photochemistry in Organic Synthesis*, Coyle, J. D., ed. The Royal Society of Chemistry: London, **1986**, p. 196.

(34) Kondo, K. and Matsumo, M. *Chem. Lett.* **1974**, 701.

(35) Rio, G. and Berthelot, J. *Bull. Soc. Chim. France.* **1971**, 2438.

(36) Adam, W. and Rebollo, H. *Tetrahedron Lett.* **1981**, *22*, 3049.

(37) Adam, W. Klug, G., Peters, E.-M., Peters, K., and von Schnering, H. G. *Tetrahedron.* **1985**, *41*, 2045.

(38)(a) Rosenfeld, S. M. *J. Chem. Educ.* **1986**, *63*, 184.   (b) Schulz, M. and Kirschke, K. *Adv. Heterocyclic Chem.* **1976**, *8*, 165.   (c) Rigaudy, J. *Pure Appl. Chem.* **1968**, *16*, 169. For reviews, see (d) Gollnick, K. and Schenck, G. O. In *1,4-Cycloaddition Reactions*, Hamer, J., ed. Academic Press: New York, 1967, p. 255.   (e) Schönberg, A. *Preparative Organic Photochemistry*. Springer Verlag: New York, 1968, p. 389, and references therein.

(39) (a) Rigaudy, J., Deletang, C., and Basselier, J.-J. *C. R. Hebd. Seances Acad. Sci.*, *Ser. C.* **1969**, *268*, 344.   (b) For the role of ISC steps in singlet-oxygen chemistry, see Turro, N. J. *Tetrahedron.* **1985**, *41*, 2089.

(40) Catalani, L. H. and Wilson, T. *J. Am. Chem. Soc.* **1989**, *111*, 2633, and references therein.

(41)(a) Matsumoto, M. and Kuroda, K. *Synth. Commun.* **1981**, *11*, 987.   (b) For $^1O_2$ lifetimes in different solvents, see Ref. 2(a), p. 233.   (c) For [4+2] cycloaddition of $^1O_2$ to *trans*-stilbene, see Kwon, V.-M., Foote, C. S., and Khan, S. I. *J. Org. Chem.* **1989**, *54*, 3378, and references therein.

(42) For a review, see (a) Foote, C. S., Wuesthoff, M. F., Wexler, S., Burstain, I. G., Denny, R., Schenck, G. O., and Schulte-Elte, K.-H. *Tetrahedron.* **1967**, *23*, 2583.   (b) Wasserman, H. H., McCarthy, K. E., and Prowse, K. S. *Chem. Rev.* **1986**, *86*, 845.

(43) Gollnick, K. and Koegler, S. *Tetrahedron Lett.* **1988**, *29*, 1003, and references therein.

(44)(a) For a review, see Feringa, B. L. *Rec. Trav. Chim. Pays-Bas.* **1987**, *106*, 469.   (b) Kernan, M. R. and Faulkner, D. J. *J. Org. Chem.* **1985**, *53*, 2773. (c) Grazino, M. L. and Iesce, M. R. *Synthesis.* **1985**, 1151.

(45) For preparative procedure of photooxygenation of 2,5-dimethylfuran, see Foote, C. S. and Uhde, G. *Org. Photochem. Synth.* **1971**, *1*, 70.

(46)(a) Giraud, M., Valla, A., Bazin, M., Santus, R. and Momzikoff, A. *J. Chem. Soc., Chem. Commun.* **1982**, 1147.   (b) See also Ref. 15.

(47) For a review, see Foote, C. S. *Acc. Chem. Res.* **1968**, *1*, 104.   (b) For the preparative procedure of photooxygenation of 2,3-dimethyl-2-butene, see Foote,

C. S. and Uhde, G. *Org. Photochem. Synth.* **1971**, *1*, 60.

(48)(a) For a review, see Stephenson, L. M., Grdina, M. J., and Orfanopoulos, M. *Acc. Chem. Res.* **1980**, *13*, 419. (b) Orfanopoulos, M., Smonou, I., and Foote, C. S. *J. Am. Chem. Soc.* **1990**, *112*, 3607, and references therein. (c) Kwon, B.-M. and Foote, C. S. *J. Org. Chem.* **1989**, *54*, 3878.

(49)(a) Fenical, W., Kearns, D. R., and Radlick, P. *J. Am. Chem. Soc.* **1969**, *91*, 3396, 7771. (b) Jefford, C. W., Kohmoto, S., Boukouvalas, J., and Burger, U. *J. Am. Chem. Soc.* **1983**, *105*, 6498. (c) Jefford, E. W., Kohmoto, S. and Boukouvalas, J. *J. Photochem.* **1984**, *25*, 537.

(50)(a) Harding, L. B. and Goddard, W. A., III, *J. Am. Chem. Soc.* **1980**, *102*, 439. (b) Semiempirical MINDO/3 calculations: Dewar, M. J. S. and Thiel, W. *J. Am. Chem. Soc.* **1977**, *99*, 2338. (c) CASSCF calculations: Hotoka, M., Roos, B., and Siegbahn, P. *J. Am. Chem. Soc.* **1983**, *105*, 5263. (d) STO-3G and unrestricted NIMDO/3 (UM3) calculations: Yamaguchi, K., Yabushita, S., Fueno, T., and Houk, K. N. *J. Am. Chem. Soc.* **1981**, *103*, 5043. (e) A recent theoretical study supports the concerted mechanism. See Davies, G. A. and Schiesser, C. H. *Tetrahedron Lett.* **1989**, *30*, 7099.

(51)(a) Clennan, E. L., Chen, X., and Koola, J. J. *J. Am. Chem. Soc.* **1990**, *112*, 5193. (b) For geminal selectivity, see Clennan, E. L., and Chen, K. *J. Org. Chem.* **1988**, *53*, 3124. (c) Orfanopoulos, M. G., Grdina, M. G., and Stephenson, L. M. *J. Am. Chem. Soc.* **1979**, *101*, 275. (d) Orfanopoulos, M. G., Stratakis, M., and Elemes, Y. *J. Am. Chem. Soc.* **1990**, *112*, 6417, and references therein.

(52)(a) Foote, C. S., Wexler, S., and Ando, W. *Tetrahedron Lett.* **1965**, 4111. (b) Schenck, G. O., Gollnick, K., Buchwald, G., Schroeter, S., and Ohloff, G. *Liebigs Ann. Chem.* **1964**, *674*, 93.

(53)(a) Litt, F. A. and Nickon, A. *Adv. Chem. Ser.* **1968**, *77*, 118. (b) Adam, W., Griesbeck, A., and Staab, E. *Tetrahedron Lett.* **1986**, *27*, 2839

(54) Adam, W. and Bakker, B. H. *Tetrahedron Lett.* **1979**, 4171.

(55)(a) Adam, W. and Liu, J. C. *J. Am. Chem. Soc.* **1972**, *94*, 2894. (b) Adam, W. and Steinmetzer, H.-C. *Angew. Chem., Int. Ed. Engl.* **1972**, *11*, 540. (c) Iwata, C. *Tetrahedron Lett.* **1985**, *26*, 3227, 3231.

(56)(a) Jaroszewski, J. W. and Ettlinger, M. G. *J. Org. Chem.* **1988**, *53*, 4335. (b) Adam, W., Hasemann, L., and Prechtl, F. *Angew. Chem., Int. Ed. Engl.* **1988**, *27*, 1536. (c) Adam, W. and Griesbeck,

A. *Angew. Chem., Int. Ed. Engl.* **1985**, *24*, 1070. (d) Orfanopoulos, M. and Foote, C. S. *Tetrahedron Lett.* **1985**, *26*, 5991.

(57) Ohloff, G., Klein, E., and Schenck, G. O. *Angew. Chem.,* **1961**, *73*, 578.

(58) von Wittenau, M. S. *J. Org. Chem.* **1964**, *29*, 2746.

(59) Kornblum, N. and DeLaMare, H. E. *J. Am. Chem. Soc.* **1951**, *73*, 880.

(60) Mihelich, E. D. and Eickhoff, D. J. *J. Org. Chem.* **1983**, *48*, 4135.

(61)(a) Adam, W., Braun, M., Griesbeck, A., Lucchini, U., Staab, E., and Will, B. *J. Am. Chem. Soc.* **1989**, *111*, 203. (b) Adam, W., Griesbeck, A., and Staab, E. *Tetrahedron Lett.*, **1986**, *27*, 2839. (c) Adam, W. and Pasquato, L. *Tetrahedron Lett.* **1987**, *28*, 311.

(62) For reviews, see (a) Adam, W. and Yang, F. In *Small Ring Heterocycles*, Hassner, A., ed. Wiley: New York, 1985, Part 3, Chapter 4. (b) Adam, W., Baader, W. J., Babatsikos, C., and Schmidt, E. *Bull. Soc. Chim. Belg.* **1984**, *93*, 605. (c) Adam, W. In *The Chemistry of Functional Groups, Peroxides*, Patai, S., ed. Wiley: New York, 1983. (d) Numerous 1,2-dioxetanes have been isolated and characterized. For recent studies, see Adam, W., Griesbeck, A. G., Gollnick, K., and Knutzen-Mies, K. *J. Org. Chem.* **1988**, *53*, 1492. (e) Akasaka, T., Nomura, Y., and Ando, W. *J. Org. Chem.* **1988**, *53*, 1670.

(63)(a) Bartlett, P. D. and Schaap, A. P. *J. Am. Chem. Soc.* **1970**, *92*, 3223. (b) For the preparative procedure for the photooxygenation of *cis*-dimethoxyethylene, see Schaap, A. P., Thayer, A. L., and Kees, K. *Org. Photochem. Synth.* **1976**, *2*, 49. (c) 1,2-dioxetanes are also formed in the singlet oxygenation of *s-cis* fixed dienes, see Clennan, E. L. and Nagraba, K. *J. Org. Chem.* **1987**, *52*, 294.

(64)(a) Vinyl ethers: Asveld, E. W. H. and Kellog, R. M. *J. Am. Chem. Soc.* **1980**, *102*, 3644. (b) Enamines: Saito, I., Matsugo, S., and Matsuura, T. *J. Am. Chem. Soc.* **1979**, *101*, 7332. (c) Vinylsulfides: Ando, W., Watanabe, K., Suzuki, J., and Migita, T. *J. Am. Chem. Soc.* **1974**, *96*, 6766.

(65)(a) Kopecký, K. R. and Mumford, C. *Can. J. Chem.* **1969**, *47*, 7093. (b) By oxidation of an alkene and trimethylsilyl hydroperoxide, see Posner, G. H., Webb, K. S., Nelson, W. M., Kishimoto, T., and Seliger, H. H. *J. Org. Chem.* **1989**, *54*, 3253. See also Ref. 18(b).

(66)(a) Richardson, W. H., Burns, J. H., Price, M. E., Cranford, R., Foster, M., and Slusser, P. *J. Am.*

Chem. Soc. **1970**, *100*, 7596. (b) Adam, W. In *Chemical and Biological Generation of Electronically Excited States*, Adam, W., Cilento, G., eds. Academic Press: New York, 1982. (c) Richardson, W. H., Sitggal-Estberg, D. L., Chen, Z., Baker, J. C., Burns, D. M., and Sherman D. G. *J. Org. Chem.* **1987**, *52*, 3143.

(67) Richardson, W. H. and Sitggal-Estberg, D. L. *J. Am. Chem. Soc.* **1982**, *104*, 4173.

(68) Richardson, W. H. and Thomson, S. A. *J. Org. Chem.* **1985**, *50*, 1803.

(69)(a) Schaap, A. P. and Zaklika, K. A., In *Singlet Oxygen*, Wasserman, H. H., and Murray, R. W., eds, Academic Press: New York, 1979, Chapter 6. (b) Bartlett, P. D. and Landis, M. E., In *Singlet Oxygen*, Wasserman, H. H., and Murray, R. W., eds, Academic Press: New York, 1979, Chapter 7. (c) Wasserman, H. H. and Lipschutz, B. H., In *Singlet Oxygen*, Wasserman, H. H., and Murray, R. W., eds, Academic Press: New York, 1979, Chapter 9. (d) as above.

(70)(a) Tonachini, G., Schlegel, H. B., Bernardi, F., and Robb, M. A. *J. Am. Chem. Soc.* **1990**, *112*, 483 and references therein. (b) The major products of singlet oxygenation of 1,4-dialkoxy-1,3-butadienes are 1,2-dioxetanes. See Clennan, E. L. and Nagraba, K. *J. Am. Chem. Soc.* **1988**, *110*, 4312, and references therein. (c) For evidence for the perepoxide mechanism, see Akasaka, T. and Ando, W. *Tetrahedron Lett.* **1987**, *28*, 217.

(71)(a) For the preparative procedure for the dioxetane, see Adam, W., Liu, J.-C., Strating, J., Wieringa, J. H., and Wynberg, H. *Org. Photochem. Synth.* **1967**, *2*, 10. (b) The results can, however, be discounted on several grounds. For details, see Frimer, A. A. *Chem. Rev.* **1979**, *79*, 359.

(72)(a) Turro, N. J. and Lechtken, P. *Tetrahedron Lett.* **1973**, 565. (b) Adam, W. *Pure Appl. Chem.* **1980**, *52*, 2591. (c) Horn, K. A., Koo, J., Schmidt, S. P., and Schuster, G. B. *Mol. Photochem.* **1978–79**, *9*, 1. (d) Turro, N. J. *Acc. Chem. Res.* **1974**, *7*, 97. (e) For steric and electronic effects altering the efficiency of triplet carbonyl production, see Adam, W. and Paader, W. J. *J. Am. Chem. Soc.* **1985**, *107*, 410, ref. 67, and references therein. (f) For enzymatic triggering of 1,2-dioxetanes, see Schaap, A. P., Handley, R. S., and Giri, B. P. *Tetrahedron Lett.* **1987**, *28*, 935.

(73)(a) Richardson, W. H. *J. Org. Chem.* **1989**, *54*, 4677. (b) For a review of oxygen diradicals derived from cyclic perepoxides, see Adam, W. *Acc. Chem. Res.* **1979**, *12*, 390.

(74) For a review, see White, E. H., Miano, J. D., Watkins, C. J., and Breux, E. J. *Angew. Chem.*, **1974**, *86*, 292.

(75) Zaklika, K., Burns, P. A., and Schaap, A. P. *J. Am. Chem. Soc.* **1978**, *100*, 318.

(76) Adam, W., Bhushan, V., Fuchs, A., and Kirchgäsner, U. *J. Org. Chem.* **1987**, *53*, 3059.

(77) Adam, W., Epe, B., Shiffmann, D., Vargas, F., and Wild, D. *Angew. Chem.*, *Int. Ed. Engl.* **1988**, *27*, 429.

(78) Jefford, C. W., Boukouvalas, J., and Kohmoto, S. *Tetrahedron.* **1985**, *41*, 2081.

(79) For reviews, see (a) Ando, W. *Sulfur Rep.* **1981**, *1*, 143. (b) Ando, W. and Takata, T. In *Singlet O$_2$*. CRC: Boca Raton, Florida, 1984, Vol. 3, Chapter 1, p. 1.

(80) Gu, C.-L. and Foote, C. S. *J. Am. Chem. Soc.* **1982**, *104*, 6060.

(81) Akasaka, T., Kako, M., Sonobe, H., and Ando, W. *J. Am. Chem. Soc.* **1988**, *110*, 494, and references therein.

(82) Akasaka, T., Yabe, A., and Ando, W. *J. Am. Chem. Soc.* **1987**, *109*, 8085.

(83) Jensen, F. and Foote, C. S. *J. Am. Chem. Soc.* **1988**, *110*, 2368.

(84) Clennan, E. L. and Yang, K. *J. Am. Chem. Soc.* **1990**, *112*, 4044.

(85) For recent studies, see (a) Clennan, E. L. and Chen, X. *J. Am. Chem. Soc.* **1989**, *111*, 5787 (diastereoselective oxidation). (b) Clennan, E. L. and Chen, X. *J. Am. Chem. Soc.* **1989**, *111*, 8212 (temperature, solvent, and substituent effects). (c) Nahm, K. and Foote, C. S. *J. Am. Chem. Soc.* **1989**, *111*, 1909 (intermediate trap by trimethyl phosphite).

(86)(a) Shiguro, K., Hirano, Y., and Sawaki, Y. *J. Org. Chem.* **1988**, *53*, 5397. (b) Hamilton, G. A. In *Molecular Mechanism of Oxygen Activation.* Hayaishi, O., ed. Academic Press: New York, 1974, p. 405.

(87)(a) For an excellent review of the electronic structure of carbonyl oxides, see Sander, W. *Angew. Chem.*, *Int. Ed. Engl.* **1990**, *29*, 344. (b) For rigorous *ab initio* and semiempirical calculations of carbonyl oxides, see Cremer, D., Schmidt, T., Sander, W., and Bischop, P. *J. Org. Chem.* **1989**, *54*, 2115, and references therein. (c) For generation and transient spectroscopy of substituted diaryl carbonly oxides, see Scaiano, J. C., McGimpsey, W. G., and Casal, H. L. *J. Org. Chem.* **1989**, *54*, 1612.

(88) Saito, I., Matsuura, T., Nakagawa, M., and Hino, T. *Acc. Chem. Res.* **1971,** *10,* 346, and references therein.

(89) For chalcogenapyrylium dyes as potential photodynamic therapeutic agents, see Detty, M. R. and Merkel, P. B. *J. Am. Chem. Soc.* **1990,** *112,* 3845.

(90) For example, 2,2′:5′,2″-terthiophene isolated from marigold, see Benny, J.-P. *J. Org. Chem.* **1982,** *47,* 2201.

(91)(a) Gollnick, K. and Paulmann, U. *J. Org. Chem.* **1990,** *55,* 5954. (b) For the oxygenation of vinylcyclopropanes, see Feldman, K. S. and Simpson, R. E. *J. Am. Chem. Soc.* **1989,** *111,* 4878.

(92)(a) Joseph, J., Gosavi, R. K., Otter, A., Kotovych, G., Lown, E. M., and Strausz, O. P. *J. Am. Chem. Soc.* **1990,** *112,* 8670. (b) For the first stable aliphatic thioaldehyde and its photochemistry, see Okazaki, R., Ishii, A., and Inamoto, N. *J. Am. Chem. Soc.* **1987,** *109,* 279.

(93) Steliou, K., Gareau, Y., Milot, G., and Salama, P. *J. Am. Chem. Soc.* **1990,** *112,* 7819, and references therein.

Chapter 14
PREPARATIVE TECHNIQUES

PREPARATIVE TECHNIQUES

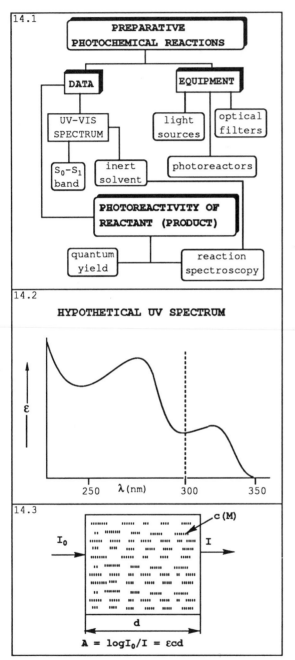

14.1

PREPARATIVE PHOTOCHEMICAL REACTIONS

DATA

EQUIPMENT

UV–VIS SPECTRUM

light sources

optical filters

S_0–S_1 band

inert solvent

photoreactors

PHOTOREACTIVITY OF REACTANT (PRODUCT)

quantum yield

reaction spectroscopy

14.2

HYPOTHETICAL UV SPECTRUM

ε

250   λ (nm)   300   350

14.3

I_0 I c (M)   d

$A = \log I_0/I = \varepsilon c d$

14.1. Preparative photochemical reactions require specific information about the reactant(s) and product(s), in addition to the known criteria for thermal reactions in organic synthesis. There are two requirements for a successful photochemical reaction: the formation of a sufficient number of electronically excited molecules, and a reasonably short irradiation time. It is also important to note that photochemical reactions are performed in special equipment that is not always readily available in organic chemical laboratories.[1]

14.2. The first step in planning a photochemical reaction is to record the electronic (UV-VIS) spectra of the reactant(s) and, whenever possible, the product(s) in the solvent to be used for the irradiation.[2] UV-VIS absorption spectra of organic molecules generally show several broad overlapping bands. Since most photochemical reactions in fluid solution proceed through the S_1 (or T_1) state, radiation corresponding to the longest wavelength band in the electronic absorption spectrum of one of the reactants should be used. If this band lies above 300 nm, either a RUL-3000 discharge lamp (see 14.9) or a high-pressure mercury lamp with a Pyrex filter (see 14.20) should be used for the irradiation. In a photosensitized reaction, radiation that is exclusively, or at least predominantly, absorbed by the photosensitizer is required.

14.3. The appropriate concentration (c) for the photoreactive component can be calculated from the *Beer–Lambert law*. This requires a knowledge of the molar absorption coefficient (ϵ) of the reactant in the wavelength region selected for the irradiation.[3]

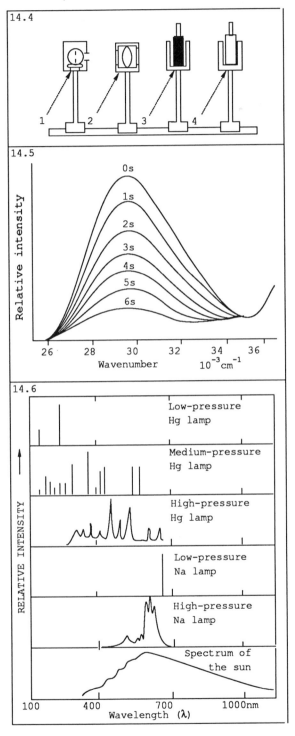

14.4

14.5

Relative intensity

0s
1s
2s
3s
4s
5s
6s

26   28   30   32   34   36
Wavenumber $10^{-3} cm^{-1}$

14.6

RELATIVE INTENSITY

Low-pressure Hg lamp

Medium-pressure Hg lamp

High-pressure Hg lamp

Low-pressure Na lamp

High-pressure Na lamp

Spectrum of the sun

100   400   700   1000nm
Wavelength (λ)

14.4. Information about the efficiency of a photochemical reaction is obtained readily by *reaction spectroscopy*. In this technique the sample is irradiated in an apparatus consisting of a lamp (1), a quartz lens (2), and a filter (3), all mounted on an optical bench. This arrangement allows a narrow beam of radiation to be directed onto the radiation cell (4).[4] More accurate information about the efficiency of a photochemical reaction can be obtained by measuring its quantum yield.[5] However, many reports of photochemical reactions do not include quantum yield measurements. It can be estimated from the reaction conditions, such as sample size, concentration, light source, and irradiation time.

14.5. A solution of known concentration of reactant is irradiated and its UV spectrum is recorded at regular time intervals. The compound whose spectrum is shown was extremely photoreactive since its longest wavelength band almost disappeared after 6 seconds of irradiation.[6] The situation is more complicated if the photoproduct absorbs at the same wavelength as the starting material. In this case the photoproduct can act as an internal filter and inhibit the desired reaction; is photoreactive and can undergo a further photochemical transformation, resulting in a mixture of products; or can revert to the starting material and result in a photostationary state. It is desirable in the investigation of a new photochemical reaction to affect the conversion only to about 10–20%, isolate the photoproduct, record its spectrum, and determine its photostability.

14.6. Preparative photochemical reactions require an intense light source that emits radiation for absorption by the reactant.[1(a),(e),7] The figure shows the emission spectra of the sun and the most commonly used radiation sources.[8]

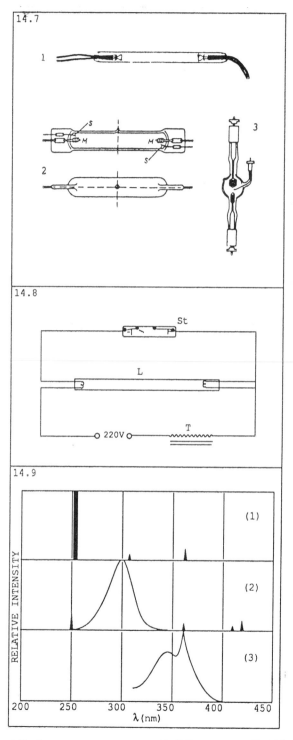

14.7. The most commonly used discharge lamps for preparative photochemistry fall into three categories: (1) *low-pressure* (resonance) *mercury lamps*, with or without a phosphor coating, (2) *medium-pressure mercury lamps*, with or without a dopant (see 16.5), and (3) *high-pressure mercury or xenon-mercury lamps*.

14.8. *Low-pressure mercury lamps* (L) are miniature U-shaped or straight-tube fluorescent lamps. A discharge of mercury vapor (10^{-5} atmosphere) converts electrical energy into UV radiative energy with an efficiency of about 60% (95% of this energy is emitted at 253.7 nm). The remaining energy is converted into visible radiation and heat. The arc length of the lamp is a function of power rating. A typical low-pressure lamp has about a 40 cm arc length with a power rating of 16 W, as compared to a 10 cm arc length with a power rating of 400 W for a high-pressure lamp. The lamp is initiated by a starter (St) that warms the electrodes. When operated at 40–50°C, low-pressure mercury lamps produce the required vapor pressure of mercury and the maximum photon flux. Overheating causes emission to occur at longer wavelengths at the expense of the 253.7 nm resonance line. The lifetime of these lamps is about 6000 hours.[9]

14.9. In a quartz envelope these lamps, e.g., RUL-2537 (1), emit radiation primarily at 253.7 nm. An envelope consisting of ultrapure quartz emits radiation at 185 nm that can be efficiently used for mercury-sensitized photoreactions. When the internal surface of the envelope is coated with a phosphor, as in lamps RUL-3000 (2) and RUL-3500 (3),[10] the primary emission is replaced by a secondary fluorescence emission at longer wavelengths. These lamps provide radiation in the wavelength region from 250–380 nm without a need for filters.

14.10

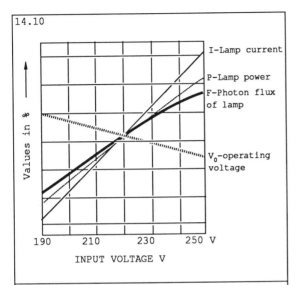

I-Lamp current

P-Lamp power

F-Photon flux of lamp

V_0-operating voltage

INPUT VOLTAGE V

14.10. One disadvantage of low-pressure mercury lamps is their low radiation flux density. To ensure a sufficient photon flux for a photochemical reaction, they are used generally in reactors with external radiation sources (see 14.26 and 14.27) in which six to 12 lamps are positioned around the reaction flask. The figure shows the primary characteristics of these lamps as a function of the input voltage.

14.11

SPECTRAL DISTRIBUTIONS OF UNDOPED MEDIUM-PRESSURE MERCURY LAMP HPK 125W[a]

Spectral region, nm	Major lines	Output watts	einstein/h
240–270	248,254, 265,270	4.3	0.03
270–300	280,289, 297	1.1	0.01
300–330	302,313	4.5	0.04
330–360	334	0.4	0.004
360–390	365.5	5.1	0.06
390–420	405–408	1.8	0.02
420–450	436	3.5	0.05
540–570	546	4.4	0.07
570–600	577–579	3.5	0.06

[a]Manufactured by North American Philips Lighting Co., Somerset, NJ.

14.11. *Medium-pressure mercury lamps* for photochemical reactions on a laboratory scale typically have a power rating of 100–500 W, an arc length of 5–15 cm, and a tube diameter of 1–2.5 cm. The mercury pressure in these lamps varies from 1–10 atm and the spectral distribution of the radiation is mainly in the form of sharp discrete lines.[11] Metal (e.g., Ga, Mg, Fe, and Tl) halide doped lamps provide additional lines characteristic of the metal dopant.[12] The 30-nm segments of the spectrum shown in the table[13] correspond to the minimum wavelength band-pass normally required in preparative photochemistry. Although absorption bands of photoreactive components in a reaction mixture are usually considerably broader than 30 nm, absorption anywhere within the band generally leads to the same photochemical reaction with the same quantum efficiency.

14.12

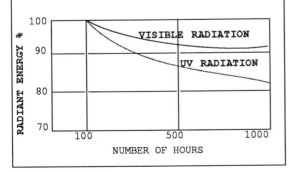

VISIBLE RADIATION

UV RADIATION

NUMBER OF HOURS

14.12. These discharge lamps convert only a small part (30–40%) of the electrical energy into radiation (about 60% UV, 40% VIS for a typical 400 W Hg lamp); the remainder is converted into heat. The normal operating temperature of these lamps are 600–800°C. Thus, photoreactors utilizing medium-pressure lamps require efficient cooling (see 14.29–14.33). During their lifetime (about 1000 hours), the emitted radiation gradually shifts toward the red because output in the UV region is dissipated faster than that in the VIS region.

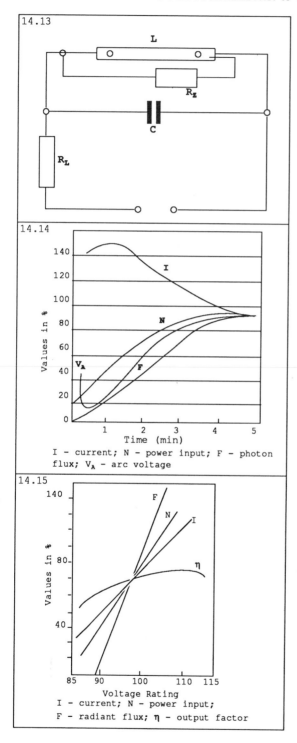

14.13

14.14

I – current; N – power input; F – photon flux; V_A – arc voltage

14.15

I – current; N – power input;
F – radiant flux; η – output factor

14.13. A medium-pressure mercury lamp (L) must be operated with an appropriate *power supply*[14] consisting of an impedance (R_Z), usually an inductance, in series with the lamp to provide stabilization of the voltage-current emissions of the discharge. When a coil (R_L) is connected in series with the lamp, the current will automatically reach the correct value and be maintained. To achieve a stable discharge it is necessary to divide the input voltage between the discharge and the power supply, approximately in a ratio of $1:1$. The use of a coil as a power supply decreases the output of the radiation by about 10%. Lamps of different sizes require different power supplies, which are commercially available.

14.14. When a medium-pressure mercury lamp is operated with a power supply, mercury condensed on the walls of the envelope evaporates and the mercury vapor pressure gradually increases to its maximum value. In the process of initiating the lamp, the current (I) and voltage (V_A), as well as the input (N) and the radiant energy (F) (photon flux), change. Initially, the current is twice the final stationary working current and the voltage of the lamp rises during warm-up from an initial value of 15–30 V to its final operating voltage. Since about 3–5 minutes is required for all of the mercury to evaporate, only then does the discharge reach its stationary parameters with full radiative output. When the lamp is powered off, it cannot be restarted until it has cooled.[15]

14.15. Variations in input voltage considerably alter the characteristics of medium-pressure mercury lamps. The dependence of some of these characteristics on the nominal voltage for a typical medium-pressure lamp is shown in the figure.

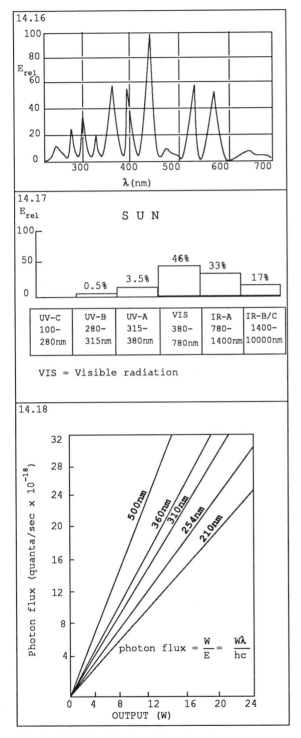

14.16

E_{rel}

λ (nm)

14.17

E_{rel}

S U N

0.5%   3.5%   46%   33%   17%

UV-C 100– 280nm	UV-B 280– 315nm	UV-A 315– 380nm	VIS 380– 780nm	IR-A 780– 1400nm	IR-B/C 1400– 10000nm

VIS = Visible radiation

14.18

Photon flux (quanta/sec × 10^{-18})

500nm   360nm   310nm   254nm   210nm

$$\text{photon flux} = \frac{W}{E} = \frac{W\lambda}{hc}$$

OUTPUT (W)

14.16. *High-pressure mercury lamps* have power ratings ranging from tenths of watts to several kilowatts and operating pressures of mercury and/or xenon from 10 to several hundred atmospheres. Therefore, they must be maintained in suitable housings to be protected from fractures, explosions, and intense UV emissions. Full output is reached about 15 minutes after ignition and the lifetime varies from 200–400 hours using DC power or 100–200 hours with AC power. They are constructed with a capillary or a compact arc [see 14.7(3)]. The latter has an arc length of only a few millimeters, essentially a point source. The emission occurs primarily in the VIS region of the spectrum. A specific region can be selected by using a monochromator or a suitable combination of filters.[16]

14.17. Medium-pressure mercury lamps of high wattage, high-pressure mercury lamps, ordinary incandescent lamps with a tungsten filament, and halogen lamps are common sources of visible radiation. Sunlight is also a valuable source of visible radiation; almost half of the sun's intensity lies at wavelengths between 315 and 780 nm with its strongest intensity at 480 nm.[17]

14.18. To determine the irradiation time for a photochemical reaction it is necessary to know the spectral output of the lamp at the appropriate wavelength. Manufacturers normally provide the output in watts. However, photochemists find it more convenient to express the *photon flux* in einsteins (see 1.4) per hour. The unit einstein/h [watts × wavelength (nm) × 3.0 × 10^{-5}][18] can be related more readily to the photochemical reaction rate, i.e., the rate (mol/h) = einstein/h absorbed × quantum yield of the reaction. The figure demonstrates the determination of the photon flux of a discharge arc for a specific output and wavelength.[19]

14.19

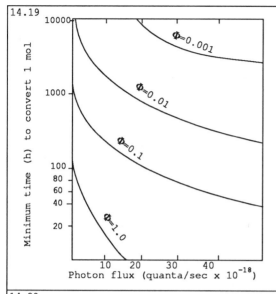

14.19. Using a value for the photon flux and the quantum yield Φ of the reaction, the minimum radiation time t for the transformation of starting material of mass g can be obtained. A more accurate value can be calculated from:

$$t = \frac{g}{M} \frac{hc}{W\Phi\lambda} 6.023 \times 10^{23}$$

where t is the time in seconds, M is the molecular weight of the reactant, W is the output in watts at the wavelength λ absorbed by the starting material, and Φ is the quantum yield of the reaction. It is clear from the figure[20] that many photochemical reactions require very long irradiation times.

14.20

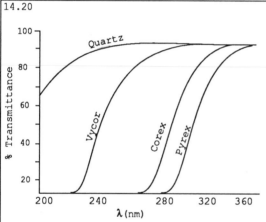

14.20. The light source must emit radiation of a wavelength corresponding to the longest wavelength band in the electronic spectrum of the starting material (photoreactive compound or sensitizer), except when the reactant is in a higher excited state. Medium-pressure mercury lamps emit radiation with a wide range of energies. If the desired product is decomposed by the radiation, an *optical filter* must be incorporated to absorb unwanted wavelengths. It is usually necessary to remove the short wavelength radiation (high energy) because it often leads to secondary (usually destructive) reactions. In preparative photochemistry, *glass* and *chemical filters* are the most commonly used optical filters. A wide range of different glasses with a variety of cut-off wavelengths are available.[21]

14.21

TRANSMITTANCE OF SOME GLASSES (2mm) AT SELECTED WAVELENGTHS

GLASS	% TRANSMITTANCE AT WAVELENGTH (nm)						
	254	302	313	334	366	405	436
QUARTZ	90	90	90	90	90	90	90
SIMAX	0	50	70	85	90	90	90
PYREX	0	50	70	90	90	90	90
SOLIDEX	0	50	65	80	90	90	90
JENA GLASS	0	10	30	70	85		
WINDOW GLASS	0	5			90	90	90
GWCa	0	0	0	5	80	90	90
GWV	0	0	0	0	0	65	80

14.21. In reactors with external light sources, the vessels containing the mixture to be irradiated are manufactured from a glass that serves as an optical filter. In immersion reactors either the immersion wells are made from these glasses or absorption sleeves are used as filters.[22]

14.22

LIQUID OPTICA1 FILTERS
FOR SELECTED SPECTRAL AREAS

TRANSMITTANCE (nm)	COMPOSITION OF THE SOLUTION
300–340	$1.75M\ NiSO_4 \cdot 6H_2O +$ $0.5M\ CoSO_4 \cdot 7H_2O$
320–370	$0.16M\ CuSO_4 \cdot 5H_2O +$ $1.2M\ NiSO_4 \cdot 6H_2O +$ $2.13M\ CoSO_4 \cdot 7H_2O +$ $0.42M\ KNO_3$
400	$1M\ NaNO_2$ (2cm)
500	$1M\ K_2CrO_4$ (2cm)

14.23

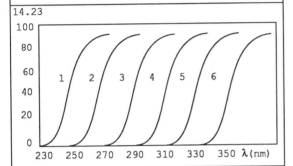

14.24

λ (nm)	Filter					
	A	B	C	D	E	F
248	9	–	–	8	–	–
254	55	–	–	52	–	–
265	26	–	–	21	–	–
276	3	–	–	1	–	–
280	8	–	–	2	–	–
289	7	1	–	–	–	–
297	19	6	1	–	–	–
302	30	14	3	–	–	–
313	64	40	22	–	–	–
334	12	10	1	–	–	–
365	100	94	–	98	47	92
405	41	41	6	41	6	41
436	62	62	28	62	6	62
546	68	68	68	68	62	68
578	56	56	56	56	55	56

14.22. Filter solutions (*chemical filters*), in combination with or in place of glass filters, can be used to remove unwanted short wavelength radiation. Typical chemical filters are aqueous solutions of inorganic salts. Numerous liquid filter systems have been described in the literature. Those shown in the figure are photostable and suitable for use in preparative photochemistry.[23]

14.23. *There are two types of chemical filters* that differ in their spectral properties: *cut-off filters*, which remove the short wavelength region of radiation (the glass filters mentioned above and the last two examples in 14.22), and *band-pass filters*, which select a particular band from the spectrum (the first two examples in 14.22). The chemical filters whose transmission curves are shown in the figure are cut-off filters. They consist of: (A) Ag_2SO_4, (B) $CaCl_2$, (C) $NaBr \cdot 2H_2O$, (D) $Hg(NO_3)_2$, (E) $Pb(NO_3)_2$, and the individual curves correspond to the following composition of these salts in g/l:[24]

1. 0.05 A + 500 B
2. 0.16 A + 400 C
3. 6.40 A + 500 C
4. 0.30 A + 375 B + 80 C + 0.10 D + 0.05 E
5. 400 C + 1.00 D
6. 650 C + 3.00 E

14.24. The table provides the relative intensities (I_{rel}) of spectral lines of a typical discharge lamp alone (A) and in combination with optical filters (B–F).[25] The line at 365 nm in column A represents an optical output of 6 W. The symbols in the table represent: A—no optical filter; B—Pyrex, 2 mm; C—Pyrex, 2 mm + K_2CrO_4 (0.4 g/l in water), 5 mm, D–5,7-dimethyl-1,4-diazacycloheptadiene perchlorate (DDCP, 0.4 g/l in water), 5 mm; E—Pyrex, 2 mm + 2,3-diphenylindenone (0,.25 g/l in EtOH), 5 mm; and F—Pyrex, 2 mm + DDCP (0.4 g/l in water), 5 mm.

14.25

PHOTOCHEMICAL REACTORS

1. *Photoreactors with external sources of radiation*

   The radiation source [one or more lamp(s)] is located outside the reaction mixture.

2. *Immersion well photoreactors*

   The radiation source (lamp) is "immersed" in the reaction mixture

14.26

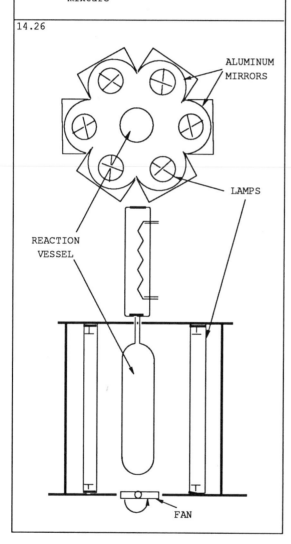

ALUMINUM MIRRORS

LAMPS

REACTION VESSEL

FAN

14.25. Preparative photochemical reactions are performed in *photoreactors*[26] consisting of a reaction vessel, a source of radiation with a power supply, an arrangement for filtering the radiation, and a means of cooling (water or air) and stirring the reaction mixture. Preparative photoreactors are divided into two types depending on the location of the radiation source with respect to the reaction vessel. In *reactors with external radiation*, several low-pressure mercury lamps (assuring a sufficient photon flux) that emit narrow spectral bands without the need for filters are used. A disadvantage of this type of reactor is that the intensity of lamp radiation decreases as a function of the square of the distance from the source. Also, the radiation can be reflected away from the reaction mixture. In *immersion well reactors*, utilization of the radiation is more efficient since the source is surrounded by the solution of the reactant and none of the radiation is lost. However, the use of medium-pressure mercury lamps requires efficient cooling and the use of liquid or glass optical filters.

14.26. In *reactors with external light sources*, the reaction mixture is placed in a reaction vessel made of suitable glass or quartz (transparent to the radiation) and positioned in the center of a circular array of about 12 low-pressure mercury discharge lamps. Curved reflectors made from aluminum, which reflects down to 200 nm, are placed behind the lamps to maximize the emitted radiation, and a fan is located under the reaction vessel to avoid overheating of the solution. In this apparatus the irradiation mixture should be contained in a vessel having the same length as the radiating length of the lamps.[27,28] Reactors with external lamps permit the use of large volumes of reaction mixtures, which is especially important when the reaction requires the use of very dilute solutions or gases.

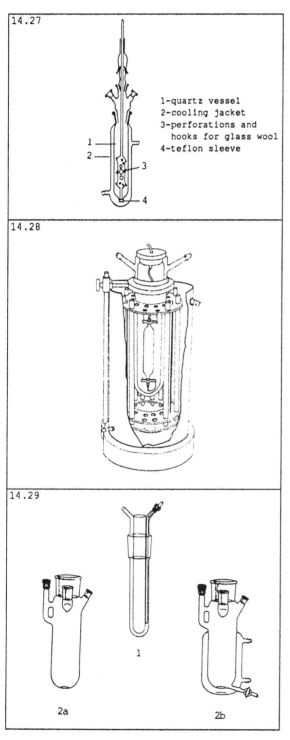

14.27
1-quartz vessel
2-cooling jacket
3-perforations and
   hooks for glass wool
4-teflon sleeve

14.28

14.29

2a

2b

14.27. Photochemical reactions are frequently accompanied by the formation of polymeric byproducts, insoluble in the solvent and which deposit on the cooler walls of the reaction vessel. These products absorb some of the radiation and, thus, prevent its efficient utilization for the desired reaction. The use of commercially available reactors with external light sources, in which the reaction vessel is equipped with a rotating brush to continuously remove the deposit from the walls of the vessel, can increase the product yield 30-fold.[29,30]

14.28. Another reactor using external radiation is the *merry-go-round*[31] or *carousel reactor*,[32] generally employed for comparative studies and quantum yield determinations. In this apparatus small-scale photolyses can be accomplished. This is especially important in the initial stages of investigation of a new photochemical reaction.

14.29. *Immersion-well reactors* are the most commonly used type in preparative photochemistry, particularly for photochemical syntheses on a laboratory scale since reaction volumes of 100–1000 ml and up to 1 mol of starting material can be irradiated efficiently. An immersion reactor consists of two parts: an *immersion well* (1) and a *reaction vessel* (2a) or (2b) with external cooling.[33] The immersion well in which a medium-pressure mercury lamp is located also serves as a means of cooling. If necessary, a filter solution can be circulated using an external pump and the lamp can be surrounded by a glass filter sleeve.[21] The reaction vessel has several outlets that serve to hold a reflux condenser, thermometer, or for the introduction of a gas, removal of samples, etc. For reactions that require temperatures below $-20°C$, an evacuated immersion well must be used in place of a water-cooled one (see 14.31).

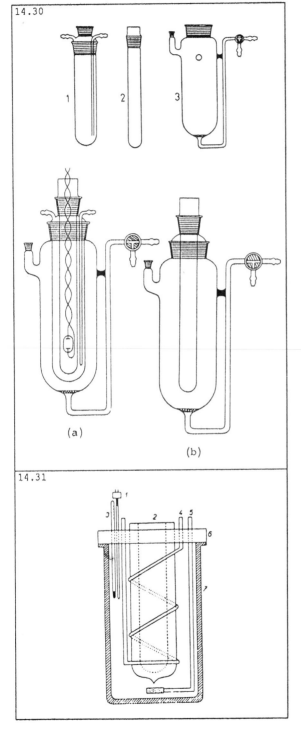

14.30

(a)

(b)

14.31

14.30. To make the immersion well as universally useful as possible,[34] it is appropriate that the two coaxial tubes (1 and 2), when placed in the reactor (3), are separated, as shown in (a). By changing the internal tube (2), which serves as an optical filter, it is possible to cut off the spectral region of the lamp emissions. The outside of the immersion well (1) is manufactured from quartz; (2) is either quartz or a suitable glass. Arrangement (b) is employed for reactions requiring high temperatures, e.g., at the boiling point of the solvent. In this design the internal tube replaces the entire immersion well. Since oxygen is an efficient quencher of photoexcited states and also reacts with free radicals, it should be removed from the solution. This is usually accomplished by flushing the solution with oxygen-free nitrogen[35] or argon, both prior to (preferably under reflux with subsequent cooling to room temperature) and during the irradiation. Degassing by sonification has also been utilized effectively.[36]

14.31. For photochemical reactions at very low temperatures, a reactor proposed by Owsley and Bloomfield[37] permits photoreactions to be accomplished with large volumes and at temperatures from $-80-+80°C$. In this photochemical reactor the immersed unit is represented by a Dewar vessel (2) evacuated at $100°C$ to 10^{-7} Pa, equipped with a cooling coil (4) through which liquid nitrogen is circulated. The plastic reaction vessel (7) has a volume of 4000 ml and is equipped with a teflon lid (6) having openings for an immersion well, a cooling coil, a thermocouple (1), a thermometer (3), and a fritted glass tube (5) for the introduction of a gas. A modification[38] of this reactor has been used successfully for the photochemically initiated $[2+2]$ cycloaddition reactions of ethylene or acetylene to conjugated carbonyl compounds.

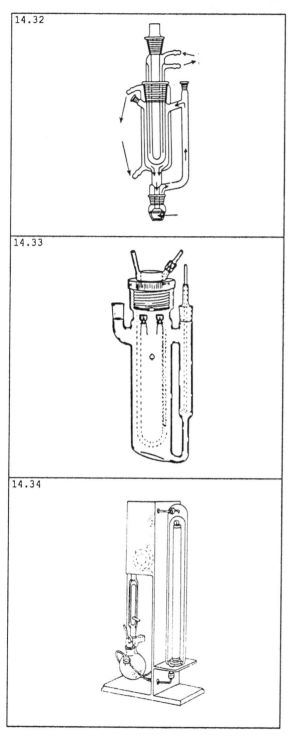

14.32

14.33

14.34

14.32. As a consequence of the concentrations used in preparative photochemistry, most of the emitted light is absorbed by a solution layer only a few millimeters thick.[3(c)] This short path length for absorption results in the formation of relatively high concentrations of excited states and other intermediates close to the wall of the immersion well, and subsequently leads to the accumulation of products in this region of the photoreactor. This can produce unwanted secondary reactions and result in the deposition of highly absorbing polymeric films on the immersion well wall (see 15.28, 15.50, 15.53, and 15.54). These problems can be at least partially avoided by the use of a commercial reactor in which vigorous stirring is accomplished by a magnet-driven high rpm turbine that sucks the reaction solution from the irradiated region through a side arm and returns it for further irradiation.[39]

14.33. An elegant solution for circulating the irradiated solution is embodied in a commercially available immersion reactor equipped with a chamber in which impeller blades pump the reactant solution from the lower to the upper part of the reactor, while stirring.[40] Circulation is so rapid that it ensures very efficient mixing of the reaction solution and, as a result, the irradiated layer located immediately next to the lamp is continuously renewed.

14.34. A *falling film photoreactor* is particularly useful for the irradiation of concentrated solutions. In this apparatus the reaction mixture is pumped from a reservoir to a nozzle, positioned such that the jet of liquid forms a thin film that covers the outer surface of a quartz or glass tube. By constantly replenishing the layer of liquid near the lamp, the buildup of deposits on the wall can be avoided.[41]

14.35

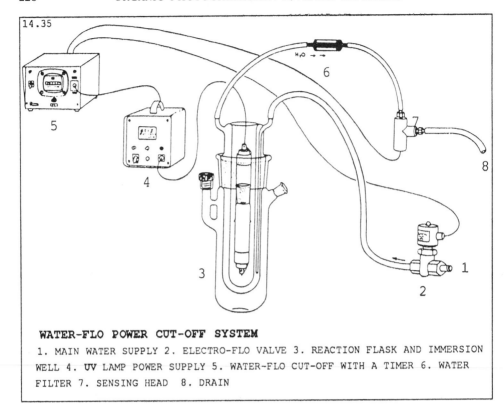

WATER-FLO POWER CUT-OFF SYSTEM

1. MAIN WATER SUPPLY 2. ELECTRO-FLO VALVE 3. REACTION FLASK AND IMMERSION WELL 4. UV LAMP POWER SUPPLY 5. WATER-FLO CUT-OFF WITH A TIMER 6. WATER FILTER 7. SENSING HEAD 8. DRAIN

14.35. Although many photochemical reactions require long radiation times, it is always desirable to perform them without interruption. This requires the use of a *control system* to guarantee the safe course of the irradiation. A commercially available[42] water flow monitoring system shuts off power to the lamp in the event of a water or power failure. The system consists of a micro-switch sensor that detects when the water flow falls below a pre-set value and a timer to record the shutdown time. A water filter prevents dirt and rust in the water supply from reaching the sensor and unnecessarily shutting down the system. Long-term experiments result in the formation of yellow-brown iron III salts from the cooling water on the inside walls of the cooling vessel. This strongly suppresses the intensity of radiation available for the desired photochemical reaction. For this reason it is necessary to periodically clean the cooling vessel with aqueous hydrochloric acid.

Safety precautions must be strictly employed during irradiation experiments. Short wavelength UV irradiation causes an inflammation of the cornea, even in small doses. In larger amounts UV light can cause permanent vision damage. It is also necessary to protect skin from UV light (it causes erytherma) by wearing suitable gloves and appropriate clothing when conducting photochemical reactions. Other hazards associated with short wavelength UV radiation include ozone production (proper ventilation is required) and the fragility of lamps (proper standards of electrical safety and careful handling of lamps should be employed). Protective glasses and gloves should be worn if it is necessary to handle a burning lamp.[43] It is strongly recommended that the reaction vessel be wrapped in aluminum foil or placed in a photochemical safety reaction cabinet during irradiation.[44]

REFERENCES

(1)(a) Rabek, J. F. *Experimental Methods in Photochemistry and Photophysics*. John Wiley: Chichester, United Kingdom, 1982. (b) Scharf, H.-D., Fleishhauer, J., and Aretz, J. In *Methoden der Organischen Chemi/Houben-Weyl*, Muller, E., ed. G. Thieme: Stuttgart, 1975, Vol. 4/5a: Photochemie, pp. 41–89. (c) Murov, S. L. *Handbook of Photochemistry*. Marcel Dekker: New York, 1973. (d) Calvert, J. G. and Pitts, J. N. *Photochemistry*, Wiley: New York, 1966. (e) ACE Glass Catalog No. 1000, pp. 228–242. (f) Kopecký, J. and Liska, F. *Chem. Listy*. **1978**, *72*, 577.

(2) Jaffe, H. H. and Orchin, M. *Theory and Applications of Ultraviolet Spectroscopy*. Wiley: New York, 1962, p. 2.

(3)(a) See Ref. 1(c), p. 123. (b) Srinivasan, R. *Org. Photochem. Synth.* **1971**, *1*, 6. (c) For the calculation of the concentration of the photoactive component, see Hutchinson, J. In *Photochemistry in Organic Synthesis*, Coyle, J. D., ed. The Royal Society of Chemistry: London, 1986, p. 29.

(4)(a) Adapted from Suppan, P. *Principles of Photochemistry*. The Royal Society of Chemistry: London, 1972, p. 64. (b) Cox, A. and Kemp, T. J. *Introductory Photochemistry*. McGraw-Hill: London, 1971, p. 157. (c) See Ref. 1(b), p. 84.

(5)(a) See Ref. 1(b), pp. 84–89. (b) Evans, T. R. In *Techniques of Organic Chemistry*. Leermakers, P. A. and Weissberger, A., eds. Interscience: New York, 1969, Vol. 14: *Energy Transfer and Organic Photochemistry*, pp. 324–340. (c) Barltrop, J. A. and Coyle, J. D. *Excited States in Organic Chemistry*. Wiley: New York, 1975, pp. 146–157.

(6) Kopecký, J., Šmejkal, J., Tureček, F., Jirkovský, J., Fojtík, A., and Hanuš, V. *Tetrahedron Lett*. **1984**, *25*, 2613.

(7) Phillips, R. *Sources and Applications of UV Radiation*. Academic Press: London, 1983, pp. 185–363.

(8) Figure adapted from Schönberg, A. *Preparative Organic Photochemistry*. Springer-Verlag: New York, 1968, p. 473.

(9)(a) See Ref. 8, p. 475. (b) See Ref. 4(c), p. 43. (c) See Ref. 7, p. 185.

(10) Adapted from Ref. 3(b), p. 8.

(11)(a) See Ref. 7, p. 206. (b) See Ref. 8, p. 475.

(12) See Ref. 7, p. 264.

(13) See Ref. 3(c), p. 23.

(14) See Ref. 1(e), p. 229.

(15) See Ref. 7, p. 227.

(16)(a) See Ref. 8, p. 477. (b) See Ref. 1(e), p. 240.

(17) Figure adapted from Wiskemann, A. *Betriebsartzliches*. **1971**, *No. 1*, 39.

(18) See Ref. 1(b), p. 9.

(19) Adapted from Ref. 3(b), p. 3.

(20) Adapted from Ref. 3(b), p. 5.

(21)(a) See Ref. 8, p. 490. (b) See Ref. 3(c), p. 23.

(22)(a) See Ref. 1(e), p. 230. (b) See also Andrist, A. M. and Baldwin, J. E. *Org. Synth*. **1988**, *Coll. Vol. 6*, 146.

(23)(a) See Ref. 1(b), p. 66. (b) See Ref. 1(c) and references therein.

(24)(a) Figure adapted from Ref. 8, p. 491. (b) For band-pass chemical filters, see Muel, B. and Malpiece, C. *Photochem. Photobiol.* **1969**, *10*, 283. (c) For chemical filters used in the *Wisconsin Black Box*, see Zimmerman, H. E. *Mol. Photochem.* **1971**, *3*, 281.

(25) Kuzmic, P. *Unpublished results*.

(26) See Ref. 1(b), pp. 69–84.

(27) See Ref. 1(b), pp. 73, 76–79.

(28)(a) Photoreactors with external radiation are available from Southern New England Ultraviolet Co., Middletown, CT, and Applied Photophysics, Ltd., England. See also 14.27. (b) For an apparatus for preparative scale mercury photosensitized dehydrodimerization, see Brown, S. H. and Crabtree, R. H. *J. Am. Chem. Soc.* **1989**, *111*, 2935.

(29)(a) Bryce-Smith, D., Forst, J. A., and Gilbert, A. *Nature*, **1967**, *213*, 1121. (b) Blair, J. M., Bryce-Smith, D., and Pengilly, B. W. *J. Chem. Soc.* **1959**, 3174. (c) Available from Engelhard, Hanovia Lamps, England.

(30) The *Wisconsin Black Box*, which allows large-scale photolysis with light of greater than 50 nm band width selected from a 1000 W AH-6 mercury lamp by means of chemical filters, has been successfully used for decades [See Ref. 23(c).]

(31) Manufactured by Southern New England Ultraviolet Co., Middletown, CT.

(32) Reproduced with the permission of Applied Photophysics, Ltd. England.

(33)(a) Reproduced with the permission of ACE Glass, Inc., Vineland, N.J. (b) For a photooxygenation apparatus, see Adam, W., Klug, G., Peters, E.-M., Peters, K., and von Schnering, H. G. *Tetrahedron*. **1985**, *41*, 2045.

(34) Figure adapted from Liška, F., Dědek, V., Kopecký, J., Mostecký, J. and Dočkal, A. *Chem. Listy* **1978,** *72*, 637 (in Czech).

(35) Meites, L. and Meites, T. *Anal. Chem.* **1948,** *20*, 984.

(36) Uyehara, T., Furuta, T., Kabawawa, Y. Yamada, J., Kato, T., and Yamamoto, Y. *J. Org. Chem.* **1988,** *53*, 3669.

(37) Adapted from Owsley, D. C. and Bloomfield, J. J. *Org. Prep. Proc. Int.* **1970,** *3*, 61.

(38) Bloomfield, J. J. and Owsley, D. C. *Org. Photochem. Synth.* **1976,** *2*, 37.

(39) Reproduced with the permission of Otto Fritz, GmBH, Germany.

(40) Super Mix reaction vessel: ACE Glass Catalog 1000, p. 232. Reproduced with the permission of ACE Glass Inc., Vineland, N.J.

(41)(a) Reproduced with permission of Applied Photophysics Ltd., England. (b) See also Davidson, R. S. *Chem. Ind.* **1978,** 180.

(42) Reproduced with the permission of ACE Glass, Inc., Vineland, N.J.

(43)(a) See Ref. 3(c), p. 36, and references therein. (b) See Ref. 7, p. 406.

(44) Photochemical safety reaction cabinet: ACE Glass Catalog 1000, p. 230.

Chapter 15

FACTORS INFLUENCING THE COURSE OF PHOTOCHEMICAL REACTIONS

FACTORS INFLUENCING THE COURSE OF PHOTOCHEMICAL REACTIONS

15.1

Reactions caused by light are so many that it should not be difficult to find some which are of practical value. I do not believe that the industries should do any longer before taking advantage of the chemical effects produced by light.　　　　**G. Ciamician** (1912)[1]

In the last decade photochemistry has grown from a minor to a major activity in organic chemical laboratories. Many striking and (thermally) unusual processes can be observed.. However, in spite of the large number of new photochemical reactions which have been discovered in recent years, rather few have yet found use in organic synthesis and fewer still have been of economic value in industry.　　**D. H. R. Barton** (1969)[2]

15.1. In spite of their undeniable advantages, photochemical reactions were not used extensively in organic synthesis until the mid-1960s because they were considered too complex and nonspecific. However, the development of new photochemical experimental techniques, separation methods, and the concurrent understanding of the mechanisms of photochemical processes have gradually improved their selectivity and control over their pathways. This has occurred to such an extent that organic chemists have acquired the understanding necessary for the successful use of photoreactions. Many photochemical reactions have been employed recently for the synthesis of various complex structures, e.g., cage compounds, natural products, etc.[3]

15.2

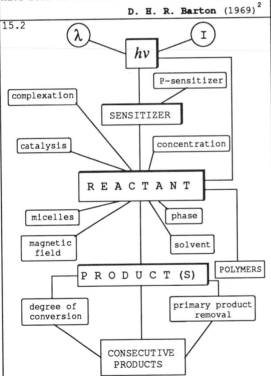

15.2. The successful application of a photochemical reaction depends on various factors that are related either to the "reagent" (the radiation source, with or without a sensitizer) or to the reactant. Some of these factors follow photochemical laws, e.g., the spectral characteristics of the starting material or sensitizer, which determine the wavelength of the radiation needed, the quantum yield of the photoreaction, and the required irradiation time, can, to some extent, be controlled by the intensity of the irradiation source. Other factors are a consequence of the type of photochemical reaction, e.g., the effect of concentration in dimerization reactions in competition with isomerization. The path of a photochemical reaction can be influenced by the solvent, the use of a catalyst, the formation of complexes, secondary reactions, etc. Photochemical reactions can follow different pathways in solution, the solid phase, or inclusion complexes.

15.3

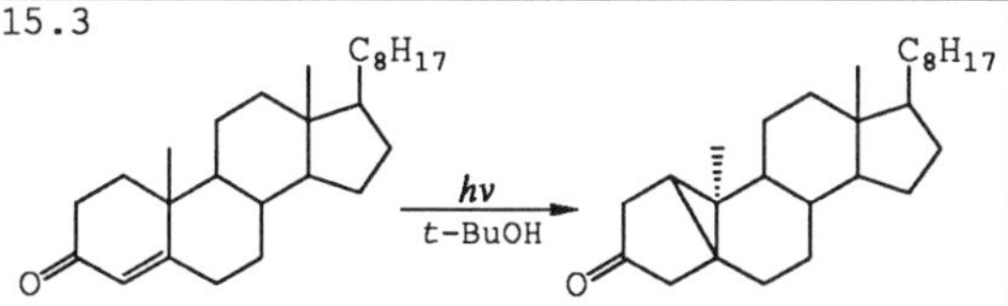

power rating	irradiation time
200 W	170 h
450 W	75 h
1000 W	24 h

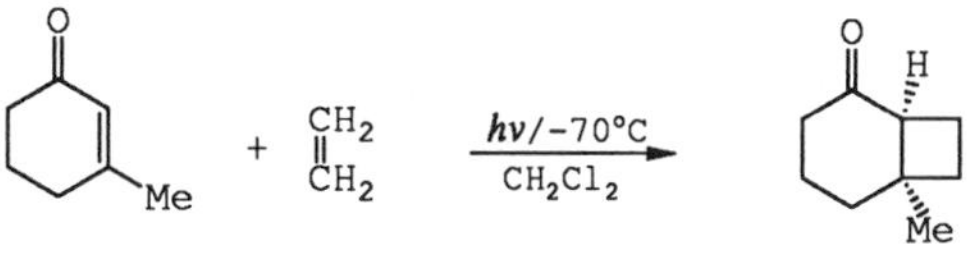

quantity (g)	power rating	irradiation time (h)	yield (%)
20	450 W	48	80
94	1000 W	8	91

15.4

LUMINESCENCE RULES

KASHA	**VAVILOV**
Emission occurs from the S_1 or T_1 state, independent of the state which was initially excited.	*The quantum yield of fluorescence is independent of the excitation radiation.*

15.5

15.3. The most important require-ment in preparative photochemistry is the delivery of a sufficient number of photons of appropriate energy to the re-action system. The irradiation time is determined by the quantum yield of the reaction and the output of the lamp in the spectral region used in the photo-chemical reaction (see 14.18 and 14.19). The photoisomerization of 4-cholesten-3-one to lumicholestenone[4] and the cycloaddition of ethylene to 3-methylcyclohexenone illustrate this dependence.[5] Replacement of a 450 W lamp with a 1000 W lamp in the reac-tion shown in 5.30 resulted in a de-crease in the irradiation time by a factor of 6–10. Irradiation times were deter-mined using relatively fresh lamps whose output had not yet significantly decreased (see 14.12).

15.4. Photochemical reactions in solution occur from vibrationally re-laxed molecules in their S_1 or T_1 states.[6] Nonradiative transitions from higher excited states are usually much faster than radiative transitions or photo-chemical transformations. Conse-quently, some photoreactions are rela-tively wavelength independent. However, when the rates of nonradia-tive transitions are unusually small or the rates of photochemical transforma-tions unusually large from higher ex-cited states, these rules and the conclu-sions that follow become invalid.

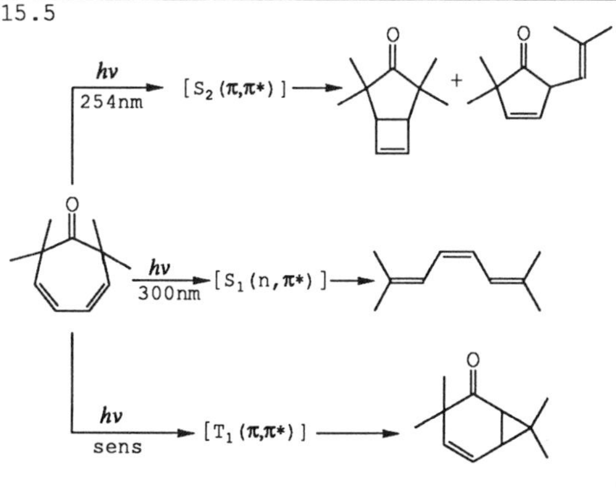

15.5. The effect of radiation wave-length has been mentioned previously (see 7.5 and 9.20). The participation of higher excited states has been estab-lished in a number of photochemical reactions.[7] For example, cyclohepta-dienone, which possesses two mutually noninteracting chromophores, yields different photoproducts depending on the energy of the radiation (or the use of a triplet sensitizer).[8] Excitation to the S_1 state of cycloheptadienone is local-ized primarily on the carbonyl group; excitation to the S_2 state occurs in the diene system.

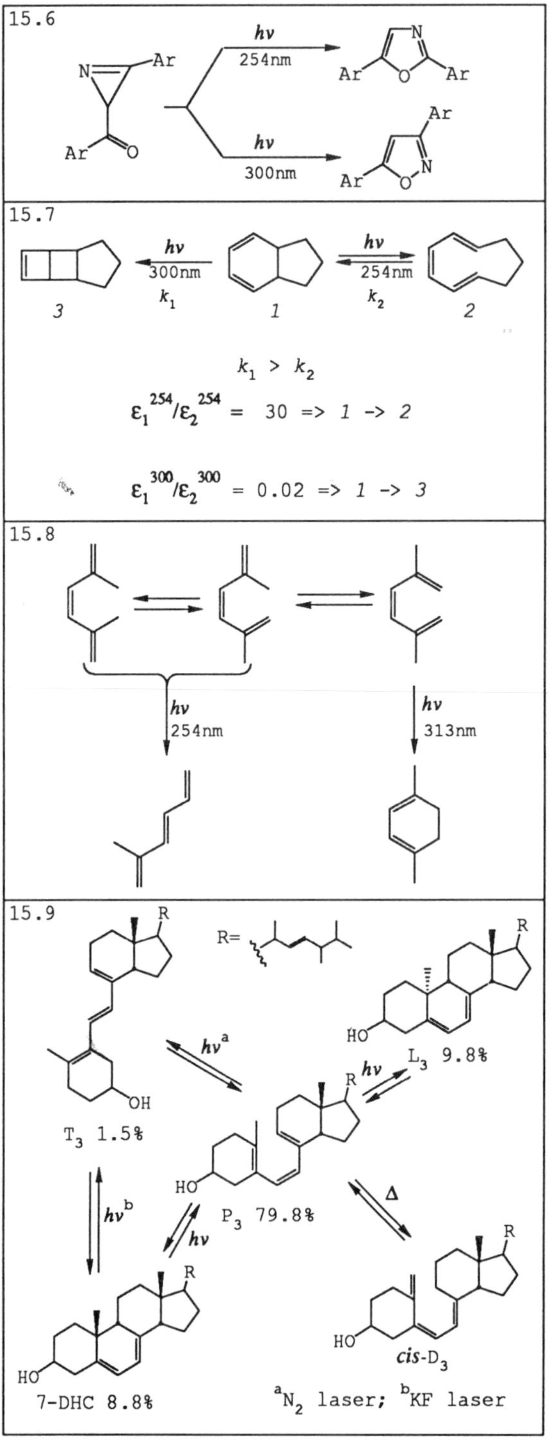

15.6

15.7

$$k_1 > k_2$$

$$\epsilon_1^{254}/\epsilon_2^{254} = 30 \Rightarrow 1 \rightarrow 2$$

$$\epsilon_1^{300}/\epsilon_2^{300} = 0.02 \Rightarrow 1 \rightarrow 3$$

15.8

15.9

$R=$

T_3 1.5%

L_3 9.8%

P_3 79.8%

cis-D_3

7-DHC 8.8%

$^a N_2$ laser; ^{b}KF laser

15.6. Excitation of acylated azirines into the $S_0 \rightarrow S_1$ band ($n \rightarrow \pi^*$, carbonyl chromophore) or triplet sensitization results in the formation of isoxazoles, while irradiation into the $S_0 \rightarrow S_2$ band ($n \rightarrow \pi^*$, azirine ring) leads to the formation of oxazoles.[9] In the former reaction, rapid ISC to T_1 apparently takes place.

15.7. Participation of higher excited states is not the only reason for wavelength dependent photoreactivity. The two reaction paths of *1*, which forms a photostationary state with *2*, result from a difference in the values of ϵ_1 and ϵ_2 at 254 and 300 nm (see the ratio $\epsilon_1 : \epsilon_2$). At 254 nm *1* absorbs much more radiation than *2*, which, therefore does not undergo the reverse reaction to yield *1*. On the other hand, *2* absorbs much more radiation than *1* at 300 nm, reverts back to *1*, and subsequently undergoes a 4π electrocyclic reaction to yield *3*, in spite of the fact that $k_2 < k_1$.[10]

15.8. Although the conformers of 1,3,5-hexatriene in the S_0 state are in rapid equilibrium, in the S_1 state there is an increased barrier against rotation about the original single bonds. Because the S_1 lifetimes are short, the conformers cannot interconvert within this lifetime and, therefore, each conformer produces its own photoproducts (the principle of *non-equilibration of excited rotamers = NEER principle*).[11] The photochemistry of hexatriene, therefore, is dependent exclusively on the ratio of conformers on the S_0 hypersurface and their molar extinction coefficients at the radiation wavelength used.

15.9. The use of two-step laser photolysis with KrF and N_2 lasers emitting radiation at 248 and 337 nm, respectively, has made it possible to reduce or eliminate the competing photochemical reactions that lead to complex photoequilibrium mixtures in the photolysis of 7-dehydrocholesterol. Previtamin D_3 (P_3) is obtained in high yield using this technique.[12]

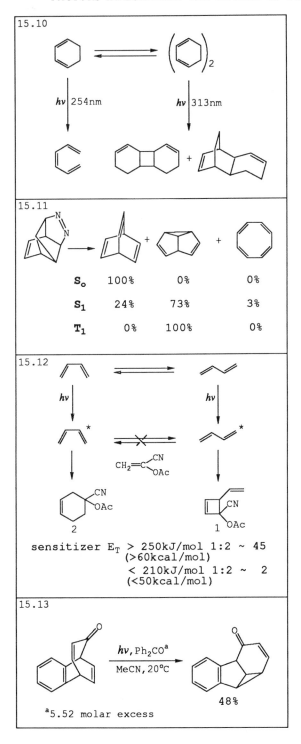

15.10

hv | 254nm   *hv* | 313nm

+

15.11

	S$_0$	100%	0%	0%
	S$_1$	24%	73%	3%
	T$_1$	0%	100%	0%

15.12

hv   *hv*

*   *

CH$_2$=C<CN,OAc

2   1

sensitizer E$_T$ > 250kJ/mol 1:2 ~ 45
                (>60kcal/mol)
            < 210kJ/mol 1:2 ~ 2
                (<50kcal/mol)

15.13

hv, Ph$_2$CO^a
MeCN, 20°C

48%

^{a}5.52 molar excess

15.10. The effect of radiation wavelength on the pathway of a photochemical reaction is also illustrated by the complexation of a reactant in the ground state, as shown for neat 1,3-cyclohexadiene. Only the monomer absorbs at 254 nm and 1,3,5-hexatriene is the sole product. At 313 nm the "dimer" absorbs most of the radiation and results in a mixture of [2+2] and [4+2] photodimers.[13]

15.11. The path of a photochemical reaction, e.g., that of azo compounds, is affected by the spin multiplicity and triplet energy of the sensitizer (see 5.18). The preparative importance of this photolysis is based on N$_2$-elimination from cyclic azo compounds with the subsequent formation of cage compounds.[14] However, some cyclic azo compounds do not undergo photolytic N$_2$-elimination[15] (*reluctant azo alkanes*, see 11.10).

15.12. The ratio of products in sensitized photodimerization and photocycloaddition reactions is influenced by the triplet energy (E$_T$) of the sensitizer. If E$_T$ is >250 kJ/mol, the *s-trans*-diene conformer, dominant in the S$_0$ thermal equilibrium, is excited and undergoes reaction with the alkene to produce predominantly a cyclobutene derivative. If the sensitizer has a E$_T$ of 210 kJ/mol, energy transfer to the *s-cis* conformer occurs. Similarly, in the dimerizations of butadiene and isoprene, a reduction in the triplet excitation energy of the sensitizer causes increased formation of [4+2], at the expense of [2+2], dimers.[16]

15.13. In sensitized photochemical reactions the sensitizer must absorb a maximum amount of the incident radiation. Difficulties are encountered if the sensitizer absorbs in the same region of the spectrum as the reactant. In the example shown, selective excitation of benzophenone is achieved by employing very high concentrations relative to the acceptor.[17]

15.14

SUPPORT ANCHORED SENSITIZERS

(P-sensitizers)

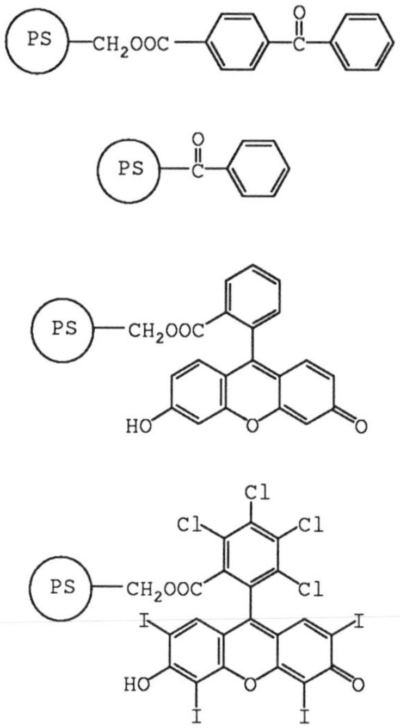

15.14. Separation of a sensitizer from the photoproduct(s) is sometimes difficult. It can be facilitated by the use of sensitizers anchored to a polymeric support that is transparent to the radiation and insoluble in the solvent used.[18] In addition to the ease of separation from products by filtration, these P-sensitizers can be regenerated and reused. P-sensitizers prepared by the condensation of chloromethylated polystyrene (PS) with benzophenone-4-carboxylic acid (*1*) or benzoylated polystyrene (*2*)[19] have been used in the photosensitization of cycloaddition and dimerization reactions of alkenes. Dye-sensitized photooxidation reactions are performed using the condensation products of chloromethylated polystyrene with fluorescein or Rose Bengal (commercially available as *Sensitox*).[20] The use of Rose Bengal attached to silylated silica gel has also been reported.[21]

15.15

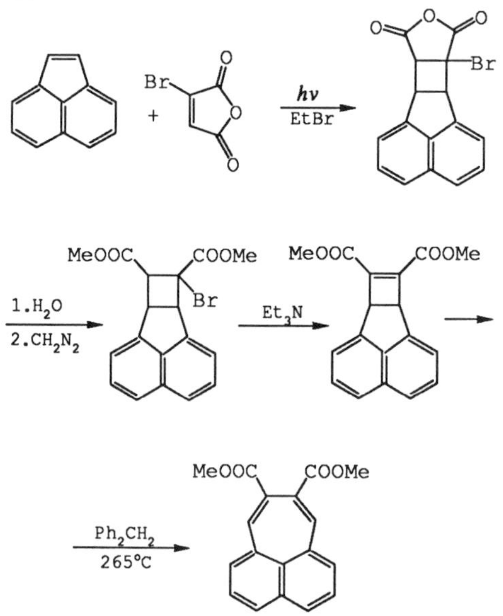

15.15. A heavy-atom environment increases the rate of the nonradiative S-T transition ("heavy-atom effect," see 5.22). The irradiation of compounds, particularly those whose lowest excited singlet state has a π,π^* configuration, in solvents having a heavy atom (especially iodine) leads to a population of their T_1 states. This allows photochemical reactions to occur from this excited state even for compounds with a large S_1-T_1 splitting, such as alkenes that have very inefficient S-T crossing. The heavy-atom effect has been utilized in the cycloaddition of acrylonitrile,[22] cyclopentadiene,[23] and others to acenaphthylene and in the synthesis of pleiadiene and its derivatives.[24] It is significant that, of all the derivatives of maleic anhydride that have been investigated, the cycloaddition of the bromo derivatives proceeds most efficiently—an example of an internal heavy-atom effect.

15.16

X=O, S, Se

$hv \updownarrow hv'$

15.16. Heavy-atom solvents also affect the course of photochemical Z-E isomerizations. The photoisomerization of the cyanine shown is extremely sensitive to both external and internal heavy-atom effects. The rate of isomerization increases in the order O < S < Se.[25,26]

15.17

CUT-OFF LIMITS FOR SELECTED SOLVENTS

Solvent	Cut-off wavelength (nm)	
	10 mm Cell	0.1 mm Cell
acetonitrile	190	180
hexane	195	173
ethanol (95%)	204	187
water	205	172
EPA[a]	212	190
chloroform	245	235
carbon tetrachloride	265	255
benzene	280	265
acetone	330	325

[a] 5:5:2 by volume mixture of ethyl ether, isopentane, and ethanol.

15.17. The choice of solvent is extremely important in preparative photochemical reactions. The solvent must be photochemically inert (unless reactivity is desired) and transparent to the incident radiation, and it must not contain absorbing impurities. The table[27] indicates the cutoff points for various solvents. These points are the wavelengths at which the absorbance is approximately one in a 10 mm and 0.1 mm cell, respectively. For the irradiation of a 1% solution of starting material, the ϵ of the reactant should be ~5000 times larger than that of the solvent. Since O_2 acts as a quencher of excited states, causes peroxide formation, etc., it must be removed from the solvent by sonification or, more commonly, by flowing oxygen-free nitrogen or argon through the reaction mixture prior to irradiation.

15.18

Solvent	$k_{rel/ml}$[a]
2-propanol	12.7
2-butylamine	12.5
2-octanol	6.5
cumene	2.4
cyclohexane	2.0
cyclohexene	2.0
2,2,4-trimethylpentane	1.9
n-hexane	1.8
2,3-dimethylbutane	1.5
tert.-butylamine	1.5
triethylamine	1.3
tert.-butylalcohol	1.1
toluene	0.94
anisole	0.49

[a] expresses the reactivity of the solvent

15.18. Solvents that are hydrogen-atom donors can lead to undesirable photoreduction products, particularly in the presence of carbonyl-containing sensitizers. If free radicals are involved in the reaction, solvents, such as alcohols, saturated hydrocarbons, and ethers, must be avoided. The solvent-derived radical resulting from hydrogen abstraction can also react with the starting material or the product to form undesired products. The table[28] shows the relative rates of hydrogen abstraction from various solvents by excited benzophenone. Benzene, acetonitrile, and acetic acid are considered photochemically inert toward carbonyl compounds.

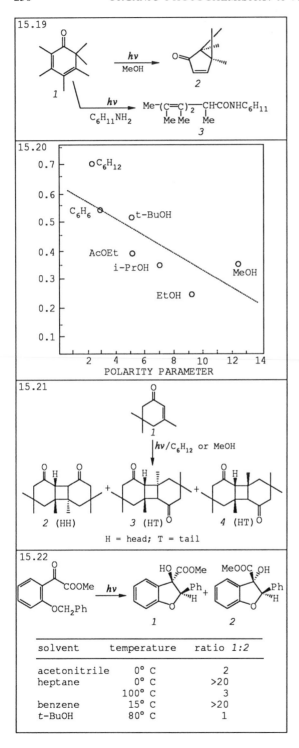

15.19. The primary step in the photolysis of 2,4-cyclohexadienones (*1*) is an α-cleavage, which leads to the formation of the corresponding dienylketene. In the presence of a weak nucleophile, e.g., MeOH, an efficient cyclization to *2* occurs. If a strong nucleophile, such as an alkylamine is present, a derivative of a dienoic acid (*3*) is formed.[29] The effect of solvent polarity on the photochemistry of 2,4-cycloheptadienones was also reported.

15.20. Polarity of the solvent also affects the regioselectivity of some photoreactions, e.g., the dimerization of cyclopentenone. A correlation has been observed between the log of the ratio of the two regioisomers (*1*) and (*2*) (see 9.22), and the *Kirkwood–Onsager polarity parameter* (some of the scatter is due to the high concentration of cyclopentenone in the solvent used).[30] However, it appears that the change in the ratio of the HH/HT dimers reflects differences in the polarity of the transition states leading to the HH and HT dimers, i.e., to either (*1*) or (*2*) from 9.22.

15.21. The [2+2] photodimerization of isophorone (*1*) is highly solvent dependent.[31(a)] The ratio of HH dimer (*2*) to total HT dimers (*3*) and (*4*) is 1:4 and 4:1 in cyclohexane and methanol, respectively. The less polar cyclohexane favors the formation of the less polar dimers (*3*) and (*4*), while the more polar methanol favors the more polar (*2*).[31(b),(c)]

15.22. Both the polarity of the solvent and the reaction temperature often have a significant effect on the stereochemical course of a photoreaction.[32] Interestingly, in heavy-atom solvents, e.g., bromoform or bromobenzene, the formation of stereoisomer (*1*) is preferred. Spin inversion ($T_1 \rightarrow S_0$) is apparently very rapid in these solvents and leads to cyclization of the intermediate biradical.[33]

solvent	temperature	ratio *1:2*
acetonitrile	0° C	2
heptane	0° C	>20
	100° C	3
benzene	15° C	>20
t-BuOH	80° C	1

15.23

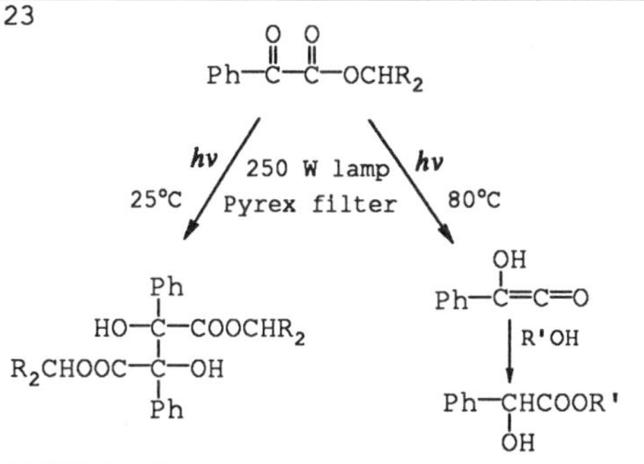

15.23. Photochemical reactions generally have a low *activation energy* (E_{act}) and, therefore, exhibit little temperature dependence. However, when different reaction paths from a given excited state have significantly different E_{act}, the dependence on temperature can be significant.[34]

15.24

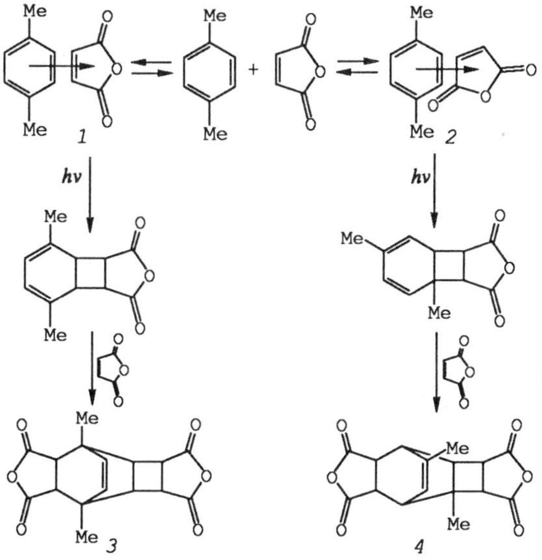

$$20°C \quad 3:4 = 20:1$$
$$100°C \quad 3:4 = 1:10$$

15.24. This [2+2] photocycloaddition reaction is another example of the effect of temperature on regioselectivity. The result is probably due to the existence of an equilibrium between the CT complexes (*1*) and (*2*) whose position depends on temperature.[35]

15.25

1. $R \xrightarrow{h\nu} R^* \longrightarrow P$

Concentration as low as possible

2. $R \xrightarrow{h\nu} R^* \xrightarrow{+S} P$

Concentration as high as possible

3. $R \xrightarrow{h\nu} R^* \xrightarrow{+S} P_1$
$\downarrow +S$
$S^* \xrightarrow{+S} P_2$

Concentration must be estimated by trial

15.25. The effect of the concentration of starting material in a photochemical reaction must be considered with respect to its molecularity. (1) *Unimolecular reactions* tend to be concentration independent and, therefore, are best performed in solutions as dilute as possible to avoid unwanted bimolecular reactions. (2) The rate of *bimolecular reactions* is enhanced with increasing concentration of the components (at least one of the components). Photocycloadditions of alkenes to cyclic enones are first-order reactions with respect to the alkene and, therefore, are critically dependent on the concentration. Even if the enone absorbs all of the radiation, the rate of cycloaddition is independent of its concentration, and it is advantageous to use the highest possible concentration at which the enone does not dimerize. (3) The most complex case is the photocycloaddition of sensitizers, e.g., the addition of carbonyl compounds to alkenes. The yield of oxetane increases with increasing concentration of the alkene. However, if the excitation of the alkene is sensitized, its dimerization rate is further enhanced. Then, it is necessary to experimentally find suitable concentrations of both components.

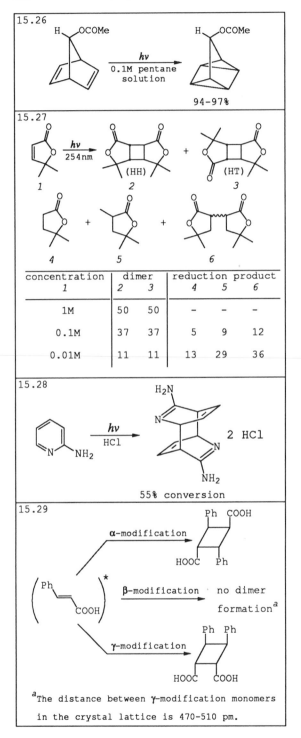

15.26

hv
0.1M pentane
solution

94–97%

15.27

hv
254nm

(HH)   +   (HT)

1   *2*   *3*

+   +

4   *5*   *6*

concentration *1*	dimer		reduction product		
	2	*3*	*4*	*5*	*6*
1M	50	50	–	–	–
0.1M	37	37	5	9	12
0.01M	11	11	13	29	36

15.28

hv
HCl

2 HCl

55% conversion

15.29

α-modification

Ph COOH

HOOC Ph

β-modification · no dimer
formation[a]

γ-modification

Ph Ph

HOOC COOH

[a]The distance between γ-modification monomers in the crystal lattice is 470–510 pm.

15.26. The product yield from unimolecular photoreactions is often limited by a specific maximum concentration of the starting material. In the example shown, the use of high concentrations of the reactant reduces the yield and purity of the product.[36] An example of a photorearrangement that requires the use of highly dilute solutions to prevent photodimerization is the photoisomerization of 4-cholesten-2-one to lumicholestenone (see 15.3).

15.27. The effect of the concentration of starting material on the course of a bimolecular photochemical reaction is illustrated.[37] At high concentrations the HH (*2*) and HT (*3*) dimers (in a ratio of 1:1) are the exclusive products. As the concentration is decreased, the amounts of other products (*4–6*) increase.

15.28. In bimolecular photochemical reactions, the efficiency decreases significantly during the irradiation as the concentration of the starting material diminishes. Therefore, it is advantageous to stop the irradiation after partial conversion has been attained.[38]

15.29. Photochemical reactions in organized systems (organic crystals, molecules adsorbed at interfaces, micelles, and inclusion complexes) in which reactants often have an enhanced degree of organization differ from the corresponding reaction in homogeneous solution. Because the electronic interactions in these assemblies are generally strong, the excitation energy transfer is very fast (10^{12} s^{-1}). As a result of the physical restraint to motion present in the crystal lattice,[39] the photodimerization of, e.g., cinnamic acid,[40] styryl derivatives,[41] and others[42] have been shown to be topochemically controlled with the formation of cyclobutane dimers. The configurations of these dimers correlate with X-ray crystallographic studies. In fluid solution only *cis-trans* isomerization occurs.

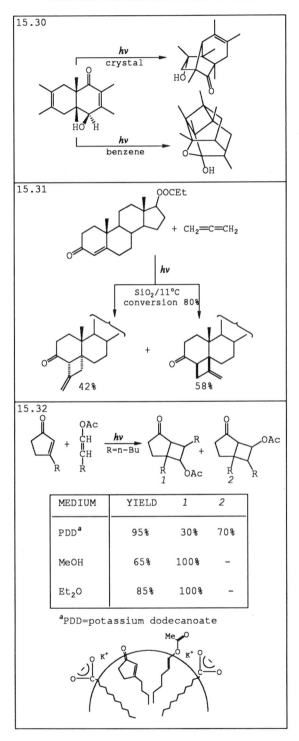

15.30

15.31

15.32

MEDIUM	YIELD	1	2
PDD[a]	95%	30%	70%
MeOH	65%	100%	–
Et$_2$O	85%	100%	–

[a]PDD=potassium dodecanoate

15.30. Although both bimolecular and unimolecular photochemical processes occur in organic crystals, the topochemical postulate applies particularly to the former. Unimolecular photochemical reactions are controlled more by the conformation of the molecules in the crystal than by their mutual orientation. On the other hand, bimolecular photoreactions in the solid phase[43] seem to follow the principle of least motion (*topological principle*[44]).[45] This occurs even if there is a high activation barrier, so that this reaction path has no significance in solution photochemistry.[46]

15.31. Molecules absorbed on various surfaces are also in forced orientations and their photoproducts are frequently different than those obtained in solution. The photocycloaddition of allene to testosterone propionate in methanol occurs primarily at the more accessible α-face of the steroidal skeleton, i.e., the α/β-cycloadduct ratio is 5:1 at 11°C and 9:1 at −78°C. Irradiation of the steriod absorbed by its α-face on a silica gel surface results in an increase in the proportion of β-cycloadducts.[47] The topochemical control of photochemical reactions in zeolite cavities has recently been studied in several laboratories.[48]

15.32. Micelles, another example of organized systems, are used to increase the effective concentration of reactants and to control the regioselectivity of photochemical reactions,[49] as shown in the figure.[50] In addition, undesired hydrogen transfer can be eliminated in this system. It is apparent that further studies of micelles will significantly contribute to a better understanding of the primary photochemical events of photosynthesis and vision, and will give rise to new electronic devices for information storage, switching, and energy conversion.

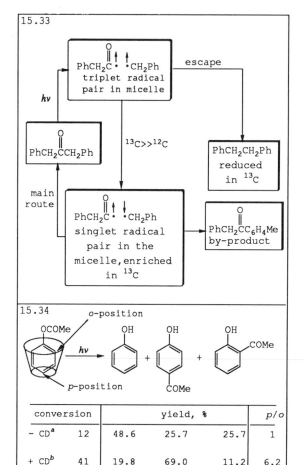

15.33

PhCH₂C–•CH₂Ph
triplet radical pair in micelle → escape

hv

PhCH₂CCH₂Ph ← main route

¹³C >> ¹²C

PhCH₂CH₂Ph reduced in ¹³C

PhCH₂C• •CH₂Ph singlet radical pair in the micelle, enriched in ¹³C

PhCH₂CC₆H₄Me by-product

15.34

o-position

OCOMe → hv → OH + OH (COMe) + OH (COMe)

p-position   COMe

conversion		yield, %			p/o
– CD[a]	12	48.6	25.7	25.7	1
+ CD[b]	41	19.8	69.0	11.2	6.2

[a] photolysis without β-cyclodextrin
[b] photolysis with β-cyclodextrin

15.35

X, X → hv₁ / hv₂ → Y

X, X → hv₁ / Δ → Z

X = [anthracene-COO⁻]

Y = [COO⁻ photodimer]

Z = [COO⁻ COO⁻ photodimer]

15.33. Micelles have been employed as a photochemical method for magnetic isotope enrichment.[51] According to the CIDNP theory, triplet radical pairs containing ¹³C undergo T_1–S_0 intersystem crossing much faster than radical pairs formed, e.g., by photolysis of dibenzyl ketone (DBK) with nonmagnetic nuclei (¹²C), because the lifetimes of the ¹²C triplet states are sufficient for the escape of these radicals from the micelle. After decarbonylation and recombination, the resulting 1,2-diphenylethane is depleted in the ¹³C isotope. The ¹³C radical pairs rapidly reach the S_0 state in which they recombine with the formation of either DBK or 1-phenyl-4-methylacetophenone, both of which are enriched in ¹³C.[52]

15.34. Cyclodextrins (CD$_x$, hosts) are naturally occurring cyclic oligosaccharides that form inclusion complexes (β-CD$_x$ 1:1; γ-CD$_x$ 1:2 host-guest complexes) with a variety of organic molecules (guests). They represent a specific subarea of organized assemblies.[53] Because of the orientation of the guest, e.g., the phenyl carboxylate shown, in the cavity of the host (β-CD$_x$), it is possible to perform regio- or stereoselective reactions on the guest.[54] The results of this photo-Fries rearrangement (see also 10.6 through 10.9) reflect the greater accessibility of the *para*-position to the acyl radical. A similar effect has been observed in the photolysis of acid anilides with β-CD$_x$.[55]

15.35. Because of its larger cavity γ-CD$_x$ can accommodate two guest molecules. This facilitates excimer formation[56] and, thus, enhances the photodimerization.[57] The anthracene moieties in the regioisometric *bis*(anthracene-9-carbonyl)-γ-CD$_x$ undergo dimerization to the *cis*-photodimer,[58] although irradiation of 9-substituted anthracenes gives only the *trans*-photodimer (see 7.15).

15.36

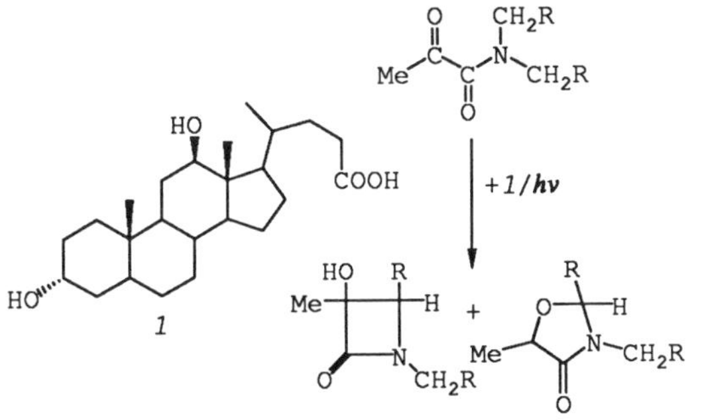

R = H; yield   42%   10%

R = Me; yield   74%

15.37

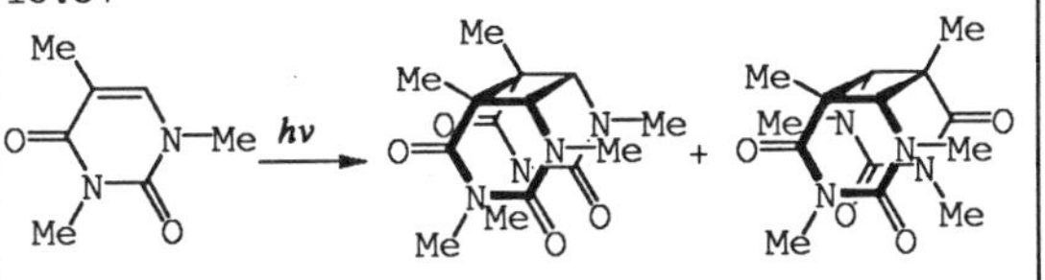

PHOTODIMERIZATION OF *1* IN CHOLESTERYL
OLEYL CARBONATE

TEMPERATURE (°C)	MESOMORPHIC PHASE	CONVERSION (%)	*2* (%)	*3* (%)
15	SMECTIC	46	96	4
30	CHOLESTERIC	27	92	8
50	ISOTROPIC	11	65	35
-70	SOLID (DMSO)	28	40	49[a]
2	SOLID (DIOXANE)	3	32	68

[a]The remainder are *anti* dimers.

15.38

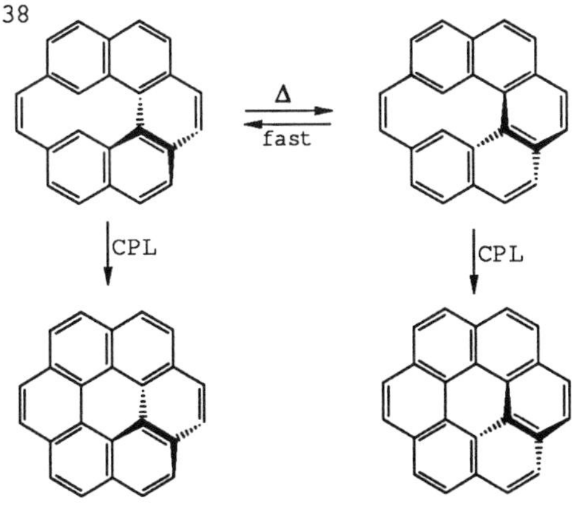

15.36. Deoxycholic acid (*1*) also forms channel-type inclusion compounds in a 4:1 ratio, e.g., with pyruvamides. Irradiation of these complexes in the solid phase produces chiral β-lactams in high yield (optical yield 10–15%), along with an oxazolidinone derivative. Chiral induction appears to be a consequence of the chirality of the channels in the deoxycholic acid crystal.[59]

15.37. Anisotropic solvents can increase both the rate and the stereoselectivity of a photodimerization, e.g., that of 1,3-dimethylthymine (*1*).[60] Irradiation of a solution of *1* in cholesteryl oleyl carbonate in its smectic (−10–19°C) or cholesteric phase (19–37°C) leads almost exclusively to the formation of the *cis-syn* dimer. Analogous results have been obtained with tetramethyluracil.[61] The stereoselectivity is rapidly lost in isotropic solutions. A fairly high degree of conversion can be attained in a mesomorphic or solid state. In isotropic solution the photolysis of the dimer to a monomer occurs faster by an order of magnitude than in an anisotropic medium.

15.38. It was shown that an organized medium can cause chiral induction in a photochemical reaction (see 15.36). Recently, considerable attention has been given to enantioselective photochemical reactions affected by circular polarized light (CPL) and chiral compounds or substituents.[62] The first method is based on the different absorption of CPL by the enantiomers leading to the formation of a larger number of excited molecules of one of the enantiomers. Optical yields are relatively low and an intense radiation source is required, since most of the energy is lost in its conversion to CPL with a narrow bandwidth. The source of CPL can be a xenon arc, e.g., a 1600 W lamp, or a suitable laser. The figure illustrates the synthesis of the chiral hexahelicene using CPL.[63]

15.39

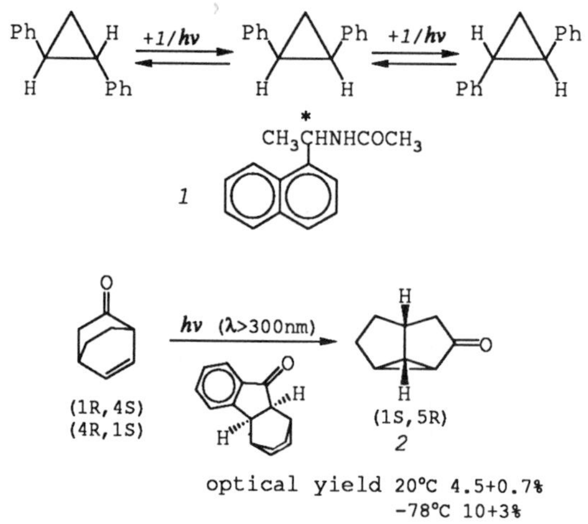

CH₃CHNHCOCH₃

1

hv (λ>300nm)

(1R,4S)
(4R,1S)

(1S,5R)
2

optical yield 20°C 4.5+0.7%
-78°C 10+3%

15.40

R =

15.41

a-c

AcO

1   *2*

a - *hv*, PhICl₂; b - KOH/MeOH; c - Ac₂O

15.42

hv ———▶

43%

1   *2*

15.39. The photochemical *chiral induction method* consists of irradiating the starting material in the presence of: (1) a chiral environment (see 15.36), (2) a chiral sensitizer, or (3) a chiral substituent in the reactant. Method (2) represents one of the classic methods for enantioselective photoreactions, first illustratred by the asymmetic *cis-trans* isomerization in which the sensitizer was the chiral acetamide *1*.[64] More recently this method has been used to effect the synthesis of tricyclic ketone *2* by an asymmetric oxadi-π-methane rearrangement.[65]

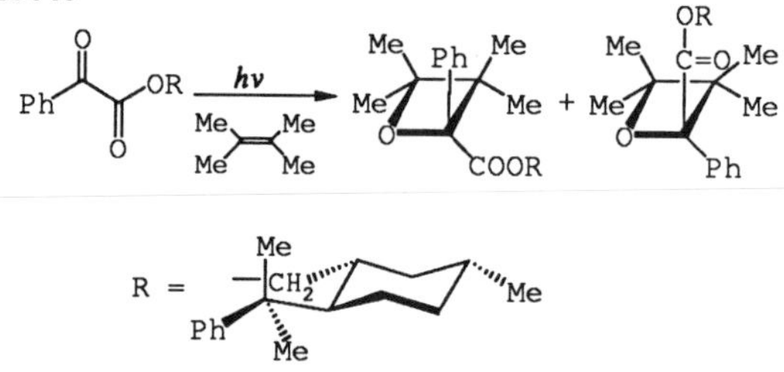

15.40. Method 3 was used for chiral induction in mixed [2+2] photocycloadditions to form cyclobutanes[66] and oxetanes.[67] In the latter, the degree of induction depends not only on the structure of the alkyl groups in the α-keto and ester moieties, but also on the nature of the alkene and the reaction conditions, especially the reaction temperature.[67]

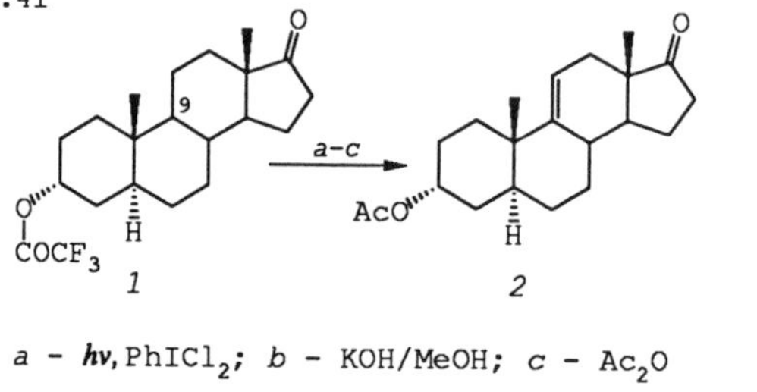

15.41. Steroidal skeletons have a number of positions of comparable reactivity, making it difficult to selectively functionalize any one position, e.g., radical halogenation can occur at three different positions to approximately the same degree. The selective functionalization of an unactivated position has been achieved by a number of photochemical methods, e.g., by the introduction of an electronegative group. For example, irradiation of *1* in the presence of phenyl-iodo-dichloride[68] leads to alkene *2* by the selective chlorination of position 9.[69]

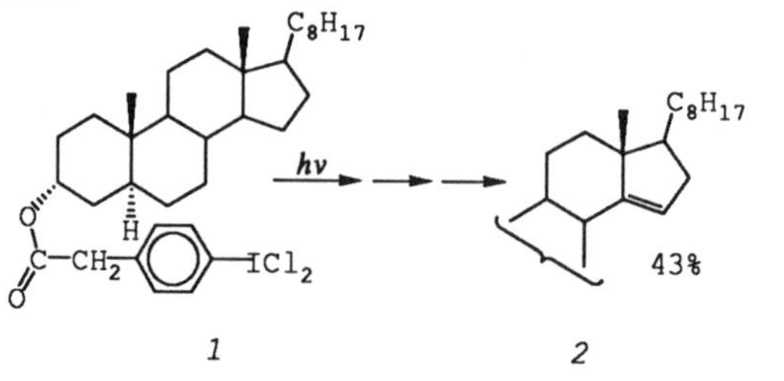

15.42. A more geneal approach is the use of steroid *remote functionalization*[70] based on accessibility of the α-face of the steroid molecule. In the iododichloride *1* the chlorine atoms interact most strongly with the hydrogen atom on C-14. As a result, its irradiation yields 14(15)-cholestenol (*2*) via the 14-chloro derivative.[69]

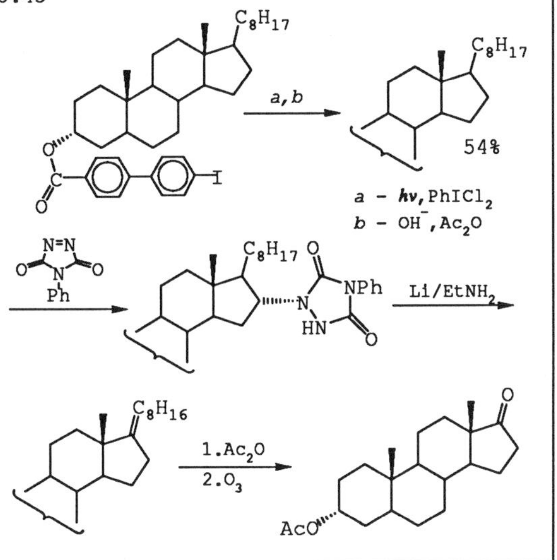

15.43. This method was employed in an elegant synthesis of aldosterone by the functionalization of position 16.[69,71] The key step is the photochlorination using $PhICl_2$ as a chlorine atom transfer agent and an esterified 4'-iodobiphenyl-4-carboxylic acid as the carrier of these atoms (radical relay mechanism). The selectivity of the attack in these *biomimetic reactions* is comparable to that of enzyme-catalyzed reactions.

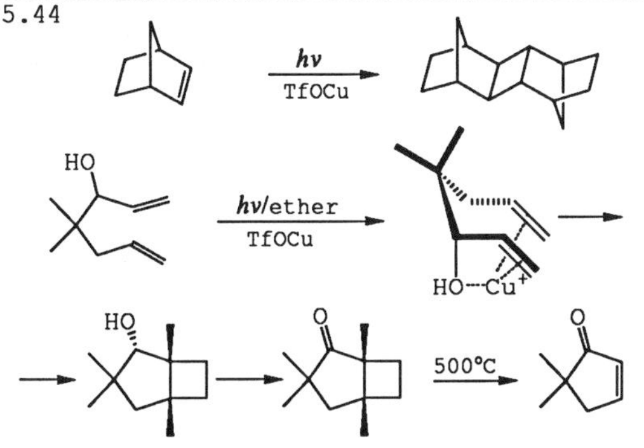

15.44. The course of photochemical reactions is strongly influenced by the use of complexes formed with the reactants. Alkenes are often photolyzed in the form of their copper complexes, especially those with *copper triflate* (trifluoromethanesulfonate, CuOTf).[72] This salt, an excellent catalyst in the photodimerization of alkenes, is prepared in the form of a benzene complex. Since the complex is weakly coordinated, benzene is easily substituted by an alkene that results in the formation of a soluble complex. Irradiation of a solution of $CuOTf (C_6H_6)_{0.5}$ in norbornene produces the dimer with a yield of 88%.[73] Using CuBr, only 38% is obtained.[74]

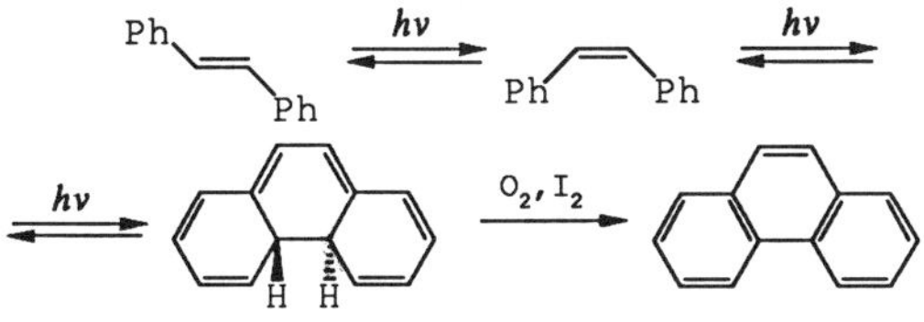

15.45. In certain substituted stilbenes the addition of one equivalent of $CuBr_2$ or $CuCl_2$ greatly enhances phenanthrene formation.[75] In addition, the simultaneous use of oxygen and iodine in the photocyclization of stilbene further increases the yields.

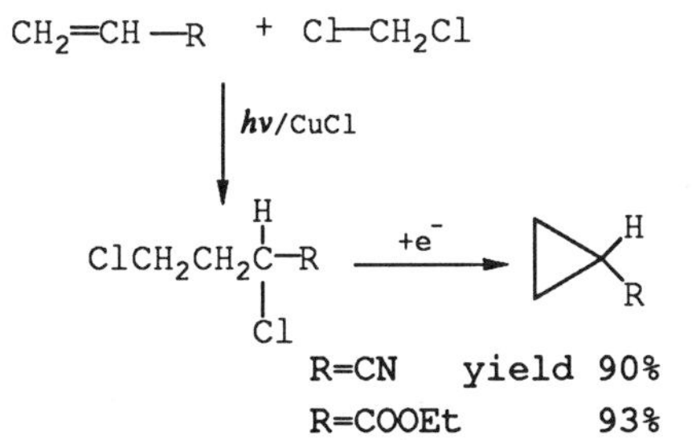

15.46. Cuprous chloride catalyzes the photochemical addition of alkyl chlorides to electron-deficient alkenes. The addition of dichloromethane leads to the corresponding 1,3-dichloro compounds in good yield. Subsequent reduction of the products provides high yields of cyclopropane carboxylic acid derivatives. This method complements the Simmons–Smith reaction, which is ineffective with electron-deficient alkenes.[76] (See also 12.10 and 12.11).

15.47

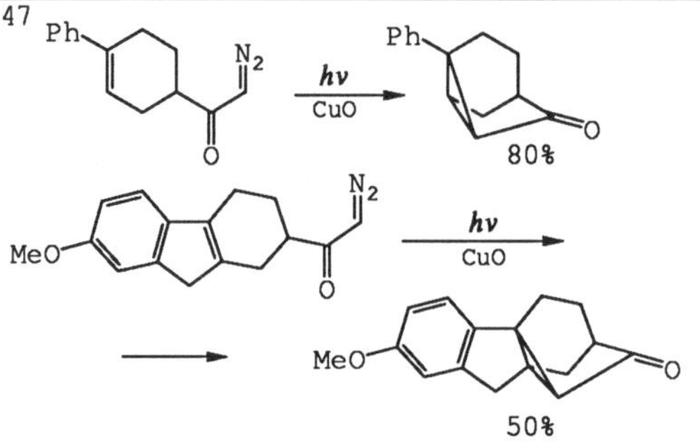

80%

50%

15.48

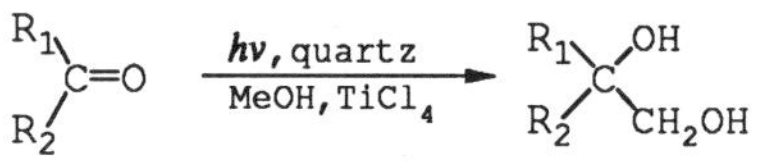

	yield		yield
R_1=Me,R_2=Et	54%	R_1=Me,R_2=n-Pr	68%
R_1=R_2=Et	55%	R_1=R_2=-(CH$_2$)$_5$-	76%

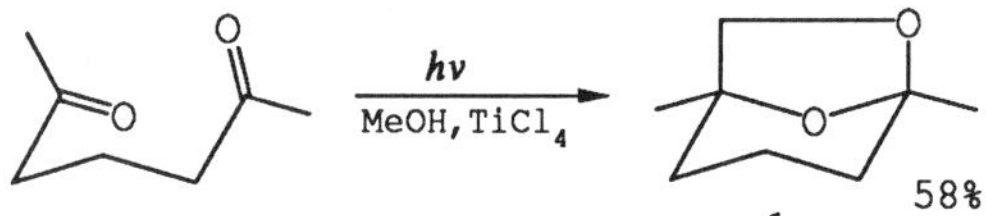

58%

1

15.49

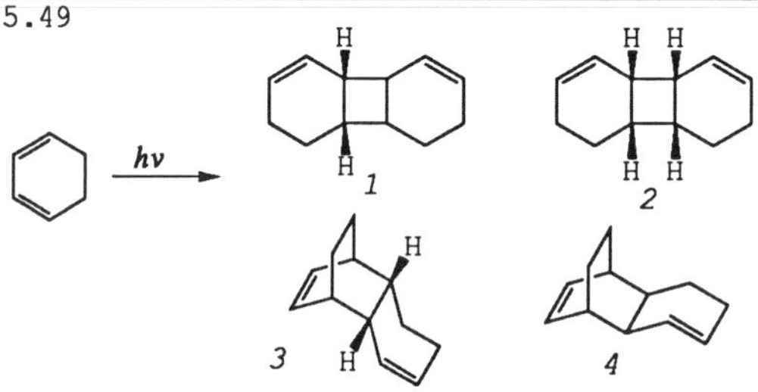

Reaction conditions	Total yield,%	Relative yield,%			
		1	*2*	*3*	*4*
hv, quartz		0.5	0.5	–	0.5
hv, 330nm		44	20	–	36
hv, ^{3}sens	92	60	20	trace	20
thermal	32	–	–	80	
hv,330nm/DCN^a	60	–	–	77	23

^{a}DCN = 1,4-dicyanonaphthalene

15.50

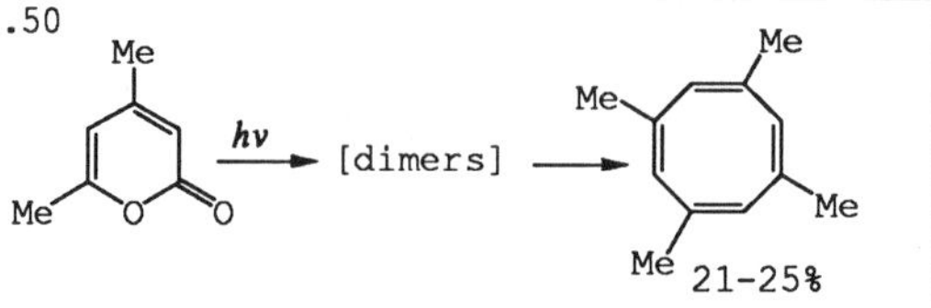

21-25%

15.47 An active CuO catalyst enhances the carbenoid addition in the photolysis of γ,δ-unsaturated α-diazoketones. This method has been used for the synthesis of tricyclic and polycyclic ketones.[77]

15.48 The photolysis of methanolic solutions of ketones in the presence of TiCl$_4$ gives high yields of 1,2-diols, along with minor amounts of ketals. In the absence of TiCl$_4$, several products in low yield are observed. This reaction was used in an elegant synthesis of pheromone *frontalin (1)*.[76]

15.49 [4+2] cycloaddition reactions with electron-rich dienophiles are generally unsuccessful. They may, however, proceed by irradiation of the diene and dienophile in the presence of a sensitizer (catalyst). The reaction occurs via an exciplex, formed from the sensitizer in the S$_1$ state and the dienophile. The exciplex, trapped by the diene, forms a ternary excited state complex or triplex (*triplex Diels–Alder reaction*).[79,80] Cyanoarenes (1,4-dicyanonaphthalene, DCN; 9,10-dicyanoanthracene, DCA) or "reluctant" azoalkanes (2,3-diazabicyclo[2.2.2]-oct-2-ene, DBO) are good sensitizers. The exciplexes have strong C–T character and tend to dissociate into radical ions[81] that generally undergo transformations different from typical photoreactions.

15.50. If the photoproduct precipitates during irradiation, as in the example shown, a common phenomenon in photodimerization reactions, it settles primarily on the cooler parts of the reactor, e.g., the immersion well in immersion reactors (see 14.29). The radiation is then absorbed by the dimer and the reaction ceases. Unless a reactor with a "brush" (see 14.27) is used, the precipitate must be removed from the immersion well and separated from the reaction mixture by filtration.[82]

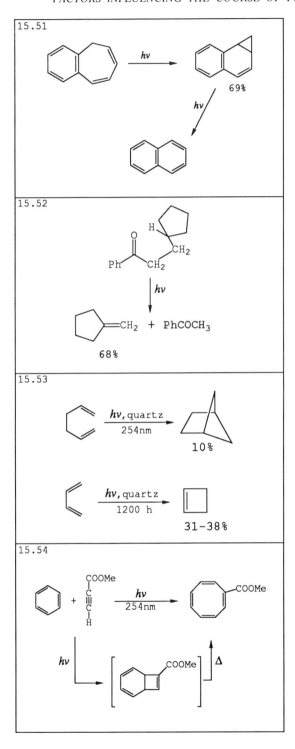

15.51 When the desired photoproduct absorbs in the same wavelength region as the starting material, the efficiency of the photoreaction is decreased and the reaction may cease altogether. This is also true if byproducts absorb the incident radiation. In either case it is advisable to stop the irradiation after partial conversion of the starting material. In the photoisomerization of 1,2-benzotropylidene to benzonorcaradiene, long irradiation times led to the formation of naphthalene.[83]

15.52. If the photoproducts are sufficiently volatile it is possible to prevent the formation of secondary products by performing the photolysis at an elevated temperature and/or reduced pressure. The formation of an oxetane can be suppressed in the Norrish Type II reaction shown by distillation of methylenecyclopentene as it is formed.[84]

15.53. If a polymer film forms on the wall of the reactor and a special reactor (see 14.27) does not succeed in its elimination, the irradiation must be interrupted periodically to remove it from the walls. Although polymers are sparingly soluble in most solvents, a 10% aqueous solution of HF will usually dissolve them. In the mercury-sensitized photolysis of 1,5-pentadiene, which requires six days of irradiation, it was necessary to clean the quartz reactor every two days.[85] A similar procedure is required for the electrocyclic ring closure of butadiene to cyclobutene.[86]

15.54. The photocycloaddition of methyl propiolate to benzene, which is accompanied by polymer formation, was performed in 15–25 cycles by distilling the starting materials from the reaction mixture after every 20 hours of irradiation. The distilled material was then irradiated again for an additional 20 hours in a clean reactor.[87]

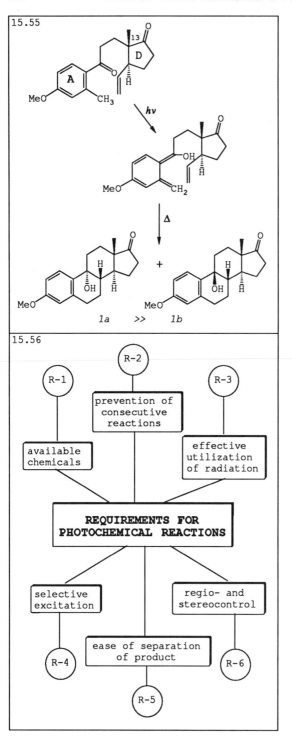

15.55

MeO — A — CH₃

13 / D / O / O / H

hv

MeO — OH — CH₂

Δ

MeO — H — ŌH — H + MeO — H — ŌH — H

1a   >>   1b

15.56

R-1   R-2   R-3

prevention of consecutive reactions

available chemicals

effective utilization of radiation

REQUIREMENTS FOR PHOTOCHEMICAL REACTIONS

selective excitation

regio- and stereocontrol

R-4

ease of separation of product

R-6

R-5

15.55. Some of the obstacles encountered in photochemical reactions are demonstrated by the synthesis of the hormone estrone. The successful execution of the photoenolization step (including the proper stereochemistry) was effected by irradiating: (1) in methylcyclohexane (optically transparent, b.p. 100.9°C) at 95°C to overcome the activation barrier of the intramolecular diene addition; (2) at 340 nm to prevent excitation of the unconjugated ketone group and, thus, loss of stereochemical integrity at C-13; (3) in the presence of pyridine, which increases both the activation barrier of the reketonization reaction and the lifetime of the Z-photoenol; (4) in the presence of 2,6-dimethylphenol or 2,4,6-trimethylphenol to protect the primary ABCD-cycloadducts (*1a*) and (*1b*) from subsequent photochemical reactions.[88]

15.56. To successfully utilize photochemical reactions in organic synthesis, certain general rules[89] should be followed. (1) The photochemical reaction should be the first, or one of the first, steps in the synthesis. (2) Preference should be given to photochemical reactions that decrease the degree of unsaturation or the continuity of unsaturated bonds in the molecule (photobleaching reactions). (3) A reasonable degree of conversion (50–70%) of the starting materials should be expected, particularly in long irradiations (photoreactions with a low Φ). (4) The excitation should be applied to a chromophore of the reactant that absorbs relatively long wavelength radiation (>300 nm) to protect other functional groups. (5) Preference should be given to dimerizations and fragmentation reactions over rearrangements. (6) Use should be made of those photochemical reactions and experimental conditions (see 15.55) offering regio- and stereoselective control.

REFERENCES

(1) Ciamician, G. *Science.* **1912**, *36*, 385.

(2) Barton, D. H. R. *Pure Appl. Chem.* **1968**, *16*, 1.

(3) A number of books and review articles concerned with preparative photochemistry have been published. Some of the more recent general ones are: (a) Srinivasan, R., ed. *Organic Photochemical Synthesis.* Wiley–Interscience: New York, 1971, Vol. 1; 1976, Vol. 2. (b) Schönberg, A. *Preparative Organic Photochemistry.* Springer Verlag: New York, 1968. (c) Muller E., ed., *Methoden der Organischen Chemie*/(Houben-Weyl). G. Thieme: Stuttgart, 1975, Vol. 4/5a: Photochemie. (d) Horspool, W. M., ed. *Synthetic Organic Photochemistry.* Plenum: New York, 1924. (e) Coyle, J. E. *Photochemistry in Organic Synthesis.* The Royal Society of Chemistry: London, 1986. (f) Carless, H. A. J. *Chem. Brit.* **1980**, *16*, 456–459. (g) Davidson, R. S. *Chem. Ind.* **1978**, 180. (h) Kossanyi, J. *J. Pure Appl. Chem.* **1979**, *51*, 181. (i) Schreiber, S. L. *Science.* **1985**, *225*, 857.

(4) Schoulders, B. A., Kwie, W. W., and Gardner, P. D. *Org. Photochem Synth.* **1971**, *1*, 42.

(5)(a) Yamada, Y., Uhda, H., and Nakanishi, I. *Chem. Commun.* **1966**, 423. (b) Zurfluh, R. L., Dunham, L., Spain, V. L., and Sidal, J. B. *J. Am. Chem. Soc.* **1970**, *92*, 425.

(6)(a) Kasha, M. *Disc. Faraday Soc.* **1950**, *9*, 14. (b) Becker, R. *Theory and Interpretation of Fluorescence and Phosphorescence.* Wiley: New York, 1969.

(7) For an excellent review of wavelength effects on organic photoreactions in solution, see Turro, N. J., Ramamurthy, V., Cherry, W., and Farneth, W. *Chem. Rev.* **1978**, *78*, 125.

(8) Schuster, D. I. *J. Org. Chem.* **1979**, *44*, 4254, and references therein.

(9) Ullman, E. G. and Singh, B. *J. Am. Chem Soc.* **1966**, *88*, 1844; *89*, 6911.

(10) Dauben, W. G. and Kellog, M. S. *J. Am. Chem. Soc.* **1971**, *93*, 3805.

(11)(a) Whitesell, J. K., Minton, M. A., and Tran, V. D. *J. Am. Chem. Soc.* **1989**, *111*, 1473, and references therein. (b) For an exhaustive review, see, Jacobs, H. J. C. and Havinga, E. *Adv. Photochem.* **1979**, *11*, 321.

(12)(a) Dauben, W. G. and Phillips, R. G. *J. Am. Chem. Soc.* **1982**, *104*, 5780. (b) Dauben, W. G., Share, P. E., and Ollmann, R. R., Jr. *J. Am. Chem. Soc.* **1988**, *110*, 2548, and references therein.

(13)(a) Bahurel, Y. L., MacGregor, D. J., Penner, T. L., and Hammond, G. S. *J. Am. Chem. Soc.* **1972**, *94*, 637. (b) Schenck, G. O., Mannsfeld, S. P., Schomburg, G., and Krauch, C. H. *Z. Naturforsch.* **1964**, *19B*, 18.

(14)(a) Bartlett, P. D. and Porter, N. A. *J. Am. Chem. Soc.* **1968**, *90*, 5317. (b) Fox, J. R. and Hammond, G. S. *J. Am. Chem. Soc.* **1964**, *86*, 4031. (c) Zimmerman, H. E., Boettcher, R. J., Buehler, N. E., and Keck, G. E. *J. Am. Chem. Soc.* **1975**, *97*, 5635. (d) For an example of the multiplicity-dependent pericyclic rearrangement, see Goldschmidt, Z. and Genizi, E. *Tetrahedron Lett.* **1987**, *28*, 4867.

(15)(a) Dilling, W. L. and Kroening, R. D. *Tetrahedron Lett.* **1968**, 5101. (b) For cyclopentadiene, see Dilling, W. L. and Droening, R. D. *Tetrahedron Lett.* **1968**, 5601.

(16) Ciabattoni, J., Crowley, J. E., and Kende, A. S. *J. Am. Chem. Soc.* **1967**, *89*, 2778.

(17) Kende, A. S. and Goldschmidt, Z. *Org. Photochem. Synth.* **1971**, *1*, 27.

(18) Akelah, A. and Sherrington, D. C. *Chem. Rev.* **1981**, *81*, 557, and references therein.

(19) Blossey, E. C. and Neckers, D. C. *Tetrahedron Lett.* **1974**, 323.

(20)(a) Blossey, E. C., Neckers, D. C., Thayer, A. L., and Schaap, A. P. *J. Am. Chem. Soc.* **1973**, *95*, 1520. (b) Schaap, A. P., Thayer, A. L., Blossey, E. C., and Neckers, D. C. *J. Am. Chem. Soc.* **1975**, *97*, 3741.

(21) Tamagaki, S., Liesner, C. E., and Neckers, D. C. *J. Org. Chem.* **1980**, *45*, 1573.

(22) Plummer, B. F. and Hall, R. A. *J. Chem. Soc. D.* **1970**, 44.

(23) Plummer, B. F. and Chihal, D. M. *J. Am. Chem. Soc.* **1971**, *93*, 2071.

(24)(a) Shields, J. E., Gavrilovic, D., and Kopecký, J. *Tetrahedron Lett.* **1971**, 271. (b) Shields, J. E., Gavrilovic, D., Kopecký, J., Hartmann, W., and Heine, H.-G. *J. Org. Chem.* **1974**, *39*, 515. (c) For additional references see, Kubo, Y., Tobawa, S., Yamane, K., Takuwa, A., and Araki, T. *J. Org. Chem.* **1989**, *54*, 4929.

(25) Cooper, W. and Rome, K. A. *J. Phys. Chem.* **1974**, *78*, 16.

(26) For heavy-atom effect in *cis-trans* photoisomerization of stilbenes, see (a) Saltiel, J., Chang, E.,

and Megarity, E. D. *J. Am. Chem. Soc.* **1974,** *96,* 6521.   (b) Fischer, G., Muszkat, K. A., and Fischer, E. *Isr. J. Chem.* **1968,** *6,* 965.

(27) Lambert, J. B., Shurvell, H. F., Lightner, D. A. and Cooks, R. G. *Introduction to Organic Spectroscopy,* Macmillan: New York, 1987, p. 257.

(28) Adapted from Evans, T. R. In *Techniques of Organic Chemistry,* Leermakers, P. A. and Weissberger, A., eds. Interscience Publication: New York, 1969, Vol. 14: *Energy Transfer and Organic Photochemistry,* p. 312.

(29) Dickinson, D. A., Hardy, T. A., and Hart, H. *Org. Photochem. Synth.* **1976,** *2,* 62.

(30) Adapted from Ref. 28, p. 319.

(31)(a) Chapman, O. L., Nelson, P. J., King, R. W., Trecker, D. J., and Griswold, A. *Rec. Chem. Progr.* **1967,** *28,* 167.   (b) For the preparative procedure of the dimerization of isophorone in acetic acid or *n*-hexane, see Trecker, D. J., Griswold, A. A., and Chapman, O. L. *Org. Photochem. Synth.* **1971,** *1,* 62. (c) For the photodimerization of isophorone in supercritical CHF_3 and CO_2, see Hrnjez, B. J., Mehta, A. J., Fox, M. A., and Johnston, K. P. *J. Am. Chem. Soc.* **1989,** *111,* 2662, and references therein.

(32) Pappas, S. P., Pappas, B. C., and Blackwell, J. E., Jr. *J. Org. Chem.* **1967,** *32,* 3066.

(33) Pappas, S. P. and Zehr, R. D. *J. Am. Chem. Soc.* **1971,** *93,* 7112.

(34)(a) Huyser, E. S. and Neckers, D. C. *J. Org. Chem.* **1964,** *29,* 276.   (b) Pappas, S. P., Pappas, B. C., Okamoto, Y., and Sakamoto, H. *J. Org. Chem.* **1988,** 4404, and references therein.

(35)(a) Bryce-Smith, D. and Gilbert, A. *J. Chem. Soc., Chem. Commun.* **1968,** 19.   (b) Adapted from Ref. 28, p. 322.

(36) Gassman, P. G. and Patton, D. S. *Org. Photochem. Synth.* **1971,** *1,* 21.

(37)(a) Anklam, E. and Margaretha, P. *Helv. Chim. Acta.* **1983,** *66,* 1466.   (b) For a recent study of the photochemistry of 2(3*H*)- and 2(5*H*)-furanones, see Fillol, L., Miranda, M. A., Morera, I. M., and Sheikh, H. *Heterocycles.* **1990,** *31,* 751.

(38) Taylor, E. C. and Spence, G. G. *Org. Photochem. Synth.* **1971,** *1,* 46.

(39) For reviews and photochemistry in the solid state, see: (a) Ramamurthy, V. and Venkatesan, K. *Chem. Rev.* **1986,** *86,* 433.   (b) Ramamurthy, V. *Tetrahedron.* **1986,** *42,* 5753.   (c) Hasegawa, M. *Chem. Rev.* **1983,** *83,* 507.   (d) Scheffer, J., ed. *Organic Chem-*

istry in Anisotropic Media, Tetrahedron Symposia-in-Print No. 29. Tetrahedron. **1987,** *43,* 1197–1746. (e) Byrn, S. R. *The Solid State Chemistry of Drugs.* Academic Press: New York, 1982.

(40)(a) Schmidt, G. M. J. *Pure Appl. Chem.* **1971,** *27,* 647.   (b) Cohen, M. D., Schmidt, G. M. J., and Sonntag, F. I. *J. Chem. Soc.* **1964,** 2000. (c) Schmidt, G. M. J. *et al. Solid State Photochemistry.* Verlag Chemie: Weinheim, 1976.   (d) For the solid-state dimerization of ethyl E-cyano-2-methoxy-cinnamate, see Chimichi, S., Sarti-Fantoni, P., Coppini, G., Perhem, F., and Renzi, G. *J. Org. Chem.* **1987,** *52,* 5124.   (e) For the solid-state photoreactivity of an ene-dione, see Scheffer, J. R. and Gudmundsdottir, A. D. *Mol. Cryst. Liq. Cryst.* **1990,** *186,* 19.

(41) Green, B. S. and Heller, L. *J. Org. Chem.* **1974,** *39,* 196.

(42)(a) Burton, W. B. *J. Chem. Ed.* **1979,** *56,* 483. (b) Ivie, C. W. and Casida, J. E. *J. Agric. Food Chem.* **1971,** *19,* 405.

(43) For reviews, see   (a) Scheffer, J. R., Garcia-Garibay, M., and Nalamasu, O. *Org. Photochem.* **1987,** *8,* 249.   (b) Scheffer, J. R. *Acc. Chem. Res.* **1980,** *13,* 283.

(44) Cohen, M. D. and Schmidt, G. M. J. *J. Chem. Soc.* **1964,** 1966.

(45) Cohen, M. D. *Tetrahedron.* **1987,** *43,* 1211.

(46) For recent examples, see: (a) Pokkuluri, P. R., Scheffer, J. R., and Trotter, J. *J. Am. Chem. Soc.* **1990,** *112,* 3676.   (b) Ariel, S., Askari, S., Scheffer, F. R., and Trotter, J. *J. Org. Chem.* **1989,** *54,* 4324. (c) Zimmerman, H. E. and Zuraw, M. J. *J. Am. Chem. Soc.* **1989,** *111,* 7974.   (d) Evans, S. V., Garcia-Garibay, M., Omkaram, N., Scheffer, J. R., Trotter, J., and Wireko, F. *J. Am. Chem. Soc.* **1986,** *108,* 5648. (e) Appel, W. K., Greenhough, T. J., Scheffer, J. R., Trotter, J., and Walsh, L. *J. Am. Chem. Soc.* **1980,** *102,* 1158, 1160.

(47)(a) Farwaha, R., De Mayo, P., and Toong, Y. C. *J. Chem. Soc., Chem. Commun.* **1983,** 739. (b) Dave, V., Farwaha, R., De Mayo, P., and Strother, J. B. *Can. J. Chem.* **1985,** *63,* 2401.

(48)(a) Corbin, D. R., Eaton, D. F., and Ramamurthy, V. *J. Org. Chem.* **1988,** *53,* 5384.   (b) Turro, N. J. *Tetrahedron.* **1987,** *43,* 1589.   (c) Turro, N. J. *Pure Appl. Chem.* **1986,** *58,* 1219.   (d) Ramamurthy, V. *Tetrahedron.* **1986,** *42,* 121.   (e) Turro, N. J., Fehlner, J. R., Hessler, D. P., Welsh, K. M., Ruderman, W., Firnberg, D., and Braun, A. M. *J. Org. Chem.* **1988,** *53,* 3731.   (f) Gessner, F., Olea, A., Lobaugh, J. H., Johnson, L. J., and Scaiano, J. C. *J. Org. Chem.*

1989, *54*, 259, and references therein. (g) For a review of photophysical and photochemical processes on clay surfaces, see Thomas, J. K. *Acc. Chem. Res.* **1988**, *21*, 275.

(49)(a) For a review of photochemistry in micelles, see Turro, N. J., Cox, G. S., and Paczowski, M. A. *Topics Curr. Chem.* **1985**, *57*, 129. (b) Ramamurthy, V. *Tetrahedron.* **1986**, *42*, 5753.

(50)(a) De Mayo, P. and Sydnes, L. K. *J. Chem. Soc., Chem. Commun.* **1980**, 994. (b) For the dimerization of the enone, see Lee, K. H. and De Mayo, P. *Photochem. Photobiol.* **1980**, *31*, 311. (c) For the cycloaddition of cyclohexene, see Berenjian, N., De Mayo, P., Sturgeon, M. E., Sydnes, L. K., and Weedon, A. C. *Can. J. Chem.* **1982**, *60*, 425.

(51)(a) Turro, N. J. and Kraeutler, B. *Acc. Chem. Res.* **1980**, *13*, 369–377. (b) Turro, N. J., Gratzel, M., and Braun, A. M. *J. Am. Chem. Soc.* **1980**, *92*, 712. (c) Turro, N. J. *Pure Appl. Chem.* **1981**, *53*, 259. (d) Turro, N. J. *Proc. Natl. Acad. Sci. USA.* **1983**, *80*, 609. (e) For the photochemical ^{17}O enrichment at C=O in dibenzyl ketone, deoxybenzoin, and benzoin methyl ether on porous silica, see Turro, N. J., Paczkowski, M. A., and Wan, P. *J. Org. Chem.* **1985**, *50*, 1399.

(52) For ^{13}C enrichment by photolysis of diphenylcycloalkanones, see Turro, N. J., Doubleday, D., Jr., Hwang, K.-C., Chen, C.-C., and Fehlner, J. R. *Tetrahedron Lett.* **1987**, *28*, 2929.

(53)(a) Bender, M. L. and Komiyama, M. *Cyclodextrin Chemistry*, Springer Verlag: New York, 1977. (b) For the photochemistry and photophysics within cyclodextrin cavities, see Ramamurthy, V. and Eaton, D. F. *Acc. Chem. Res.* **1988**, *21*, 300.

(54) For an example of photo-Fries rearrangement in micelles, see Singh, A. K. and Raghuraman, T. S. *Tetrahedron Lett.* **1985**, *26*, 4125.

(55) For the consequences of β-CD$_x$ complexation to Norrish Type I and/or II reactions, see: (a) Dasaratha Reddy, G., Jayasree, B., and Ramamurthy, V. *J. Org. Chem.* **1987**, *52*, 3107. (b) Nageshwer, Rao B., Syamala, M. S., Turro, N. J., and Ramamurthy, V. *J. Org. Chem.* **1987**, *52*, 5517. (c) Dasaratha Reddy, G. and Ramamurthy, V. *J. Org. Chem.* **1987**, *52*, 5521.

(56)(a) Ueno, A., Moriwaki, F., Osa, T., Hamada, F., and Murai, K. *Tetrahedron.* **1987**, *43*, 1571, and references therein. (b) Ueno, A., Moriwaki, F., Osa, T., Hamada, F., and Murai, K. *Tetrahedron.* **1987**, *43*, 1571. (c) For the photochemistry of anthraquinone-substituted β-cyclodextrins, see Aquino, A. M., Abelt, C. J., Berger, K. L., Darragh, C. M., Kelley,

S. E., and Cossette, M. V. *J. Am. Chem. Soc.* **1990**, *112*, 5819.

(57)(a) Tamaki, T. *Chem. Lett.* **1984**, 53. (b) Tamaki, T., Kokubu, T., and Ichimura, K. *Tetrahedron.* **1987**, *43*, 1485.

(58) Ueno, A., Moriwaki, F., Azuma, A., and Osa, T. *J. Org. Chem.* **1989**, *54*, 295.

(59)(a) Aoyama, H., Miyazaki, K., Sakamoto, M., and Omote, Y. *J. Chem. Soc., Chem. Commun.* **1983**, 333. (b) For solid-state photoreaction of N,N-dialkylpyruvamides, see Aoyama, H., Miyazaki, K.-I., Sakamoto, M., and Omote, Y. *Tetrahedron.* **1987**, *43*, 1513. (c) For photoreactions of 2-oxoamides and tropolones in crystalline inclusion complexes, see Kaftory, M., Toda, F., Tanaka, K., and Yagi, M. *Mol. Cryst. Liq. Cryst.* **1990**, *186*, 167. (d) For enantioselective photoreactions of tropolone alkyl ethers in a crystalline inclusion complex, see Toda, F. and Tanaka, K. *J. Chem. Soc., Chem. Commun.* **1986**, 1429. (b) For *cis-trans* isomerization of alkenes in deoxycholic acid cavities, see Guarino, A., Possagno, E., and Bassanelli, R. *Tetrahedron.* **1987**, *43*, 1541.

(60) Kunieda, T., Takahashi, T., and Hirobe, M. *Tetrahedron Lett.* **1983**, *24*, 5107.

(61) Nagamatsu, T., Kawano, C., Orita, Y., and Kunieda, T. *Tetrahedron Lett.* **1987**, *28*, 3263.

(62) Rau, H. *Chem. Rev.* **1983**, *83*, 535–547.

(63) Berstein, W. J., Calvin, M., and Buchardt, O. *J. Am. Chem. Soc.* **1972**, *94*, 494.

(64)(a) Hammond, G. S. and Cole, R. S. *J. Am. Chem. Soc.* **1965**, *87*, 3256. (b) Ouannes, C., Beugelmans, R., and Roussi, G. *J. Am. Chem. Soc.* **1973**, *95*, 8472. (c) For asymmetrically sensitized *cis-trans* isomerization of cyclooctene, see Inoue, Y., Yokoyama, T., Yamasaki, N., and Tai, A. *J. Am. Chem. Soc.* **1989**, *111*, 6480.

(65) Demuth, M., Raghavan, P. R., Carter, C., Nakano, K., and Schaffner, K. *Helv. Chim. Acta.* **1980**, *53*, 2434.

(66)(a) Lange, G. L., Decicco, C. and Lee, M. *Tetrahedron Lett.* **1987**, *28*, 2833. (b) Tolbert, L. M. and Ali, M. B. *J. Am. Chem. Soc.* **1982**, *104*, 1742.

(67) Buschmann, H., Scharf, H.-D., Hoffmann, N., Plath, M. W., and Runsink, J. *J. Am. Chem. Soc.* **1989**, *111*, 5367, and references therein.

(68) Tanner, D. D. and Van Bostelen, P. B. *J. Org. Chem.* **1967**, *32*, 1517.

(69) Breslow, R., Corcoran, R. J., Snider, B. B., Doll, R. J., Khanna, P. L., and Kaleya, R. *J. Am. Chem. Soc.* **1977,** *99,* 905.

(70)(a) For a most recent report, see Breslow, R. and Guo, T. *Tetrahedron Lett.* **1987,** *28,* 3187.   (b) For earlier works, see Breslow, R. *Acc. Chem. Res.*, **1980,** *13,* 170.   (c) For pyridine esters as templates, see Breslow, R., Brandl, M., Hunger, J., and Adams, A. D. *J. Am. Chem. Soc.* **1987,** *109,* 3799.   (d) For steroid functionalization by groups attached to the β-face, see (i) Breslow, R., Maitra, V., and Heyer, D. *Tetrahedron Lett.* **1984,** *25,* 1123.   (ii) Maitra, V. and Breslow, R. *Tetrahedron Lett.* **1986,** *27,* 3087.   (e) For a long-range intramolecular H-abstraction by alkoxyl radicals, see Orito, K., Ohto, M., and Suginome, H. *J. Chem. Soc., Chem. Commun.* **1990,** 1076.   (f) For the remote functionalization of nonactivated carbon atoms by (diacetoxyiodo)benzene and iodine, see Dorta, R. L., Hernandez, R., Salazar, J. A., and Suarez, E. *J. Chem. Res. Synop.* **1990,** 240.

(71) Breslow, R. and Maitra, V. *Tetrahedron Lett.* **1984,** *25,* 5843.

(72) For a review of homogeneous metal catalysis in organic photochemistry, see Salamon, R. G. *Tetrahedron.* **1983,** *39,* 485.

(73) Salamon, R. G. and Kochi, J. K. *Tetrahedron Lett.* **1973,** 2529.

(74) Trecker, D. J. and Foote, R. S. *Org. Photochem. Synth.* **1971,** *1,* 81,

(75) Collins, D. J. and Hobbs, J. J. *Chem. Ind.* **1965,** 1725.

(76) Mitani, M., Yamamoto, Y., and Koyama, K. *J. Chem. Soc., Chem. Commun.* **1983,** 1446.

(77)(a) Ghatak, U. R., Chakraborti, P. C., Ranu, B. C., and Sanyal, B. *J. Chem. Soc., Chem. Commun.* **1973,** 548.   (b) For the use of a Ni catalyst in this reaction, see Chakraborti, A. K., Saha, B., and Ghatak, U. R. *Ind. J. Chem., Sect. B.* **1981,** *20B,* 911.

(78)(a) Sato, T., Yamaguchi, S., and Kaneko, H. *Tetrahedron Lett.* **1979,** 1863.   (b) For the synthesis of frontalin from an azoalkane via an endoperoxide, see Kleinermanns, K. and Wolfrum, J. *Angew. Chem., Int. Ed. Engl.* **1987,** *26,* 50, ref. 73.   (c) For a review of frontalin syntheses, see Braun, M. *Nachr. Chem., Tech. Lab.* **1985,** *33(5),* 392.

(79)(a) Calhoun, G. C. and Schuster, G. B. *J. Am. Chem. Soc.* **1984,** *106,* 6870.   (b) Calhoun, G. C. and Schuster, G. B. *Tetrahedron Lett.* **1986,** *27,* 911.   (c) Calhoun, G. C. and Schuster, G. B. *J. Am. Chem. Soc.* **1986,** *108,* 8021.   (d) Jones, C. R., Allman, B. J., Mooring, A., and Spahic, B. *J. Am. Chem. Soc.* **1983,** *105,* 652.

(80) For acyclic and cyclic 1,3-dienes, alkenyl and alkynyl-benzenes, and enol ethers as dienophiles, see (a) Hartsough, D. and Schuster, G. B. *J. Org. Chem.* **1989,** *54,* 3.   (b) Akbulut, N., Hartsough, E., Kim, J.-I., and Schuster, G. B. *J. Org. Chem.* **1989,** *54,* 2549.

(81)(a) For the mechanism of electron-transfer sensitization, see Mattay, J. *Angew. Chem, Int. Ed. Engl.* **1987,** *26,* 825.   (b) Chung, W.-S., Turro, N. J., Mertes, J. and Mattay, J. *J. Org. Chem.,* **1989,** *54,* 4881.

(82) De Mayo, P., McIntosh, C. L., and Yip, R. W. *Org. Photochem. Synth.* **1971,** *1,* 99.

(83) Pomerantz, M. and Gruber, G. W. *Org. Photochem. Synth.* **1971,** *1,* 23.

(84) Neckers, D. C., Kellogg, R. M., Prins, W. L., and Schoustra, B. *Org. Photochem. Synth.* **1976,** *2,* 77.

(85) Srinivasan, R. *Org. Photochem. Synth.* **1971,** *1,* 31.

(86) Sonntag, F. I. and Srinivasan, R. *Org. Photochem. Synth.* **1971,** *1,* 39.

(87) Grunewald, G. L. and Grindel, J. M. *Org. Photochem. Synth.* **1976,** *2,* 20.

(88) Quinkert, G. and Stark, H. *Angew. Chem., Int. Ed. Engl.* **1983,** *22,* 637.

(89) Jones, G., II. *Org. Photochem.* **1981,** *5,* 1.

Chapter 16
APPLICATIONS OF ORGANIC PHOTOCHEMISTRY

APPLICATIONS OF ORGANIC PHOTOCHEMISTRY

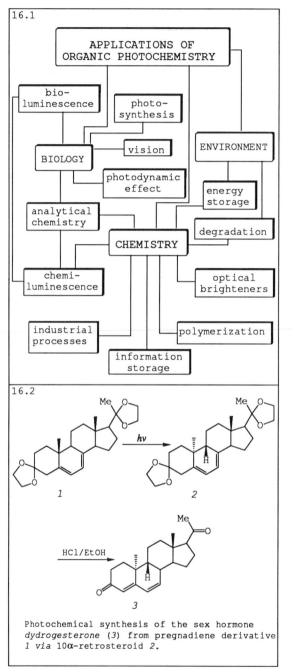

16.1

APPLICATIONS OF ORGANIC PHOTOCHEMISTRY

- bioluminescence
- photosynthesis
- vision
- BIOLOGY
- photodynamic effect
- ENVIRONMENT
- analytical chemistry
- energy storage
- CHEMISTRY
- degradation
- chemiluminescence
- optical brighteners
- industrial processes
- polymerization
- information storage

16.2

Photochemical synthesis of the sex hormone *dydrogesterone* (3) from pregnadiene derivative 1 via 10α-retrosteroid 2.

16.1. Photochemical processes as the basis of many natural occurrences are not only necessary for life on our planet, but are also utilized in many important technological procedures. Unfortunately, however, not all photochemical reactions that occur in the atmosphere or biosphere are beneficial. In fact, some of them lead to severe health damage. In addition, many activities associated with technological progress result in photochemical processes that endanger life on earth. In general, however, most photochemical processes are advantageous: they underlie many technological methods and their products are commonly used in daily life.[1]

Among the many applications of organic photochemistry, those dealing with the biochemical[2] or biological[3] effects of ultraviolet radiation or with the utilization of these effects have been selected. The criteria used to choose examples for discussion include their technical and economic significance, protection of the environment, and the solution of some practical and theoretical questions in biochemistry and biology. It should be emphasized that this selection is somewhat subjective and that the list of applications is far from complete.

16.2. The most important industrial photochemical processes are those involving the synthesis of products: (1) whose production costs are substantially less than those of comparable conventional synthetic methods,[4] especially by shortening the number of steps in the synthesis; (2) for which the cost of electric power is negligible in the total accounting; and (3) whose annual production volume does not exceed units of tons, e.g., rose oxide (see 13.27), vitamin D, and vitamin A acetate (see 6.20).[5]

16.3   INDUSTRIAL PHOTOCHEMISTRY

ECONOMICAL ASPECTS:

1. $E_{el} \longrightarrow h\nu$
   Tranformation of electrical energy into useful energy of radiation.

2. $E_{el} \longrightarrow 10\% \; h\nu + 90\% \; heat$
   Cooling of the system is necessary.

3. **Dimerization vs Isomerization**
   Recycling of solvent from very dilute solutions ($\downarrow$dimerization).

4. $\Phi_{polychromatic} < \Phi_{h\nu}$
   Irradiation time increases with the polychromatic light.

5. $\lambda = 254nm \longrightarrow 1 \; Einstein = 131 \; W$
   $546nm \qquad\qquad\qquad\qquad 61 \; W$
   The longest wavelength radiation should be used.

6. **Efficiency ~ molecular mass**
   Irradiation of a compound with the largest molecular mass.

16.4

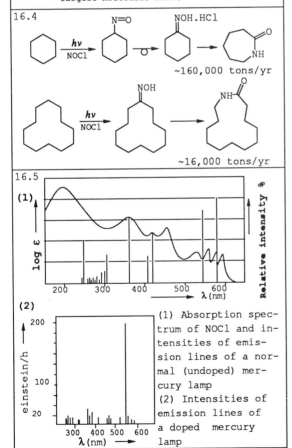

~160,000 tons/yr

~16,000 tons/yr

16.5

(1) Absorption spectrum of NOCl and intensities of emission lines of a normal (undoped) mercury lamp

(2) Intensities of emission lines of a doped mercury lamp

16.3. The use of photochemical reactions on an industrial scale presents considerable economic difficulties, such as: (1) the inefficient transformation of electrical energy into useful radiation; (2) the expense of cooling, since only about 10% of the power input is transformed into useful radiation (the remainder is converted to heat); (3) expensive solvent loss and recycling (many photochemical reactions must be performed in dilute solutions to avoid undesirable side reactions, such as dimerizations); (4) the need for long irradiation times (the Φ for polychromatic radiation is less than the Φ at a single wavelength); (5) a radiant flux of 1 einstein at 254 nm is equivalent to 131 W, but at 546 nm to only 61 W; (6) a higher molecular weight photoproduct increases the economic advantage of a photochemical reaction.[6]

16.4. One of the most technologically significant photochemical processes is the photonitrosation of cyclohexane and cyclododecane with nitrosyl chloride to produce the corresponding oximes. These are subsequently converted to lactams, which furnish the starting materials for the production of Nylon 6 and 12, respectively.[5,6] In addition to the processes mentioned in 16.2, photochemistry is used widely in the industrial preparation of various fragrant chemicals.[7]

16.5. The economic success of the photonitrosation reaction depends on the use of a special mercury discharge lamp doped with thallous iodide. This converts most of the short wavelength radiation to longer wavelength radiation of 535 nm, which is needed for photonitrosation. An annual production of 1000 tons of cyclohexanone oxime requires a reactor with 50 discharge lamps, each with an output of 60 kW (arc length 2 m, tube diameter 6 cm). Under optimal conditions the yield of this reaction is 89%.[5,6]

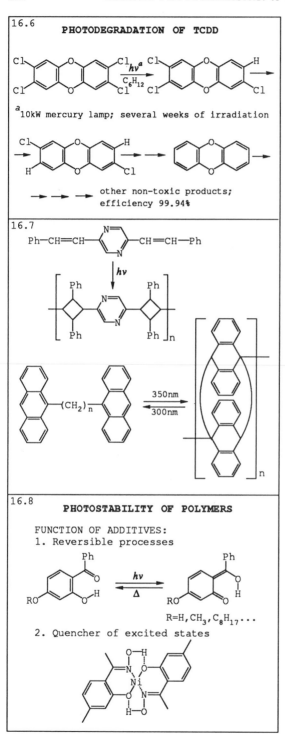

16.6

PHOTODEGRADATION OF TCDD

^{a}10kW mercury lamp; several weeks of irradiation

other non-toxic products; efficiency 99.94%

16.7

Ph—CH=CH—[pyrazine]—CH=CH—Ph

16.8

PHOTOSTABILITY OF POLYMERS

FUNCTION OF ADDITIVES:
1. Reversible processes

R=H,CH$_3$,C$_8$H$_{17}$...

2. Quencher of excited states

16.6. It can be expected that some industrial chemical processes will be replaced by ecologically less objectionable photochemical reactions. Also, the use of photochemical decomposition methods for toxic waste materials should increase. An example is the photochemical destruction of 2,3,7,8-tetrachlorodibenzo-*p*-dioxine (TCDD, a teratogen which is approximately 2000 times more toxic than strychnine). The ecological catastrophe in Soveso, Italy prompted an investigation of the degradation reactions of TCDD. Photodegradation in the presence of hydrogen donors proved to be the most successful.[8]

16.7. VIS or UV radiation can cause polymerization by one of two mechanisms: (1) the homolytic cleavage of a bond in an excited state of the monomer or initiator generates free radicals that promote a chain reaction (photoinitiated polymerization);[9] (2) the monomer reacts in an electronically excited state with another monomer in its ground state (photopolymerization). The most common processes include $[2\pi+2\pi]$ photodimerizations, e.g., the photopolymerization of 2,5-distyrylpyrazine,[10] and *photocrosslinking* of polyvinylcinnamate.[11] Other topological types, such as the $[4\pi+4\pi]$ photopolymerization of 9-substituted anthracenes (shown), are known.[12]

16.8. The effect of solar radiation and atmospheric oxygen induces polymer degradation that changes the physical properties of the polymer, e.g., artificial fibers become brittle. Photodegradation can be inhibited by various additives (photostabilizers). According to the mechanism of their action, there are two types of additives: compounds that absorb near UV radiation and whose photoproducts revert to the starting materials, e.g., 4-alkoxy-2-hydroxybenzophenones, and those compounds that quench excited states, e.g., certain Ni^{+2} complexes.[13-15]

16.9

PHOTODEGRADATION OF POLYMERS

Insertion of $>C=O$ into the polymer
molecule, e.g. $Ph-CH=CH_2 + CH_3COCH=CH_2$

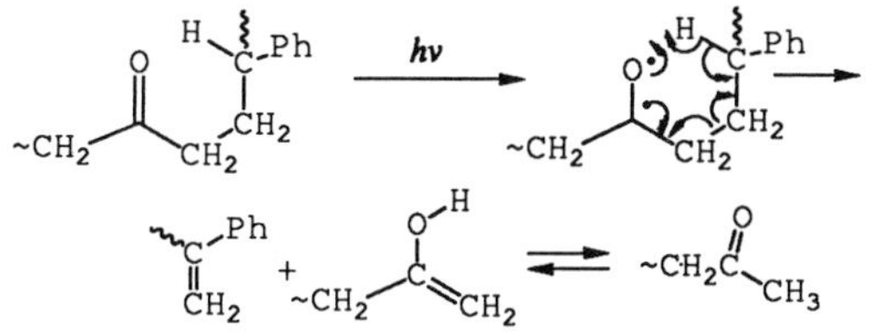

16.10

OPTICAL BRIGHTENERS

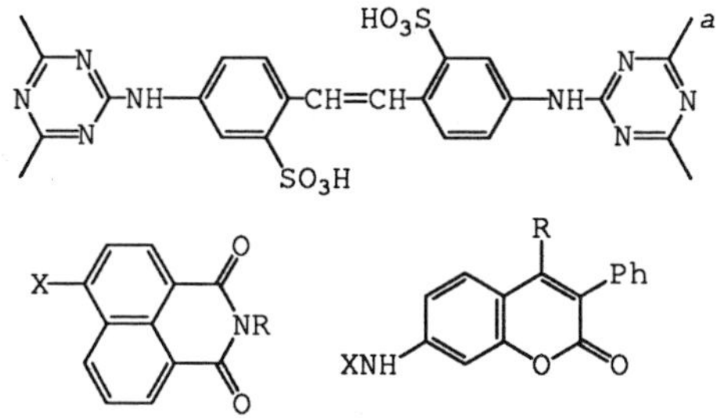

[a] Absorption at <400nm;
emission at >400 nm

16.11

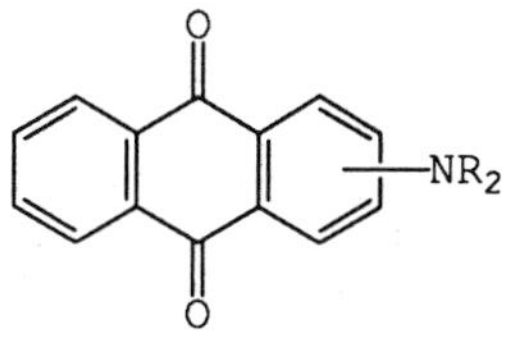

$R_2 = -CH_2CH_2CONHCH_2CH_2-$

16.12

SENSITIVITY OF PHOTOGRAPHIC PROCESSES

SENSITIVITY (cm^{-1})	PROCESS
$1-10^{-1}$	diazotype
10^{-2}	photochromism
$10^{-3}-10^{-4}$	photoresists
10^{-5}	free radicals
$10^{-6}-10^{-9}$	AgX photography

16.9. For ecological reasons it is desirable to accelerate the photodegeneration of polymers (particularly waste plastics at disposal sites). One possible method is to incorporate a carbonyl group into the polymer by the addition of a small amount of an appropriate material, such as methylvinyl ketone, to the polymerization mixture. Under the influence of solar radiation those segments of the polymer containing a carbonyl group subsequently undergo a Norrish Type II bond cleavage in the polymer chain (see 8.21).[13,16]

16.10. *Optical brighteners*, either added directly or impregnated into fabrics, are strongly fluorescent compounds. An ideal optical brightener absorbs in the near UV but not in the VIS region of the spectrum and emits light in the VIS region with a high quantum yield, giving the impression that the fabric emits more radiation than it absorbs.[17]

16.11. An important property of an ideal dye is its photostability. The energy of radiation absorbed by a dye must be rapidly dissipated in a nonradiative transition to the ground state without allowing time for the excited molecule of the dye to undergo a chemical transformation. It should be noted, however, that even a photostable dye can function as a photosensitizer, e.g., the photostable anthraquinone dye shown in the figure can sensitize the degradation of polyamide fibers.[18]

16.12. Optical recording of information is an important application of photochemistry. Nonconventional photographic processes are based on organic photochemical reactions. Although the sensitivity of these reactions is substantially lower than that of silver halide photography, they exhibit an extremely high resolving power, which is limited only by the wavelength of the radiation absorbed.[19]

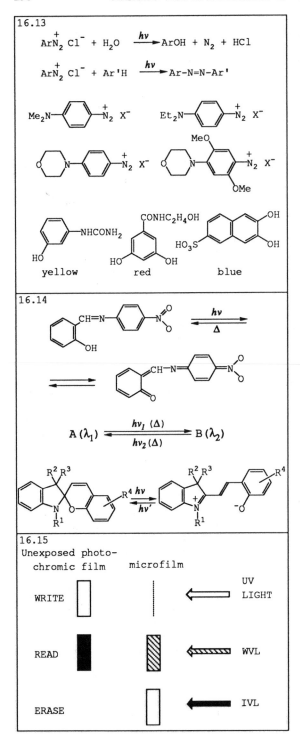

16.13. The classic *diazo type process* is still an important representative of nonconventional photographic processes. The latent image consists of the undecomposed diazonium salt and is developed by coupling it with a phenol and subsequent conversion into a dye. The color depends on the developer (phenol) used. The diazo type process is most sensitive to the UV and blue region (*blue print process*) of the spectrum. Recently, its sensitivity has been extended into the red region.[20]

16.14. *Photochromism*[19(d),21] is a reversible transformation of one molecule into another with a different electronic spectrum. The transformation must occur photochemically, at least in one direction. The simplest example is the color change caused by the absorption of radiation by a photochromic molecule incorporated in a rigid matrix, e.g., a film. There is no development process and the photochromic change can be reversed by either heating or irradiation at a different wavelength. A good photochromic system should provide a large number of reversible cycles without fatigue. Typical photochromic systems are salicylaldehyde anils.[22] Modern photochromic systems are the colorless spiropyrans, which are transformed into colored merocyanines upon irradiation.[23,24]

16.15. Since it occurs on a molecular level, the modification of a spectrum (information recording) in photochromism provides very high resolution. It was used in the development of the *PhotoChromic MicroImage Process* (PCMIP) for the production of ultramicrofilms.[25] Using this technique, a standard size book page can be reduced to a height of less than 1 mm. A 35-mm slide can be used to record a 1200-page book. The advantage of PCMIP is that, in addition to reading with weak visible light (WVL), intense visible light (IVL) can be used to erase unwanted information and replace it with new data.

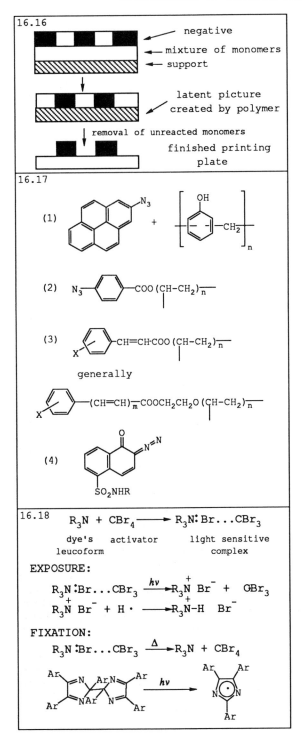

16.16

negative
← mixture of monomers
← support

latent picture
created by polymer

removal of unreacted monomers

finished printing
plate

16.17

(1)

(2)

(3)

generally

(4)

16.18

$$R_3N + CBr_4 \longrightarrow R_3N\overset{\bullet}{\cdot}Br\ldots CBr_3$$

dye's   activator   light sensitive
leucoform   complex

EXPOSURE:

$$R_3N\overset{\bullet}{\cdot}Br\ldots CBr_3 \xrightarrow{h\nu} \overset{+}{R_3N}\ Br^- + \overset{\bullet}{C}Br_3$$

$$\overset{+}{R_3N}\ Br^- + H\cdot \longrightarrow \overset{+}{R_3N}\text{-}H\ Br^-$$

FIXATION:

$$R_3N\overset{\bullet}{\cdot}Br\ldots CBr_3 \xrightarrow{\Delta} R_3N + CBr_4$$

16.16. *Photoresists* are systems that utilize light-sensitive materials dissolved in an organic solvent together with a film-forming high molecular weight polymer.[26] The light-sensitive layer in a photoresist changes its solubility when exposed to light. The irradiated regions become either insoluble (*negative photoresist*) or highly soluble (*positive photoresist*). Photoresists are employed in printed and integrated circuits, etched resistors, chemical machining, offset printing plates, reprography, etc.[27]

16.17. When a light-sensitive material is irradiated, either crosslinks between the reacting molecule and its environment are formed or its polarity is changed. Light-sensitive systems contain: (1) an azido compound in combination with a suitable polymer,[28] (2) the polymer itself possesses azido groups,[29] (3) double bonds, e.g., in the polyvinyl ester of cinnamic acid,[30] or (4) an analogous compound. Positive photoresists contain *o*-quinone diazide[31] as the light-sensitive material.

16.18. Certain free radicals from the photolysis of haloalkanes, e.g., the tribromomethyl radical or bromine atom from CBr_4, are oxidizing agents capable of initiating the formation of color.[32] CBr_4, an activator in this process, forms radical pairs with numerous aromatic amines (photosensitive complexes). The photodecomposition of these complexes leads to the transformation of amines, e.g., into triphenylmethane dyes. Typical amines are the leucoforms of triphenylmethane dyes, such as cyanines, xanthenes, etc. The fixation of the image is performed by heating or dissolving the activator in a suitable solvent. The decomposition of radical pairs can be further sensitized by the dye that enhances the image. Also, hexaarylbiimidazole can be used as an activator.[33]

16.19

$$CO_2 + H_2O \xrightarrow[x=8;\ n=6]{x\ h\nu} 1/n\ (CH_2O)_n + O_2$$

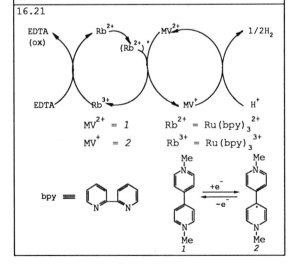

chlorophyll-a; R=Me
chlorophyll-b; R=CH=O

R' =

β-carotene

16.19. The transformation of solar radiation into chemical energy occurs during *photosynthesis*, according to the equation shown (for green plants, x = 8 and n = 6). In the process CO_2 and H_2O are consumed, while organic compounds (sugars) are synthesized and O_2 (from H_2O) is produced. *Chlorophyll-a* and *-b* are the components of green plants that absorb solar radiation. β-carotene (*2*), a component of all photosynthetic systems, protects plants from lethal oxidation.[34]

16.20

PHOTOSYNTHESIS/YEAR IN 10^{11} TONS		HUMAN ACTIVITY/YEAR IN 10^9 TONS	
C-BONDED	~1	CRUDE OIL	3.23*
BIOMASS	2.5	COAL	3.16
CO_2 bonded	3.7	*1973 data	
O_2 free	2.7	Energy content in	
Accumulated energy=		crude oil output=	
0.47-2.5x10^{15}kW		3.75x10^{13}kW	

16.20. Photosynthesis represents a primary production process without which life, as we know it on earth, would not be possible. On the average, photosynthesis fixes about 1% of the solar energy that reaches the earth.[35] The thermal part of solar radiation is used for heating water in solar collectors. The possibilities of chemical[36] and biological[37] (by increasing the yield of photosynthesis) photoconversion of solar energy into fuel are under investigation. A disadvantage of solar energy is its low energy density, requiring large surface areas of irradiation. Furthermore, it is not continuously available (day and night) and its intensity is strongly dependent on the geographical latitude and the weather.

16.21

EDTA (ox) Rb^{2+} $(Rb^{2+})^*$ MV^{2+} $1/2H_2$

EDTA Rb^{3+} MV^+ H^+

$MV^{2+} = 1$ $Rb^{2+} = Ru(bpy)_3^{2+}$
$MV^+ = 2$ $Rb^{3+} = Ru(bpy)_3^{3+}$

bpy ≡

16.21. One system for transforming solar energy into fuel is represented by photochemical oxidation-reduction systems, based on the generation of H_2 (fuel) by the decomposition of water.[38] A well-known system is the bipyridyl complex of Ru, which produces hydrogen in a homogeneous system with methylviologen (MV) in the presence of colloidal platinum. MV^{2+} is reduced by the excited ruthenium complex (Rb^{2+}) to MV^+, which, in the presence of Pt, transforms an electron to H^+, producing hydrogen. The reverse reaction is blocked by the irreversible oxidation of ethylenediaminetetraacetic acid (EDTA).[39]

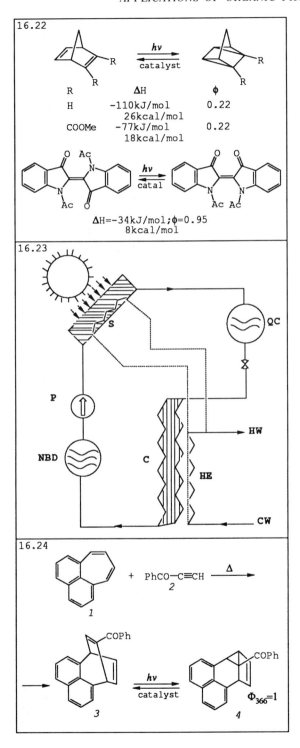

16.22

R	ΔH	φ
H	−110kJ/mol 26kcal/mol	0.22
COOMe	−77kJ/mol 18kcal/mol	0.22

ΔH=−34kJ/mol;φ=0.95
8kcal/mol

16.23

16.24

16.22. Another system is based on the synthesis of a thermodynamically less stable isomer (accumulation of solar energy in fuel) during solar irradiation and its subsequent conversion to a more stable isomer with the liberation of heat. One of the most attractive is the norbornadiene (NBD)/quadricyclane (QC) system.[40] However, because of a symmetry-related kinetic barrier for the QC → NBD isomerization, a catalyst is needed. Its advantages: NBD is relatively inexpensive; both components are liquids; the system has a long-term storage capacity; and the yields are quantitative in both directions. Its disadvantages: the high E_T of NBD (< 300 nm, therefore, a 1 : 1 complex with CuCl is used, with E_T > 300 nm); in spite of the fairly high value of ΔH, the system has a low efficiency for the accumulation of solar energy, η.[41,42]

16.23. This system has been proposed for home heating (QC, NBD = storage of QC and NBD, respectively; HE = heat exchanger; CW, HW = cold and hot water, respectively; C = catalyst bed; P = pump; S = collector with a sensitizer on a support).[37(f)] It should be emphasized that the energy required for the production of such devices for accumulating solar energy must be an order of magnitude smaller during its lifetime than the energy to be obtained from the sun.

16.24. A number of other systems, based on geometric or, more commonly, valence isomerization with the formation of a cage compound,[43] have been proposed for solar energy storage. One of the most remarkable is based on the di-π-methane rearrangement of the product of the diene addition of pleiadiene (*1*) to benzoylacetylene (*2*). The product *3*, a 1,4-diene, rearranges to *4* upon irradiation. The transformation of *3* to *4* represents a model system that fulfills most of the requirements: λ_{exc}, Φ, and the yields of (*3*) ⇌ (*4*) are quantitative.[44]

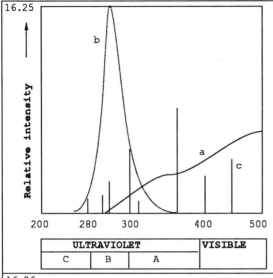

16.25

Relative intensity

b

a

c

200   280   300   400   500

ULTRAVIOLET			VISIBLE
C	B	A	

16.26   **OXYGEN FORMATION :**

$$H_2O + h\nu \text{ (short UV)} \longrightarrow 2H + O$$

$$O + O \longrightarrow O_2$$

STRATOSPHERIC OZONE BALANCE:

$$O_2 + h\nu \ (\lambda < 242.4 \text{ nm}) \longrightarrow 2 \ O \ (^3P_J)$$

$$O \ (^3P_J) + O_2 \longrightarrow O_3$$

$$O_3 + h\nu \text{ (VIS/UV)} \longrightarrow O_2 + O(^3P_J)$$

$$O_3 + h\nu \ (\lambda < 310 \text{ nm}) \longrightarrow O_2(^1\Delta) + O(^1D_2)$$

$$O_3 + O(^3P_J) \longrightarrow 2 \ O_2$$

16.27

Photolysis of Freons:

$$CFCl_3 + h\nu \longrightarrow CFCl_2 + Cl$$
$$CF_2Cl_2 + h\nu \longrightarrow CF_2Cl + Cl$$

Depletion of ozone:

$$Cl + O_3 \longrightarrow ClO + O_2$$
$$ClO + O \longrightarrow Cl + O_2$$

The role of chlorine monoxide dimer:

$$ClO + ClO \longrightarrow (ClO)_2$$
$$(ClO)_2 + h\nu \longrightarrow Cl + ClOO$$
$$ClOO + M \longrightarrow Cl + O_2$$

M = a molecule which remains unchanged
    in the reaction

16.25. Solar radiation, especially its UV component, can represent a health hazard. According to its biological effects, UV radiation is divided into three types: UVA (400–320 nm), UVB (320–280 nm), and UVC (< 280 nm). The figure shows this separation, as well as the spectral distribution of the energy of solar radiation (a), a fluorescent lamp (b), and a medium-pressure mercury discharge lamp (c). Of the total solar radiation, only wavelengths greater than 300 nm reach the surface of the earth due to the pressure of ozone in the stratosphere. The ozone layer absorbs most of the UVB radiation, which is responsible for health risks, e.g., skin cancer (by DNA damage).[45]

16.26. The survival of life on earth is dependent on the existence of the ozone layer. In the beginning phases of the development of our planet, oxygen necessary for the formation of the ozone layer resulted from the photolysis of water by far UV radiation. The maximum concentration of ozone formed at an altitude of 30 km, where an equilibrium involving the reactions shown has been established.[46]

16.27. Depletion of the ozone layer could sufficiently increase the transmittance of short wavelength UV radiation that could destroy all life (increased incidence of skin cancer, significant decrease of photosynthesis, etc.). Recently, there has been an increase in emissions, e.g., CF_2Cl_2 and $CFCl_3$, of cooling fluids from refrigerators and air conditioners (freon), blowing agents in polyurethane foams, etc.. The newly discovered *antarctic ozone hole* accelerated research into the mechanism of O_3 destruction in the stratosphere. Chlorine-catalyzed chain reactions have been shown to be responsible and several mechanisms have been proposed in which chlorine monoxide, as well as its dimer, are suspected of involvement in O_3 loss.[47]

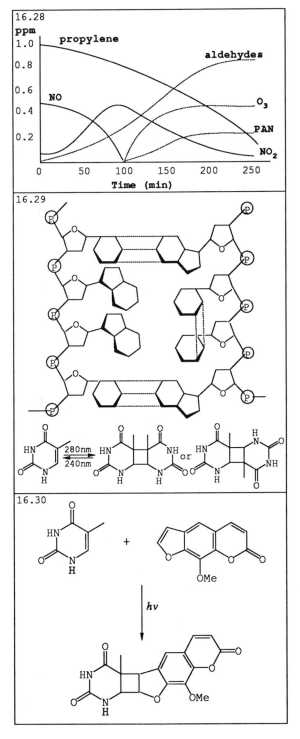

16.28. Human activities contaminate the air with pollutants that are transformed by solar radiation into *photochemical smog*, a dry fog with detrimental physiological and toxic properties. Smog results from the interaction of nitrogen oxides and hydrocarbons with ozone under the influence of solar radiation. Its most toxic components are ozone and peroxyacylnitrates, such as peroxyacetylnitrate (PAN, $CH_3COOONO_2$), which result from the action of NO_2 on acetaldehyde. Ozone is produced from oxygen by photosensitization with nitrogen oxides that absorb in the longer wavelength region (UVA). The formation of a photochemical smog (Los Angeles type) has been modeled in the laboratory.[48]

16.29. UVB radiation induces mutations and skin cancer by promoting photodimerization at the 5,6 double bond of the thymine molecules in DNA.[49] The dimerization causes cleavage of the hydrogen bonds between the purine and pyrimidine components of DNA, deformation of the DNA helix, and, thus, a perturbation of its replication. It can be repaired by DNA photolyases,[49] light-requiring enzymes that catalyze splitting of the cyclobutane ring of the dimer to regenerate thymine. If repair enzymes are in insufficient supply, replication without biological control prevails, and chronic damage, e.g., skin cancer, results.[50]

16.30. UV radiation also has a positive effect on human health, e.g., the cure of skin tuberculosis, rickets, neonatal jaundice, bactericidal effects, etc. It is used in photochemotherapy of certain severe skin diseases, e.g., psoriasis and vitiligo, associated with an overproduction of skin keranocytes. Their removal is aided by the [2+2] cycloaddition of 8-methoxypsoralen to the thymine component of DNA, reducing the rate of mitosis, and eventually inducing the death of the cell. Because of their carcinogenicity,[51] the medical use of psoralens is limited.[52]

16.31

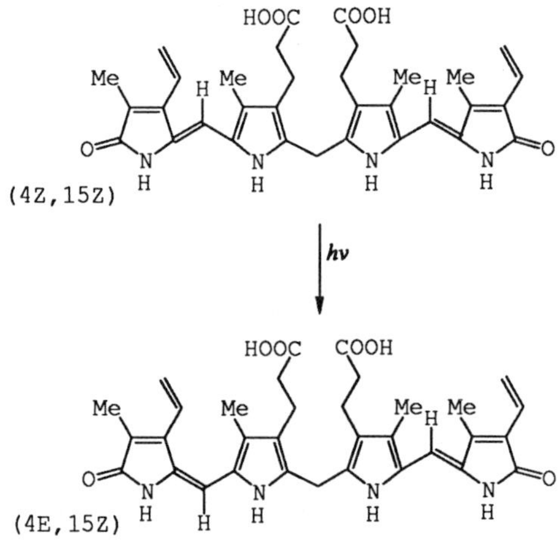

(4Z,15Z)

| *hν*

(4E,15Z)

16.32

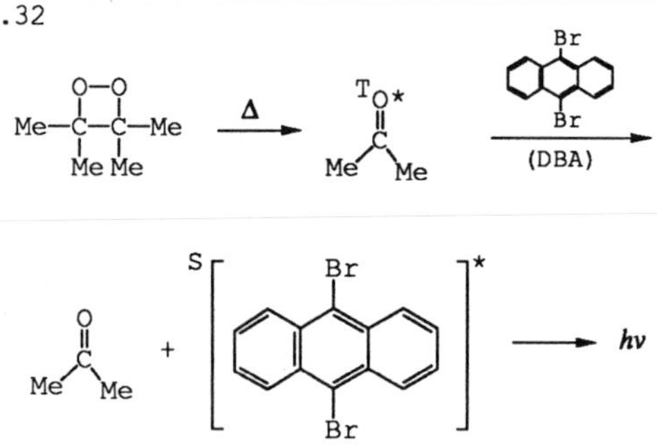

16.33

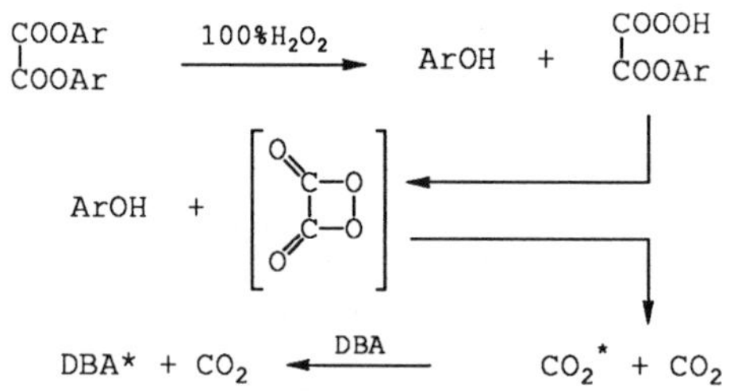

16.34

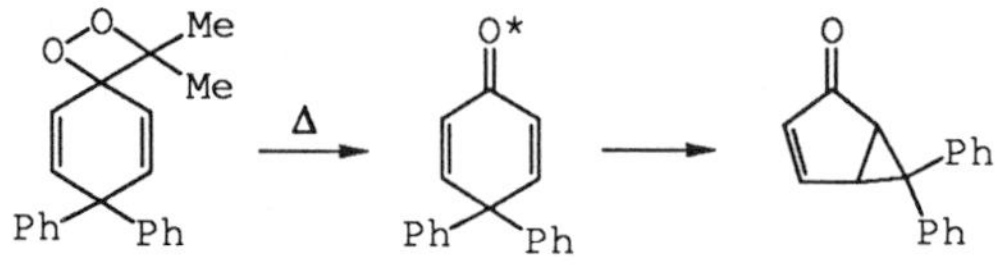

16.31. *Neonatal jaundice*, also known as methemoglobinurea of newborns, is caused by the accumulation of bilirubin in the skin and brain tissue. Treatment consists in irradiating the skin with 450 nm light (bilirubin absorption) to increase the excretion of the neurotoxic metabolite, (4Z,15Z)-bilirubin IX. The thermally unstable photoproducts of bilirubin are rapidly excreted into the bile and, to a minor extent, into urine, because of their appreciable solubility in water. This is due to the Z–E isomerization of bilirubin, i.e., of the stable 4Z,15Z isomer to a photoequilibrium mixture containing the more polar 4E,15Z; 4Z,15E; and 4E,15E isomers.[53]

16.32. Radiant energy in photochemical reactions is converted into chemical energy. The thermal decomposition of certain energy-rich compounds can induce the reverse process, i.e., a chemiluminescent or bioluminescent reaction in which radiation (usually UV) is emitted. This can be transformed into visible light by means of a suitable acceptor, e.g., dibromoanthracene (DBA). The best-known group of chemiluminescent compounds are 1,2-dioxetanes, which decompose thermally into two molecules of carbonyl compounds, one of which is formed in its $^3(n,\pi^*)$ state.[54]

16.33. The perhydrolysis of some oxalic acid derivatives causes a strong chemiluminescent reaction. These systems are available commercially as *Cyalume* and are useful as light sources in humid or explosive environments.[55] The most effective systems have chemiluminescent quantum yields as high as 0.34.[56,57]

16.34. The thermal decomposition of 1,2-dioxetanes in the presence of acceptors has been used to study the mechanism of some photoreactions. For example, it was confirmed that the cyclohexadienone rearrangement proceeds via the $^3(n,\pi^*)$ state rather than the $^3(\pi,\pi^*)$ state.[58]

16.35

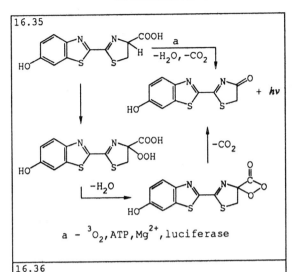

$$a - {}^3O_2, ATP, Mg^{2+}, \text{luciferase}$$

16.36

PICTORIAL REPRESENTATION OF MICROTOX™ TEST SYSTEM

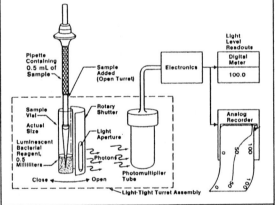

16.37

COMPOUND	mg/L	
	Microtox	Fish
	5 min EC_{50}	96 h LC_{50}
HgCl$_2$	0.065	0.01-0.9
pentachlorophenol	0.5	0.21-0.6
Aroclor 1242	0.7	0.3 -1.0
p-cresol	1.5	3.5 -19
benzene	2.0	17-32
Malathion	3.0	0.07-19.5
formaldehyde	3.0	18-185
phenol	25	9-66
1-butanol	3300	1940
urea	24000	12000
ethanol	31000	13500

16.35. The underlying mechanism of the firefly bioluminescence[59] is the luciferase-catalyzed oxidation of *luciferin*, which causes a molecule of oxygen to oxidize luciferin to the α-hydroperoxycarboxylic acid. Dehydration to α-peroxylactone and its subsequent thermal decomposition yield carbon dioxide, a carbonyl compound,[60] and a quantum of radiation. The quantum yield of bioluminescence is about 0.9.

16.36. Bioluminescence also occurs in other living organisms, e.g., bacteria, sponges, marine crustaceans, and fish. The change of Φ_{fl} luminescence of marine bacteria, *Photobacterium phosphoreum* in the presence of various chemical substances, has been used as a method for the determination of the toxicity of waste waters and water-soluble compounds. In the *Microtox* system lyophilized bacteria, suspended in a regenerative medium,[61] begin to fluoresce immediately. The fluorescence is measured before and after the addition of an aqueous solution of a test sample. Toxic materials in the sample kill some of the bacteria, resulting in a decrease in the fluorescence, which is used to determine EC_{50}, the concentration of the toxic compound at which half of the bacteria die.

16.37. The toxicity of waste water is usually determined by tests performed on fish. The disadvantages of fish testing are its long duration (24-96 hours compared with *Microtox*,[62] which requires a maximum of only 30 min) and the large error in EC_{50} because a small number of fish are used in each experiment. On the other hand, *Microtox* uses 10^6 bacteria for one analysis. The table compares the two methods. One of our important environmental problems is the pollution of stream water. In view of the accuracy and speed of the *Microtox* test, it provides a suitable alarm system for monitoring water purity.[63]

16.38

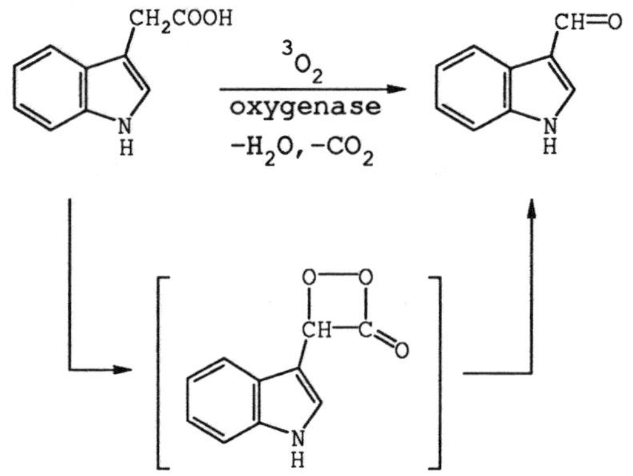

16.39

16.40

16.38. After the elucidation of chemical luminescence in certain organisms, it became obvious to investigate whether electronically excited states also form in the cells of other organisms. These electronically excited states would not necessarily be observed, since any triplet state that was formed could be transformed into the ground state in a nonradiative process. An example was actually observed in a study of the metabolism of the gland growth hormone 3-indole-3-acetic acid (IAA). Indole-3-carboxaldehyde (ICA) and carbon dioxide are formed (via a peroxylactone) by the horseradish peroxidase(HRP)-catalyzed oxidation of IAA.[64] Recently, additional enzymatic reactions of this type have been reported.[65,66]

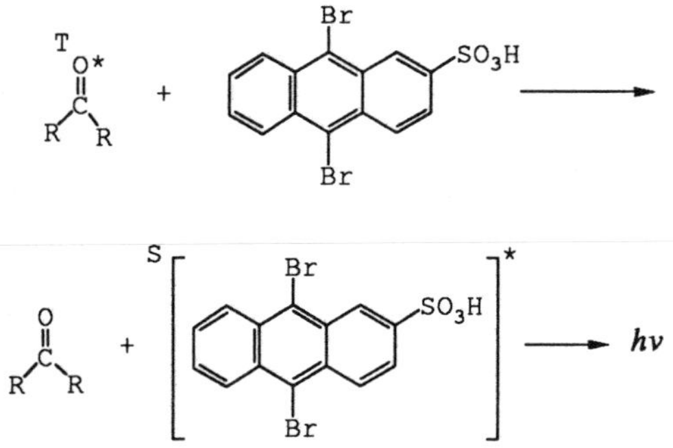

16.39. The excited triplet state of ICA, which normally reverts to its ground state by a luminescence process, can efficiently transfer its triplet excitation energy to a suitable acceptor, e.g., 9,10-dibromoanthracene-2-sulfonic acid whose fluorescence can be observed. These studies suggest that the mechanism of this conversion is analogous to the mechanism of the bioluminescence of luciferin.[67]

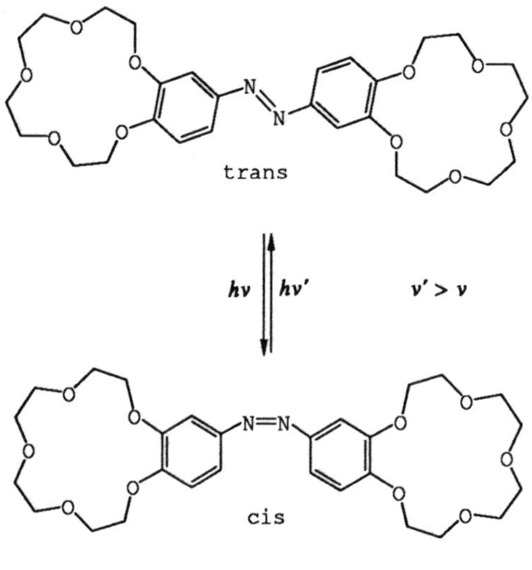

16.40. Recently, photoregularity systems that model the transport of ions through biological membranes have been constructed. These systems are based on the *cis-trans* isomerization of the azobenzene part of the *azobis*(benzocrown ether) molecule. The system shown in the figure indicates that the light-induced transport of K^+ through a membrane, such as *o*-dichlorobenzene, takes place 27 times faster than the transport of Na^+. In contrast to Na^+, K^+ forms an intramolecular 1:2 complex with the *cis* form of the crown ether. *Cis-trans* isomerization does take place in the dark, but occurs rapidly when the system is irradiated with visible light.[68]

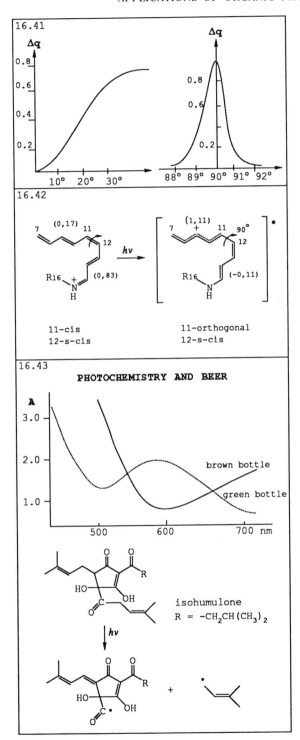

PHOTOCHEMISTRY AND BEER

isohumulone
R = -CH$_2$CH(CH$_3$)$_2$

16.41. Quantum chemical calculations indicate that the pyramidalization (sp^3 hybridization) of one of the carbon atoms in the orthogonal ethylene molecule causes a significant separation of charges in the excited Z$_1$ state (see 2.39). Rotation about the central double bond in the *s-cis-*, *s-trans-*diallyl intermediate from the photoisomerization of 1,3,5-hexatriene causes a separation of charges to occur suddenly in the narrow range of 89–90° (*sudden polarization effect*).[69,70]

16.42. The mechanism of the permeability change in the retinal rod membrane, which makes the passage of sodium ions in the vision process possible, is still an unanswered question. This prompted a theoretical study of the changes due to the absorption of a photon by the chromophore of rhodopsin,[70,71] modeled by a nonatetraenylidenemethyliminum ion in its 11-*cis*-12-*cis* configuration. After photon absorption the molecule rotates about the 11-12 bond and, in the orthogonal geometry, undergoes sudden polarization. The positive charge on the nitrogen atom disappears and a new positive charge appears in the region of carbons 7-11. It is hypothesized that this change either acts directly electrostatically on the pores of the membrane or causes conformational changes in the protein.[69(b),72]

16.43. In concluding the discussion of photochemical applications, a practical example is provided. When beer is exposed to light, especially sunlight, it acquires a repulsive taste, known as the *light-struck flavor*. The irradiation of beer causes isohumulone to undergo α-cleavage (see 8.4) with the formation of a dimethyl allyl radical, which reacts with sulfur compounds in beer to produce offensively odiferous products. The photostability of bottled beer is dependent on the phototransmission of the glass. As shown, beer drawn into brown bottles is preferable.[73]

REFERENCES

(1) For excellent reviews of the benefits and risks of photochemical processes, see (a) Schenck, G. O. *Betriebsaertzliches.* **1977**, *No. 1*, 1. (b) Wiskemann, A. *Betriebsaertzliches.* **1977**, *No. 1*, 39.

(2) For a comprehensive review of DNA photochemistry, see Wang, S. Y., ed. *Photochemistry and Photobiology of Nucleic Acids.* Academic Press: New York, 1976, Vols. I and II.

(3)(a) For a review of medical applications, see Rodighiero, G., Dall'Aqua, F., and Pathak, M. A. In *Topics in Photomedicine.* Smith, K. C., ed.; Plenum: New York, 1984. (b) For a more complete account of photoreactions in biological systems, see Smith, K. C., ed. *The Science of Photobiology.* Plenum: New York, 1977.

(4) Rappoldt, M. P. and Rix, T. R. *Rec. Trav. Chim. Pays-Bas.* **1971**, *90*, 27.

(5) Figure adapted from Fischer, M. *Angew. Chem., Int. Ed. Engl.* **1978**, *17*, 16.

(6)(a) Pfoertner, K. H. In *Photochemistry in Organic Synthesis.* Coyle, J. D., ed. The Royal Society of Chemistry: London, 1986, p. 256. (b) Pfoertner, K. H. *J. Photochem.* **1984**, *25*, 91.

(7) For the photoisomerization of (a) verbenone to chrysanthenone: Schuster, D. I. and Widman, D. *Tetrahedtron Lett.* **1971**, 3571. (b) β-ionone to pyran derivative: Ferguson, D. and Fordham, W. D. *Fr. 1,567,370; Chem. Abstr.* **1970**, *72*, 132524v. Cerfontain, H. and Geene-Vasen, J. A. *J. Tetrahedron.* **1981**, *37*, 1571.

(8) Crosby, D. G. *ACS Symp. Ser.* **1978**, 73 (Disposal Decontam. Pestic.), 1.

(9) Considerable attention has been given recently to a third mechanism, i.e., a photoinduced electron-transfer mechanism. For a review, see Timpe, H.-J. *Topics Curr. Chem.* **1990**, *156*, 167.

(10) Nakanishi, H., Suzuki, Y., Suzuki, F., and Hasegana, M. *J. Polym. Sci. Part A-1.* **1969**, *7*, 753.

(11) Mellor, J. M., Phillips, D., and Salisbury, K. *Chem. Brit.* **1974**, *10*, 160.

(12) DeSchryver, F. C., Anand, L., Smets, G., and Switten, J. *J. Polym. Sci. Part B.* **1971**, *9*, 777.

(13) Ranby, B. and Rabek, J. F., eds. *Photodegradation, Photooxidation and Photostabilization of Polymers.* Wiley: Chichester, United Kingdom, 1975.

(14) Bolliger, H. R. *Rep. Progr. Appl. Chem.* **1970**, *55*, 186, and references therein.

(15)(a) Heller, H. J. *Eur. Polym. J.* **1969**, (suppl.), 105, and references therein. (b) The excited state intramolecular proton transfer through intramolecular hydrogen bond mechanisms has been questioned recently. See Catalan, J., Fobero, F., Guijarro, M. S., Claramunt, R. M., Santa Maria, M. D., Foces-Foces, M. de la C., Cano, F. H., Elguero, J., and Sastre, R. *J. Am. Chem. Soc.* **1990**, *112*, 747, and references therein.

(16) Guillet, J. E. and Amerik, Y. *Macromolecules.* **1971**, *4*, 375.

(17) Bolliger, H. R. *Rep. Progr. Appl. Chem.* **1970**, *55*, 181.

(18)(a) Suppan, P. *Principles of Photochemistry.* The Royal Society of Chemistry: London, 1972, p. 72. (b) Considerable attention has been given recently to electron-transfer processes in imaging. For a review, see Eaton, D. F. *Topics Curr. Chem.* **1990**, *156*, 199.

(19) For reviews of unconventional photographic systems, see (a) Hanson, P. P. *Photogr. Sci. Eng.* **1970**, *14*, 438. (b) Baumann, N. *Nachr. Chem. Tech.* **1974**, *22*, 477. (c) Delzenne, G. A. *Adv. Photochem.* **1979**, *11*, 1. (d) Heller, H. G. *Chem. Ind.* **1978**, 193. (e) Jacobson, R. E. *Chem. Brit.* **1980**, *16*, 468. (f) Sahyun, M. R. V. *J. Chem. Educ.* **1973**, *50*, 88.

(20)(a) Delzenne, G. A. *Adv. Photochem.* **1979**, *11*, 42, and references therein. (b) Kosar, J. *Light Sensitive Systems.* Wiley: New York, 1965.

(21)(a) El'tsov, A. V., ed. *Organic Photochromes.* Plenum: New York, 1990. (b) Dürr, H. and Bouas-Laurent, H., eds. *Photochromism: Molecules and Systems.* Elsevier: Amsterdam, 1990. (c) Brown, G. H., ed. *Photochromism.* Wiley–Interscience: New York, 1971. (d) Dorion, G. H. and Wiebe, A. F. *Photochromism, Optical and Photographic Applications.* Focal Press: London, 1970. (e) Delzenne, G. A. *Adv. Photochem.* **1979**, *11*, 58.

(22) Cohen, M. D., Hirschberg, Y., and Schmidt, G. M. J. *J. Chem. Soc.* **1964**, 2051, 2060, and references therein.

(23)(a) Berman, E. *J. Phys. Chem.* **1962**, *66*, 2275. (b) For photochromic macrocycles, see Winkler, J. D. and Deshayes, K. *J. Am. Chem. Soc.* **1987**, *109*, 2190, and references therein.

(24)(a) For thermally irreversible photochromic systems, see Irie, M. and Mohri, M. *J. Org. Chem.* **1988**, *53*, 803, and references therein. (b) For the photomodulation of polypeptide conformation by sunlight in spiropyran-containing poly(L-glutamic acid), see Ciardelli, F., Fabbri, D., Pieroni, O., and Fissi, A. *J. Am. Chem. Soc.* **1989**, *111*, 3470. For polymer matrices, see (c) Gardlund, Z. G. *Polym. Lett.* **1968**, *6*, 57. (d) Gardlund, Z. G. and Laverty, J. J. *Polym. Lett.* **1969**, *7*, 719. (e) For a furylfulgide photochromism study, see Kurita, S., Kashiwagi, A., Kurita, Y., Miyasaka, H., and Mataga, N. *Chem. Phys. Lett.*, **1990**, 171, 553. (f) For a short review of three-dimensional optical storage memory, see Parthenopoulos, D. A. and Rentzepis, P. M. *Science.* **1989**, *245*, 843. (g) For review of photoresponsive polymers, see Irie, M. *Adv. Polymer Sci.* **1990**, *24*, 27.

(25) Figure adapted from Fischer, E. *Chem. Unser. Zeit 1975*, *9*, 85.

(26)(a) Yovits, M. C., ed. *Large Capacity Memory Techniques for Computer Systems*. Macmillan: New York, 1962, p. 385. (b) King, D. E. *Chem. Brit.* **1968**, *4*, 107.

(27) For reviews, see (a) Walker, P., Webers, V. J., and Thommes, G. A. *J. Photogr. Sci.* **1970**, *18*, 150. (b) Delzenne, G. A. *Adv. Photochem.* **1979**, *11*, 11. (c) Reichmanis, E. and Thompson, L. F. *Chem. Rev.* **1989**, *89*, 1273.

(28) For an introduction to stereolithography, see (a) Neckers, D. C. *The Spectrum.* **1989**, *2*(4), 1. (b) Israel, B. and Bailey, W. J. *Polymer Preprints.* **1989**, *29*, 1, 218. (c) Jacobine, A. *Polymeric Materials Sci. Eng.* **1989**, *60*, 211. (d) For refractive index photoresists, see Tomlinson, W. J. and Chandross, E. A. *Adv. Photochem.* **1980**, *12*, 201.

(29) Tsunoda, T., Yamaoka, T., Soabe, Y., and Hata, Y. *Photogr. Sci. Eng.* **1976**, *20*, 188.

(30)(a) Takahashi, H., Someya, K., Hirohashi, M., Asano, T., Matsunaga, M., and Nagamatsu, G. *Japan 74 23,843*; *Chem. Abstr.* **1975**, *82*, 105218y. (b) Merrill, S. H. and Unruh, C. C. *J. Appl. Polym. Sci.* **1963**, *7*, 273.

(31)(a) Kato, M. *J. Polym. Sci. Part B.* **1969**, *7*, 605. (b) Kato, M., Ichijo, T., Ishii, K., and Hasegawa, M. *J. Polym. Sci. Part A-1.* **1971**, *9*, 2109.

(32) Pacansky, J. and Lyerla, J. R. *IBM J. Res. Dev.* **1979**, *23*, 42.; *Chem. Abstr.* **1979**, *90*, 130611u.

(33) Sprague, R. H., Richter, H. L., and Wainer, E. *Photogr. Sci. Eng.* **1961**, *5*, 98.

(34)(a) Fotland, R. A. *J. Photogr. Sci.* **1970**, *18*, 33. (b) Delzenne, G. A. *Adv. Photochem.* **1979**, *11*, 76.

(35)(a) Foyer, C. H. *Photosynthesis.* Wiley: New York, 1984. (b) For a review of chemical approaches to artificial photosynthesis, see Meyer, T. *J. Acc. Chem. Res.* **1989**, *22*, 163. (c) For a recent paper in this field, see Hasharoni, K., Levanon, H., Tang, J.,

Bowman, M. K., Norris, J. R., Gust, D., Moore, T. A., and Moore, A. L. *J. Am. Chem. Soc.* **1990**, *112*, 6477.

(36) Robinson, S. P. and Walker, D. A. In *The Biochemistry of Plants*, Hatch, M. D. and Boardman, N. K., eds. Academic Press: New York, 1981, Vol. 8, p. 194.

(37) For reviews on solar energy storage, see (a) Calzaferri, G. *Chimia.* **1978**, *32*, 241; **1981**, *35*, 209. (b) Jones, G., Chiang, S.-H., and Xuan, P.-T. *J. Photochem.* **1979**, *10*, 1. (c) Kutal, C. *J. Chem. Educ.* **1983**, *60*, 802. (d) Laird, T. *Chem. Ind.* **1978**, 186. (e) Scharf, H.-D., Fleischhauer, J., Leismann, H., Ressler, I., Schleker, W., and Weitz, R. *Angew. Chem.*, *Int. Ed. Engl.* **1979**, *18*, 652. (f) Schumacher, E. *Chimia.* **1978**, *32*, 193. (g) Schwerzel, R. E. *Prog. Astronaut. Aeronaut.* **1978**, *61* (Radiat. Energy Convers. Space), 626. (h) Connolly, J. S., ed. *Photochemical Conversion and Storage of Solar Energy.* Academic Press: New York, 1981.

(38) Krasnovskii, A. A. *Photosynth. Sol. Energy Convers. Storages, Proc. Conf. SEV, 18th.* **1983**, 5.

(39)(a) Sutin, N. and Creutz, C. *Pure Appl. Chem.* **1980**, *52*, 2717. (b) Kalyanasundaram, K. *Coord. Chem. Rev.* **1982**, *46*, 159.

(40) Gratzel, M. *Acc. Chem. Res.* **1981**, *14*, 376.

(41) For recent papers on NBD/QC systems, see (a) Gassman, P. G. and Hershberger, J. W. *J. Org. Chem.* **1987**, *52*, 1337, and references therein. (b) Murray, R. W. and Pillay, M. K. *Tetrahedron Lett.* **1988**, *29*, 15.

(42)(a) For a charge transfer sensitization by an orthometalated transition metal complex, see Grutsch, P. A. and Kutal, C. *J. Am. Chem. Soc.* **1986**, *108*, 3108. (b) For the QC → NBD conversion mediated by semiconductor powders, see Ikezava, H. and Kutal, C. *J. Org. Chem.* **1987**, *52*, 3299. (c) For colored acyl-NBD's, see Hirao, K., Ando, A., Hamada, T., and Yonemitsu, O. *J. Chem. Soc., Chem. Commun.* **1984**, 300. (d) For an electrochemical *switch* for this system, see Ref. 39(a); for dimethyldioxirane as a catalyst for QC → NBD isomerization, see Ref. 39(b).

(43)(a) Fisher, D. P., Piermattie, V., and Dabrowiak, J. C. *J. Am. Chem. Soc.* **1977**, *99*, 2811. (b) Caldwell, R. A., Mizuno, K., Hansen, P. E., Vo, L. P., Frentrup, M., and Ho, C. D. *J. Am. Chem. Soc.* **1981**, *103*, 7263. (c) Mehta, G. and Srikrishna, A. *J. Chem. Soc. Chem. Commun.* **1982**, 218.

(44) Demuth, M., Burger, U., Mueller, H. W., and Schaffner, K. *J. Am. Chem. Soc.* **1979**, *101*, 6763.

(45) Adapted from Ref. 1.

(46)(a) For leading references on atmospheric chemistry, see Wiesenfeld, J. R. *Acc. Chem. Res.* **1982,** *15,* 110.   (b) Elliott, S. and Rowland, F. S. *J. Chem. Ed.* **1987,** *64,* 387, and references therein.   (c) Wayne, R. P. *Chemistry of Atmospheres.* Clarendon: Oxford, 1985.

(47)(a) *C&EN.* **1987,** *65(44),* 22.   (b) *C&EN.* **1988,** *66(22),* 16.

(48) Adapted from Leighton, P. A. *Photochemistry of Air Pollution.* Academic Press: New York, 1961.

(49) For comprehensive reviews of DNA photochemistry, see   (a) Morrison, H., ed. *Bioorganic Photochemistry. Vol. 1, Photochemistry and the Nucleic Acids.* Wiley: New York, 1990.   (b) Wang, S. Y., ed., *Photochemistry and Photobiology of Nucleic Acids.* Academic Press, New York, 1976, Vols. I and II. (c) Friedberg, E. C. *DNA Repair.* Freeman: New York, 1985.   (d) Hutchinson, F. *Photochem. Photobiol.* **1987,** *45,* 897.   (e) Saito, I. and Sugiyama, H. *Bioorg. Photochem.* **1990,** *1,* 317.   (f) For the isolation of the DNA photoadduct with an antitumor agent, see McGee, L. R. and Misra, R. *J. Am. Chem. Soc.* **1990,** *112,* 2386.

(50)(a) Hartman, R. F., Van Camp, J. R., and Rose, S. D. *J. Org. Chem.* **1987,** *52,* 2684.   (b) For a spectroscopic probe for the DNA structure, see Witmer, M. R., Altmann, E., Young, H., and Begley, T. P. *J. Am. Chem. Soc.* **1989,** *111,* 9264.   (c) For the 1,3-dimethyluracil dimer-splitting sensitized by indole, see Kim, S. T., Hartman, R. F., and Rose, S. D. *Photochem. Photobiol.* **1990,** *52,* 789.   (d) For 3-ureidoacrylonitriles as photolytic products of cytosine and 5-methylcytosine, see Shaw, A. A. and Shetlar, M. D. *J. Am. Chem. Soc.* **1990,** *112,* 7736.

(51) For recent studies in this field, see   (a) Taylor, J.-S. and O'Day, C. L. *J. Am. Chem. Soc.* **1989,** *111,* 401.   (b) Hartman, R. F., Van Camp, J. R., and Rose, S. D. *J. Org. Chem.* **1987,** *52,* 2684.   (c) Cochran, A. G., Sugasawara, R., and Schultz, P. G. *J. Am. Chem. Soc.* **1988,** *110,* 7888.   (d) Jorns, M. S. *J. Am. Chem. Soc.* **1987,** *109,* 3133.

(52) For recent studies, see   (a) Wulff, W. D., McCallum, J. S., and Kunng, F.-A. *J. Am. Chem. Soc.* **1988,** *110,* 7419.   (b) Specht, K. G., Midden, W. R., and Chedekil, M. R. *J. Org. Chem.* **1989,** *54,* 4125, and references therein.   (c) For psoralen-substituted dioxetane as DNA intercalator for photoxenotoxic studies, see Adam, W., Beinhauer, A., Fischer, R., and Haver, H. *Angew. Chem., Int. Ed. Engl.* **1987,** *26,* 796. Adam, W., Epe, B., Schiffman, D., Vargas, F., and Wild, D. *Angew. Chem., Int. Ed. Engl.* **1988,** *27,*

429.   (d) For the solid-state photodimerization of psoralen derivatives, see Courseille, C., Hospital, M., Decout, J. C., and Lhomme, J. *Tetrahedron Lett.* **1990,** *31,* 5031.

(53) For a review of the molecular mechanism of phototherapy of neonatal jaundice, see Lightner, D. A. and McDonagh, A. F. *Acc. Chem. Res.* **1984,** *17,* 417.

(54)(a) Wilson, T., Golan, D. E., Harris, M. S., and Baumstark, A. L. *J. Am. Chem. Soc.* **1976,** *98,* 1086. (b) Turro, N. J., Lechtken, P., Schuster, G., Orell, J., and Steinmetzer, H. C. *J. Am. Chem. Soc.* **1974,** *96,* 1627.   (c) Wilson, T. and Schaap, A. P. *J. Am. Chem. Soc.* **1971,** *93,* 4126.

(55) For reviews, see   (a) Rauhut, M. M. *Acc. Chem. Res.* **1969,** *2,* 80.   (b) Horn, K. A., Koo, J., Schmidt, S. P., and Schuster, G. B. *Mol. Photochem.* **1978–79,** *9,* 1.   (c) Schuster, G. B., Dixon, B., Koo, J., Schmidt, S. P., and Smith, J. P. *Photochem. Photobiol.* **1979,** *30,* 17.   (d) For the application of chemiluminescence in analytical chemistry, see Birks, J. W., ed. *Chemiluminescence and Photochemical Reaction Detection in Chromatography.* VCH Weinheim/VCH: New York, 1989.   (e) Krasovitskii, B. M. and Markovich, B. *Organic Luminescence Materials.* VCH: New York, 1988.

(56) Rauhut, M. M., Bollyky, L. J., Roberts, B. G., Loy, M., Whitman, R. H., Iannotta, A. V., Semsel, A. M., and Clarke, R. A. *J. Am. Chem. Soc.* **1967,** *89,* 6515.

(57) Tseng, S. S., Mohan, A. G., Haines, L. G., Vizcarra, L. S., and Rauhut, M. M. *J. Org. Chem.* **1979,** *44,* 4113.

(58) For recent reviews on the mechanisms of chemiluminescence of this system.   (a) Schuster, G. *Acc. Chem. Res.* **1979,** *12,* 366.   (b) Orlovic, M., Schowen, R. L., Givens, R. S., Alvarez, F., Matuszewski, B., and Parekh, N. *J. Org. Chem.* **1989,** *54,* 3606.

(59) Zimmerman, H. E., Keck, G. E., and Pflederer, J. L. *J. Am. Chem. Soc.* **1976,** *98,* 5574.

(60)(a) For an introductory review, see Adam, W. *J. Chem. Educ.* **1975,** *52,* 138.   (b) Adam, W. and Cilento, G. *Angew. Chem., Int. Ed. Engl.* **1983,** *22,* 529.   (c) Cilento, G. *Acc. Chem. Res.* **1980,** *13,* 225.

(61)(a) McCapra, F., Chang, Y. C., and Francois, V. P. *J. Chem. Soc., Chem. Commun.* **1968,** 22. (b) Hopkins, T. A., Selinger, H. H., White, E. H., and Cass, M. W. *J. Am. Chem. Soc.* **1967,** *89,* 7148. (c) Reprinted with permission of Microbics Corp., Carlsbad, California.

(62) Adapted from Beckman, Inc., *Microtox* Model 2055; Toxicity analyzer system. The current manufacturer of the *Microtox* system is Microbics Corp., Carlsbad, California.

(63)(a) Bulich, A. A. and Green, M. W. *Proc. Int. Symp. Anal. Appl. Biolumin. Chemilumin.* **1978** (publ. 1979), 193. (b) Bulich, A. A. *Process Biochem.* **1982** *(3/4)*, 45. (c) For comparison of the *Microtox* and *Daphnia* tests, see Grange, D. and Pescheux, F. *Bull. Liaison Lab. Ponts Chaussees.* **1985,** *136*, 37.

(64)(a) Vidigal, C. C. C., Faljoni-Alario, A., Duran, N., Zinner, K., Shimizu, Y., and Cilento, G. *Photochem. Photobiol.* **1979,** *30*, 145. (b) De Mello, M. P., De Toledo, S. M., Haun, M., Cilento, G., and Duran, N. *Biochemistry.* **1980,** *19*, 5270.

(65)(a) Cilento, G. *Acc. Chem. Res.* **1980, 13,** 225. (b) Adam, W. and Cilento, G. *Chemical and Biological Generation of Excited States.* Academic Press: New York, 1982. (c) Rivas-Suarez, E. and Cilento, G. *Biochemistry,* **1981,** *20*, 7329. (d) Schulte-Herbrueggen, T. and Cadenas, E. *Photobiochem. Photobiophys.* **1985,** *10*, 35.

(66) For chemical and enzymatic triggering of 1,2-dioxetanes, see (a) Schaap, A. P., Handley, R. S., and Giri, B. P. *Tetrahedron. Lett.* **1987,** *28*, 935. (b) Schaap, A. P., Chen, T. S., Handley, R. S., DeSilva, R., and Giri, B. P. *Tetrahedron Lett.* **1987,** *28*, 1155. (c) Schaap, A. P., Sandison, M. D., and Handley, R. S. *Tetrahedron Lett.* **1987,** *28*, 1159. (d) For naphthyl dioxetane phosphates as substrates for enzymatic chemiluminescent assay, see Edwards, B., Sparks, A., Voyta, J. C., Strong, R., Murphy, O., and Bronstein, I. *J. Org. Chem.* **1990,** *55*, 6225.

(67) Duran, N., Makita, Y., and Innocentini, L. H. *Biochem. Biophys. Res. Commun.* **1979,** *88*, 642.

(68)(a) Shinkai, S., Nakaji, T., Ogawa, T., Shigematsu, K., and Manabe, O. *J. Am. Chem. Soc.* **1981,** *103*, 111. (b) Shinkai, S., Minami, T., Kusano, Y.,

and Manabe, O. *J. Am. Chem. Soc.* **1983,** *105*, 1851, and references therein.

(69)(a) Adapted from Bonačić-Koutecký, V., Bruckman, P., Hiberty, P., Koutecký, J., Leforestier, C., and Salem, L. *Angew. Chem., Int. Ed. Engl.* **1975,** *14*, 575. (b) For a recent study, see Dormans, G. J. M., Groenenboom, G. C., Van Dorst, W. C. A., and Buck, H. M. *J. Am. Chem. Soc.* **1988,** *110*, 1406.

(70)(a) For a review, see Salem, L. *Acc. Chem. Res.* **1979,** *12*, 87. (a) For a MNDO/CI study of *cis-trans* isomerization of retinal-like protonated Schiff bases, see Dormans, G. J. M., Groenenboom, G. C., van Dorst, W. C. A., and Buck, H. M. *J. Am. Chem. Soc.* **1988,** *110*, 1406.

(71) For recent reviews, see (a) Ottolenghi, M. *Adv. Photochem.* **1980,** *12*, 87. (b) Birge, R. R. *Ann. Rev. Biophyl. Bioeng.* **1981,** *10*, 315. (c) Sandorfy, C. and Vocele, D. *Can. J. Chem.* **1986,** *64*, 2257. (d) Ottolenghi, M., *Adv. Photochem.* **1980,** *12*, 97. (e) For a photoaffinity labeling study of bacteriorhodopsin with [15-^{3}H]-3-diazo-4-keto-*all-trans*-retinal, see Boehm, M. F., Gawinowicz, M. A., Foucault, A., Derguini, F., and Nakanishi, K. *J. Am. Chem. Soc.* **1990,** *112*, 7779. (f) For a review of the chemistry of vitamin A and vision, see Rando, R. R. *Angew. Chem., Int. Ed. Engl.* **1990,** *29*, 461.

(72)(a) Salem, L. *Science.* **1976,** *191*, 822. (b) Bonačić-Koutecký, V. *J. Am. Chem. Soc.* **1978,** *100*, 396. For recent studies of the possible role of the sudden polarization effect in vision, see (c) Childs, R. F. and Shaw, G. S. *J. Am. Chem. Soc.* **1988,** *110*, 3013. (d) For the likely role of twisted-internal charge-transfer (TICT) and of proton translocation in the initial step of vision, see Bonačić-Koutecký, V., Koutecký, J., and Michl, J. *Angew. Chem., Int. Ed. Engl.* **1987,** *26*, 170.

(73) Adapted from Vogler, A. and Kunkely, H. *J. Chem. Educ.* **1982,** *59*, 25.

INDEX